COMMISSION OF THE EUROPEAN COMMUNITIES

AGREP

PERMANENT INVENTORY
OF
AGRICULTURAL RESEARCH
PROJECTS IN THE
EUROPEAN COMMUNITIES

VOL. II INDEXES

MAY 1980

SPRINGER-SCIENCE+BUSINESS MEDIA, B.V.

This volume has a Library of Congress Cataloging in Publication classification.

ISBN 978-94-009-8266-6   ISBN 978-94-009-8264-2 (eBook)
DOI 10.1007/978-94-009-8264-2

Publication arranged by:
Commission of the European Communities,
Directorate-General Information Market and Innovation,
Luxembourg

Data processing by
I/S Datacentralen af 1959,
Retortvej 6-8,
2500 Valby,
Denmark

EUR 5895

LEGAL NOTICE
Neither the Commission of the European Communities nor any person acting on behalf of
the Commission is responsible for the use which might be made of the following information.

Photocomposition by Special-Trykkeriet Viborg a-s

Table of Contents

EXPLANATORY NOTES

– INDEX of SUBJECTS AREAS (List of Subject Areas on page V)

Example of typical entry:

1)   **B6400 Fertilizers and water for plants in general**
2)   85, 126, 144, 304, 315, 393, 476, 518, 634, 686,
       714, 748, 852, 860, 946, 948, 1607, 1610, 1782,
       1939, 2673, 2676, 2731, 2736, 2740, 2742, 2753,

1) Code and title of subject areas
2) Reference number(s) of relevant research projects

– LIST OF RESEARCH ORGANISATIONS
Research organisation codes see vol. I for explanation.

– LIST OF SCIENTISTS

Example of typical entry:

1) Deckers, J. 2257,2270,3157,3186   2)
    4041,5769

1) Name of scientist
2) Reference number(s) of project(s) in which scientist is involved

Subject Areas

Note: What follows is a complete listing of the Classification Scheme for Subject Areas. This does not necessarily imply that there are projects assigned to all codes. To codes without assigned projects, an asterix (*) has been entered instead of the page number.

(Subject Areas are subdivided by Fields of Research, see Vol. I.)

B – SUBJECT AREAS

B – SUBJECT AREAS

B1000 Biosphere in general
4, 1177, 1793, 1835, 1864, 1865, 2104, 2113, 2114, 2115, 2136, 2167, 2202, 2218, 2253, 2311, 2405, 2471, 2504, 2539, 2604, 2687, 4392, 4584, 5117, 6163, 7828, 16180, 16416, 17179, 18225, 19580, 19581, 19703, 19754, 19773, 19793, 19836, 19994, 19997, 19998, 20018, 20096, 20119, 20356, 20535, 20537, 20674, 20711, 20750, 20757, 20826, 20828, 20992, 20999, 21088

B1100 Soil in general
1120, 1121, 1124, 1145, 1146, 1173, 1182, 1229, 1249, 1306, 1365, 1366, 1410, 1430, 1439, 1442, 1510, 1511, 1526, 1532, 1550, 1557, 1614, 1655, 1730, 1731, 1732, 1733, 1734, 1739, 1746, 1761, 1764, 1770, 1772, 1774, 1782, 1783, 1784, 1787, 1804, 1806, 1815, 1817, 1819, 1878, 1879, 1881, 1906, 1930, 2047, 2049, 2083, 2107, 2184, 2203, 2209, 2216, 2238, 2247, 2325, 2370, 2440, 2446, 2482, 2517, 2556, 2565, 2566, 2594, 2595, 2645, 2653, 2675, 2679, 2874, 2877, 2884, 3010, 3048, 3277, 3278, 3284, 3346, 3566, 3598, 3613, 3746, 3775, 3924, 4101, 4204, 4238, 4393, 4399, 4414, 4455, 4712, 4765, 4833, 4966, 5002, 5006, 5066, 5147, 5168, 5171, 5189, 5212, 5232, 5244, 5252, 5257, 5306, 5339, 5400, 5454, 5485, 5486, 5506, 5654, 5773, 5787, 5789, 5897...

Soil science (D1100)
1, 5, 6, 7, 8, 9, 98, 99, 100, 101, 102, 103, 104, 105, 106, 107, 108, 109, 110, 111, 112, 113, 114, 115, 116, 117, 118, 119, 120, 121, 122, 123, 124, 125, 126, 127, 128, 129, 130, 131, 132, 133, 134, 135, 136, 137, 138, 139, 140, 141, 142, 143, 144, 145, 146, 147, 148, 149, 150, 151, 152, 153, 154, 155, 156, 157, 158, 159, 160, 161, 162, 163, 164, 165, 166, 167, 168, 169, 170, 171, 172, 173, 174, 175, 176, 177, 178, 179, 180, 181, 182, 183, 184, 185, 186, 187, 188, 189, 190, 191, 192, 193, 194, 195, 196, 197, 198, 199, 200, 201, 202, 203, 204, 205, 206, 207, 208, 209, 210, 211, 212, 213, 214, 215, 216...

B1110 Soil composition – general
1122, 1160, 1161, 1162, 1163, 1164, 1165, 1166, 1167, 1168, 1169, 1170, 1174, 1178, 1235, 1238, 1295, 1300, 1302, 1316, 1336, 1342, 1345, 1363, 1366, 1369, 1371, 1372, 1373, 1395, 1396, 1559, 1608, 1637, 1640, 1727, 1758, 1773, 1777, 1778, 1904, 2037, 2058, 2248, 2256, 2259, 2324, 2445, 2549, 2690, 2701, 2705, 2927, 3129, 3521, 3659, 3812, 4021, 4131, 4210, 4607, 4770, 4775, 5058, 5080, 5131, 5173, 5195, 5201, 5205, 5206, 5230, 5237, 5319, 5347, 5355, 5356, 5358, 5365, 5471, 5576, 5600, 5605, 5628, 5634, 5635, 5636, 5642, 5644, 5652, 5671, 5673, 5675, 5677, 5692, 5696, 5755, 5801, 5936, 5950, 5969, 5970, 5976, 5979, 5996, 6436, 6787, 7763, 7764, 7816, 7876, 7886, 8022, 8655, 9425, 10093, 10118, 10285, 10288, 14570, 15534, 17021, 19701, 19842...

Soil science (D1100)
14, 15, 158, 171, 430, 431, 432, 433, 434, 435, 436, 437, 438, 439, 440, 441, 442, 443, 444, 445, 446, 447, 448, 449, 450, 451, 452, 453, 454, 455, 456, 457, 458, 459, 460, 461, 462, 463, 464, 465, 466, 467, 468, 469, 470, 471, 472, 473, 474, 475, 476, 477, 478, 479, 480, 481, 482, 483, 484, 485, 486, 487, 488, 489, 490, 491, 492, 493, 494, 495, 496, 497, 498, 499, 500, 501, 502, 503, 504, 505, 506, 507, 508, 509, 510, 511, 512, 513, 514, 515, 516, 517, 518, 519, 520, 521, 522, 523, 524, 525, 526, 527, 528, 529, 530, 531, 532, 533, 534, 535, 536, 537, 538, 539, 540, 541, 542, 543, 544, 545, 546, 547, 548, 549, 550...

B1111 Soil composition – inorganic
1234, 1338, 1412, 1571, 1606, 1630, 1635, 1636, 1638, 1653, 1656, 1715, 1735, 1902, 2100, 2652, 3374, 4868, 4976, 4977, 5063, 5077, 5087, 5088, 5103, 5126, 5127, 5130, 5142, 5169, 5204, 5207, 5208, 5234, 5236, 5238, 5276, 5291, 5296, 5329,

5331, 5338, 5340, 5342, 5348, 5349, 5352, 5353, 5361, 5366, 5372, 5384, 5409, 5451, 5473, 5507, 5541, 5603, 5606, 5619, 5620, 5627, 5637, 5666, 5693, 5720, 5750, 5816, 5819, 5878, 5931, 5938, 5951, 5964, 5974, 7780, 8390, 8391, 9027, 10182, 10928, 19678, 19685, 20344

Soil science (D1100)
16, 169, 569, 570, 571, 572, 573, 574, 575, 576, 577, 578, 579, 580, 581, 582, 583, 584, 585, 586, 587, 588, 589, 590, 591, 592, 593, 594, 595, 596, 597, 598, 599, 600, 601, 602, 603, 604, 605, 606, 607, 608, 609, 610, 611, 612, 613, 614, 615, 616, 617, 618, 619, 620, 621, 622, 623, 624, 625, 626, 627, 628, 629, 630, 631, 632, 633, 634, 635, 636, 637, 638, 639, 640, 641, 642, 643, 644, 645, 646, 647, 648, 649, 650, 651, 652, 653, 654, 655, 656, 657, 658, 659, 660, 661, 662, 663, 664, 665, 666, 667, 668, 669, 670, 671, 672, 673, 674, 675, 676, 677, 678, 679, 680, 681, 682, 683, 684, 685, 686, 687, 688, 689, 690, 691...

B1112 Soil composition – organic
1607, 1653, 1749, 1843, 2257, 2755, 2824, 3483, 3653, 4799, 5071, 5136, 5152, 5175, 5214, 5235, 5323, 5324, 5460, 5615, 5889, 5890, 5891, 5956, 12183, 15331

Soil science (D1100)
10, 169, 585, 600, 602, 607, 618, 637, 654, 674, 683, 687, 688, 693, 695, 700, 711, 725, 726, 727, 728, 729, 730, 731, 732, 733, 734, 735, 736, 737, 738, 739, 740, 741, 742, 743, 744, 745, 746, 747, 748, 749, 750, 751, 752, 753, 754, 755, 756, 757, 758, 759, 760, 761, 762, 763, 764, 765, 766, 767, 768, 769, 770, 771, 772, 773, 774, 775, 776, 777, 778, 779, 780, 781, 782, 783, 784, 785, 786, 787, 788, 789, 790, 791, 792, 793, 794, 795, 796, 797, 798, 799, 800, 801, 802, 803, 804, 805, 806, 807, 808, 809, 810, 811, 812, 813, 814, 815, 816, 817, 834, 1067, 1081, 1104, 2609, 2988, 4972, 5062, 5172, 5298, 5302, 5332, 5604

B1113 Soil composition – soil air, soil water
1110, 1188, 1190, 1191, 1201, 1204, 1205, 1207, 1210, 1222, 1223, 1224, 1225, 1226, 1227, 1230, 1231, 1233, 1237, 1239, 1240, 1241, 1242, 1243, 1246, 1248, 1251, 1257, 1259, 1266, 1272, 1276, 1277, 1278, 1279, 1280, 1281, 1282, 1289, 1290, 1304, 1305, 1307, 1311, 1314, 1315, 1323, 1325, 1326, 1331, 1333, 1334, 1337, 1339, 1344, 1346, 1365, 1367, 1368, 1369, 1391, 1393, 1397, 1398, 1399, 1400, 1402, 1411, 1412, 1413, 1415, 1417, 1418, 1422, 1436, 1448, 1454, 1455, 1457, 1458, 1459, 1465, 1466, 1470, 1485, 1487, 1488, 1491, 1497, 1499, 1503, 1505, 1509, 1512, 1524, 1527, 1528, 1534, 1539, 1541, 1542, 1544, 1546, 1547, 1548, 1558, 1560, 1561, 1562, 1563, 1564, 1565, 1566, 1568, 1571, 1573, 1574, 1575, 1576, 1579, 1580, 1581, 1582, 1583, 1584...

Soil science (D1100)
20, 578, 603, 628, 630, 657, 658, 703, 713, 723, 818, 819, 820, 821, 822, 823, 824, 825, 826, 827, 828, 829, 830, 831, 832, 833, 834, 835, 836, 837, 838, 839, 840, 841, 842, 843, 844, 845, 846, 847, 848, 849, 850, 851, 852, 853, 854, 855, 856, 857, 858, 859, 860, 861, 862, 863, 873, 875, 878, 880, 881, 887, 888, 917, 922, 923, 957, 958, 1009, 1095, 1179, 1180, 1332, 1403, 1404, 1421, 1428, 1429, 1431, 1443, 1444, 1496, 1504, 1523, 1543, 1551, 5298, 5421, 20116

B1119 Soil composition – other
2680, 4026, 17095

Soil science (D1100)
864, 865, 9035

B1120 Soil structure
1147, 1209, 1340, 1374, 1385, 1413, 1418, 1447, 1548, 1568, 1573, 1610, 1611, 1616, 1640, 1686, 1726, 1736, 1737, 1738, 1740, 1741, 1744, 1746, 1750, 1751, 1752, 1753, 1754, 1755, 1756, 1759, 1760, 1769, 1771, 1773, 1776, 1777, 1779, 1780, 1781, 1784, 1802, 2715, 2716, 2912, 2993, 3064, 3065, 3259,

3294, 3389, 3521, 3847, 4476, 5036, 5319, 5327, 5347, 5356, 5559, 5721, 5762, 5889, 5890, 5891, 5968, 6831, 10173, 10255, 15293, 15311, 15313, 15317, 15332, 15390, 15517, 15773, 17445, 20387

Soil science (D1100)

435, 442, 460, 480, 482, 515, 525, 552, 556, 558, 616, 645, 646, 678, 698, 699, 713, 718, 737, 781, 821, 860, 866, 867, 868, 869, 870, 871, 872, 873, 874, 875, 876, 877, 878, 879, 880, 881, 882, 883, 884, 885, 886, 887, 888, 889, 890, 891, 892, 893, 894, 895, 896, 897, 898, 899, 900, 901, 902, 903, 904, 905, 906, 907, 908, 909, 910, 911, 912, 913, 914, 915, 916, 917, 918, 919, 920, 921, 922, 923, 924, 925, 926, 927, 928, 929, 930, 931, 932, 933, 934, 935, 936, 937, 938, 939, 940, 941, 942, 943, 944, 945, 946, 947, 948, 949, 950, 951, 952, 953, 954, 955, 956, 957, 958, 959, 960, 961, 962, 963, 964, 965, 966, 967, 968...

B1130 Bio–communities in the soil

1189, 2034, 2140, 2256, 2259, 2297, 2327, 2331, 2363, 2659, 2799, 2800, 2801, 2882, 2949, 3615, 3636, 3641, 3653, 3922, 3931, 4021, 4030, 4369, 4464, 4835, 4836, 4837, 4838, 4875, 4895, 4896, 5010, 5066, 5151, 5211, 5268, 5283, 5284, 5285, 5313, 5314, 5374, 5435, 5462, 5647, 5648, 5655, 5661, 5743, 6069, 6489, 6767, 6801, 6802, 7770, 7771, 7778, 7780, 7804, 7818, 7861, 7886, 7944, 8031, 8040, 8085, 8101, 8108, 8142, 8146, 8147, 8150, 8151, 8153, 8155, 8157, 8187, 8188, 8212, 8214, 8221, 8317, 8350, 8369, 8374, 8375, 8376, 8385, 8390, 8391, 8399, 8404, 8416, 8448, 8513, 8549, 8561, 8573, 8578, 8622, 8750, 8808, 8815, 8859, 8918, 9030, 9086, 9096, 9097, 9102, 9103, 9224, 9227, 9241, 9242, 9245, 9248, 9261, 9281, 9305, 9537, 9560, 9601, 9607...

Soil science (D1100)

97, 434, 514, 516, 517, 535, 542, 544, 549, 586, 601, 669, 696, 700, 707, 727, 746, 751, 763, 779, 782, 785, 816, 980, 981, 982, 983, 984, 985, 986, 987, 988, 989, 990, 991, 992, 993, 994, 995, 996, 997, 998, 999, 1000, 1001, 1002, 1003, 1004, 1005, 1005, 1006, 1007, 1008, 1009, 1010, 1011, 1012, 1012, 1013, 1014, 1015, 1016, 1017, 1018, 1019, 1020, 1021, 1022, 1023, 1024, 1025, 1026, 1027, 1028, 1029, 1030, 1031, 1032, 1033, 1034, 1035, 1036, 1037, 1038, 1039, 1040, 1041, 1042, 1043, 1044, 1045, 1046, 1047, 1048, 1049, 1050, 1051, 1052, 1053, 1054, 1055, 1056, 1057, 1058, 1059, 1060, 1061, 1062, 1063, 1064, 1065, 1066, 1067, 1068, 1069, 1070, 1071, 1072, 1073, 1074, 1075, 1076, 1077, 1078, 1079...

B1190 Other subjects related to soil

1757, 1831, 1847, 6006, 10318, 17301, 19961

Soil science (D1100)

2, 3, 434, 1107, 1108, 1109, 2601

B1200 Water in general

185, 186, 398, 429, 442, 628, 630, 724, 832, 1027, 1111, 1127, 1128, 1136, 1141, 1143, 1184, 1209, 1213, 1215, 1220, 1244, 1252, 1256, 1263, 1264, 1265, 1269, 1270, 1294, 1301, 1332, 1357, 1358, 1376, 1378, 1379, 1401, 1403, 1404, 1407, 1428, 1429, 1432, 1442, 1466, 1503, 1541, 1567, 1580, 1625, 1641, 1644, 1646, 1647, 1670, 1673, 1674, 1675, 1679, 1693, 1694, 1695, 1698, 1701, 1711, 1722, 1723, 1724, 1787, 1868, 2083, 2107, 2226, 2341, 2653, 5051, 5141, 5194, 5426, 5575, 9846, 13877, 16193, 16270, 16407, 16408, 16429, 17087, 18094, 18819, 18892, 19186, 19714, 19753, 20682, 20945, 21089

B1210 Water composition

450, 578, 707, 836, 1100, 1107, 1112, 1113, 1114, 1115, 1132, 1196, 1197, 1198, 1199, 1219, 1224, 1260, 1293, 1295, 1296, 1303, 1310, 1319, 1320, 1321, 1330, 1341, 1359, 1375, 1380, 1381, 1386, 1387, 1389, 1419, 1445, 1446, 1448, 1453, 1555, 1569, 1571, 1578, 1600, 1601, 1602, 1603, 1604, 1605, 1606, 1608, 1630, 1631, 1638, 1648, 1652, 1655, 1656, 1658,

1660, 1669, 1681, 1687, 1696, 1705, 1706, 1715, 1716, 1719, 1832, 1834, 1836, 1975, 1976, 2060, 2248, 2251, 2257, 2267, 2277, 2278, 2279, 2280, 2281, 2282, 2284, 2285, 2286, 2289, 2343, 3072, 3945, 5122, 5196, 5673, 5675, 10112, 10336, 11451, 12752, 13419, 13498, 15120, 15608, 16357, 16436, 16584, 16622, 16664, 19265, 19271, 19570, 19578, 19592, 19726, 19728, 19729, 19730, 19731, 19732, 19733, 19734, 19737, 19738...

B1220 Bio–communities in the water

1100, 1117, 1129, 1130, 1131, 1132, 1133, 1135, 1137, 1138, 1139, 1142, 1144, 1212, 1214, 1247, 1260, 1283, 1297, 1299, 1310, 1320, 1321, 1347, 1359, 1370, 1388, 1446, 1533, 1553, 1555, 1603, 1663, 1672, 1676, 1677, 1687, 1696, 1715, 1789, 1832, 1953, 1975, 1976, 1994, 2145, 2180, 2197, 2225, 2244, 2245, 2251, 2267, 2268, 2275, 2277, 2278, 2279, 2280, 2281, 2282, 2286, 2308, 2315, 2318, 2333, 2334, 2343, 2346, 2347, 2350, 2722, 5196, 8861, 11399, 12726, 13502, 16939, 19727, 19734, 19735, 20494, 20507, 20615, 20687, 20752, 21030

B1290 Other subjects related to water

18, 19, 818, 1134, 1140, 1288, 12873, 15344

B1300 Air and climate in general

24, 216, 304, 472, 1186, 1466, 1471, 1816, 1867, 1872, 2032, 2074, 2099, 2372, 2611, 2653, 2657, 2658, 2886, 2888, 3138, 3786, 3787, 3833, 4091, 4333, 4560, 4567, 4574, 4589, 4675, 5103, 5280, 5352, 5896, 8365, 10343, 10650, 10654, 10707, 10895, 11048, 11233, 11245, 11357, 11358, 11363, 11364, 12744, 15613, 15616, 15625, 15634, 15635, 15675, 15764, 15780, 16816, 19543, 19621, 19625, 19648, 19656, 19700, 19713, 19715, 19716, 19717, 19718, 19740, 19746, 19748, 19751, 20043, 20085, 20185, 20191, 20551, 20647, 20733, 20824, 20886, 21005

B1310 External climate

16, 21, 96, 121, 133, 134, 165, 185, 202, 268, 272, 281, 291, 293, 299, 302, 303, 310, 329, 331, 333, 336, 338, 340, 341, 382, 449, 465, 466, 493, 514, 538, 611, 628, 633, 646, 827, 828, 835, 848, 910, 912, 917, 928, 963, 1008, 1158, 1176, 1179, 1180, 1183, 1189, 1222, 1248, 1249, 1258, 1259, 1260, 1274, 1275, 1291, 1304, 1309, 1312, 1315, 1325, 1326, 1327, 1338, 1345, 1371, 1373, 1392, 1397, 1398, 1399, 1400, 1420, 1422, 1423, 1474, 1490, 1496, 1498, 1504, 1505, 1508, 1517, 1521, 1534, 1556, 1568, 1572, 1577, 1591, 1611, 1624, 1629, 1633, 1643, 1654, 1670, 1673, 1674, 1679, 1688, 1720, 1721, 1880, 1957, 1962, 2032, 2083, 2183, 2224, 2238, 2323, 2324, 2325, 2420, 2492, 2540, 2655, 2667, 2672...

B1320 Internal climate

1570, 2536, 2655, 2677, 2707, 2878, 3401, 3693, 3694, 3700, 3707, 3711, 3753, 3754, 3768, 3769, 3770, 3772, 3774, 3779, 3780, 3781, 3782, 3844, 3870, 3883, 3942, 3976, 3984, 4388, 4496, 4497, 4519, 4520, 4521, 4570, 4572, 4604, 4612, 4617, 4619, 4667, 4728, 5881, 5887, 6458, 6820, 7015, 7162, 7173, 7548, 7708, 10342, 10385, 10672, 10714, 10726, 10903, 10906, 11097, 11100, 11101, 11103, 11139, 11178, 11179, 11182, 11189, 11192, 11239, 11248, 11252, 11259, 11262, 11284, 11307, 11309, 11312, 11314, 11329, 11345, 11350, 11496, 11498, 11502, 11754, 12974, 13611, 13612, 13646, 14393, 14425, 14875, 15501, 15502, 15519, 15520, 15531, 15532, 15579, 15581, 15591, 15595, 15596, 15597, 15598, 15599, 15607, 15613, 15614, 15615, 15617, 15618, 15619, 15620, 15621, 15622, 15623, 15624, 15626, 15627, 15628, 15629, 15630, 15631...

B1390 Other subjects related to air and climate

11646, 19992

B1400 Range, uncultivated land and natural vegetation in general

223, 268, 363, 368, 376, 393, 394, 400, 412, 557, 1120, 1126, 1493, 1564, 1590, 1643, 1653, 1654, 1745, 1791, 1792,

17273, 17276, 17438, 17531, 17654, 17678, 17687, 17728, 18339, 20492, 20558, 20604, 20967

Plant production general and crop husbandry (D2100)
2179, 2639, 2640, 2641, 2642, 2643, 2644, 2645, 2646, 2647, 2648, 2649, 2650, 3616, 3618, 3623, 3624, 4608, 4685, 4687, 4713, 4759, 4764, 4848, 4850, 4983, 5000, 5001, 10283

Plant nutrition and fertilization (D2200)
4608, 5925, 5991

Plant breeding (D2300)
4713, 6009, 6016, 6017, 6018, 6019, 6020, 6707, 6723, 6728, 7655, 7689, 7693, 7708, 7709, 7710, 7712, 7713, 7714

Plant protection (D2400)
2639, 7791

Pests of plants and pest control (D2410)
8025, 8026, 8027, 8028, 8429

Plant diseases and disease control (D2420)
4759, 6009, 8994, 8995, 8996, 9341, 9550, 9781, 10011, 10036

Weeds and weed control (D2430)
2031, 2646, 10072, 10073, 10165, 10179, 10283, 10311

Miscellaneous plant disorders (D2490)
10328, 10329, 10330, 10392

B1620 Arboreta and botanical gardens
2079, 2080, 2081

Plant production general and crop husbandry (D2100)
4823

B1630 Sportfields, play and camping grounds
361, 551, 806, 946, 949, 1233, 1614, 1689, 1699, 1729, 1856, 2182, 2297, 2417, 2418, 2419, 2486, 2548

Plant production general and crop husbandry (D2100)
2644, 3389, 3402, 3616, 3623, 3624

Plant nutrition and fertilization (D2200)
5086, 5087, 5088

Plant breeding (D2300)
6020, 6707, 6723, 6728, 7627

Pests of plants and pest control (D2410)
8597

Plant diseases and disease control (D2420)
9559

B1690 Other man–made recreational resources
1360, 1696, 2522, 15724, 20507

B1900 Other subjects areas related to biosphere and recreational resources
1857, 1986, 2006, 2038, 2376, 2432, 2550, 18019, 19996, 20302, 20332

B2000 Plants and animals in general
2092, 2117, 2118, 19804, 19868, 19876, 19979, 20052, 20088, 20138, 20143, 20690, 20751, 20757, 20792, 20895, 20969, 20970, 20971, 20976

Animal nutrition (D3200)
11623

B2100 Plants and parts of plants in general
8, 158, 165, 535, 576, 837, 1342, 1423, 1800, 1801, 1849, 1858, 1906, 1931, 2047, 2089, 2090, 2091, 2107, 2132, 2134, 2554, 11562, 16271, 16309, 16540, 16960, 18473, 18800, 19048, 19067, 19579, 19599, 19601, 19616, 19678, 19961, 19963, 19969, 19976, 19978, 19995, 20050, 20055, 20080, 20086, 20093, 20094, 20105, 20111, 20116, 20122, 20123, 20130, 20135, 20161, 20162, 20185, 20224, 20519, 20525, 20532, 20533, 20534, 20535, 20536, 20538, 20540, 20541, 20542, 20562, 20564, 20566, 20570, 20608, 20616, 20630, 20631, 20632, 20692, 20708, 20718, 20733, 20878, 20906, 20908, 20910, 20914, 20915, 20917, 20922, 20927, 20929, 20934, 20935, 20940, 20942, 20943, 20946, 20958, 20961, 20979, 21000, 21003, 21007, 21021, 21022, 21023, 21024, 21025, 21026, 21027, 21028, 21029, 21031, 21032, 21046,

21048

Plant production general and crop husbandry (D2100)
10, 1415, 2596, 2651, 2652, 2653, 2654, 2655, 2656, 2657, 2658, 2659, 2660, 2661, 2662, 2663, 2664, 2665, 2678, 2876

Plant nutrition and fertilization (D2200)
1087, 1422, 2659, 5064, 5089, 5090, 5091, 5092, 5093, 5094, 5095, 5096, 5097, 5098, 5099, 5100, 5101, 5102, 5103, 5104, 5105, 5106, 5107, 5108, 5109, 5110, 5111, 5112, 5113, 5114, 5115, 5231, 5951, 6032

Plant breeding (D2300)
6021, 6022, 6023, 6024, 6025, 6026, 6027, 6028, 6029, 6030, 6031, 6032, 6620, 7534

Plant protection (D2400)
104, 7772, 7773, 7774, 7775

Pests of plants and pest control (D2410)
8029, 8030, 8031

Plant diseases and disease control (D2420)
6021, 8997, 8998, 8999, 9000, 9001, 9002, 9003, 9004, 9005, 9006, 9007

Weeds and weed control (D2430)
10059, 10074, 10075, 10076, 10077, 10078, 10079, 10080

Miscellaneous plant disorders (D2490)
10318, 10331, 10332, 10333, 10334, 10335, 10336, 10337

B2200 Plant communities as ecological systems
822, 984, 1012, 1149, 1244, 1504, 1785, 1829, 1830, 1846, 1847, 1892, 1952, 1953, 2006, 2009, 2044, 2045, 2049, 2073, 2088, 2116, 2190, 20077, 20538, 20914, 20915

Plant production general and crop husbandry (D2100)
2611, 2666, 2667, 4788

Plant nutrition and fertilization (D2200)
5116, 5117, 5118, 5119, 5577, 5950

Plant breeding (D2300)
6736

Plant protection (D2400)
7745, 7746, 7776

Pests of plants and pest control (D2410)
8032, 8070

Weeds and weed control (D2430)
10081, 10082, 10083, 10084, 10114, 10185, 10239, 10250

Miscellaneous plant disorders (D2490)
10338

B2300 Animals and parts of their bodies
1873, 2223, 7757, 7988, 9388, 19867, 20039, 20048, 20137, 20139, 20172, 20179, 20180, 20582, 20602, 20622, 20661, 20735, 20736, 20741, 20742, 20744, 20747, 20748, 20749, 20783, 20814, 20815, 20816, 20817, 20818, 20928, 20974, 20990, 21050, 21051, 21052, 21053, 21077, 21078, 21080, 21081, 21082, 21083, 21084, 21085, 21086

Animal management general and animal husbandry (D3100)
10423, 10424, 10425, 10426, 10427, 10428, 10429, 10430, 10431, 10432, 10433, 10434, 10435, 10436, 10437, 10438, 11628, 11632, 11633

Animal nutrition (D3200)
11624, 11625, 11626, 11627, 11628, 11629, 11630, 11631, 11632, 11633, 11634, 11635, 11636, 11946, 13523

Animal breeding (D3300)
12882, 13528

Animal diseases, veterinary medicine (D3400)
11631, 13478, 13479, 13522, 13523, 13524, 13525, 13526, 13527, 13528, 13529, 13530, 14069

B2400 Animal communities as ecological systems
1154, 1828, 1842, 1994, 2063, 2108, 2291, 2302, 2335, 2336, 2337, 2338, 2339, 2340, 2379, 7962, 7963, 8256, 8293, 19556, 20825

Animal nutrition (D3200)

1827
Animal diseases, veterinary medicine (D3400)
13531, 13532
B2500 Animal and plant communities as ecological systems
543, 1786, 1922, 1993, 1995, 1996, 1997, 2146, 2381, 2390, 19662, 20737
Plant production general and crop husbandry (D2100)
3514
Plant breeding (D2300)
6008
Pests of plants and pest control (D2410)
8033
Animal management general and animal husbandry (D3100)
10542
B2600 Animal – plant interrelationships
1785, 1841, 2091, 20739
Plant production general and crop husbandry (D2100)
2598
Pests of plants and pest control (D2410)
2291, 8034, 8240, 8241, 8311, 8794
B2900 Other subjects related to plants and animals in general
19559
Plant protection (D2400)
7777
Plant diseases and disease control (D2420)
8997
B3000 Crops in general
15, 31, 64, 65, 67, 68, 100, 242, 301, 302, 309, 332, 336, 340, 344, 350, 353, 368, 380, 395, 439, 467, 480, 488, 496, 515, 567, 594, 610, 636, 705, 738, 751, 757, 802, 808, 826, 855, 868, 871, 873, 883, 884, 886, 920, 924, 934, 937, 940, 941, 944, 945, 948, 949, 951, 952, 954, 956, 959, 962, 963, 968, 969, 971, 972, 974, 975, 977, 978, 998, 1061, 1080, 1088, 1121, 1197, 1206, 1285, 1314, 1315, 1325, 1326, 1373, 1374, 1378, 1379, 1380, 1381, 1393, 1397, 1412, 1425, 1427, 1435, 1436, 1437, 1470, 1492, 1495, 1501, 1513, 1548, 1550, 1586, 1590, 1610, 1627, 1629, 1633, 1641, 1643, 1650, 1752, 1771, 1772, 1774, 1775, 1776, 1780, 1792, 1793, 1795, 1797, 1808, 1858, 1994...
Plant production general and crop husbandry (D2100)
10, 159, 174, 189, 203, 449, 492, 532, 878, 887, 897, 1064, 1382, 1424, 1426, 1432, 1536, 1542, 1740, 1741, 1750, 2030, 2393, 2597, 2601, 2602, 2604, 2668, 2669, 2670, 2671, 2672, 2673, 2674, 2675, 2676, 2677, 2678, 2679, 2680, 2681, 2682, 2683, 2684, 2685, 2686, 2687, 2688, 2689, 2690, 2691, 2692, 2693, 2694, 2695, 2696, 2697, 2698, 2699, 2700, 2701, 2702, 2703, 2704, 2705, 2706, 2707, 2708, 2709, 2710, 2711, 2712, 2713, 2714, 2715, 2716, 2717, 2718, 2719, 2720, 2721, 2722, 2723, 2724, 2725, 2726, 2727, 2728, 2729, 2730, 2731, 2732, 2733, 2734, 2735, 2736, 2737, 2738, 2739, 2740, 2741, 2742, 2743, 2744, 2745, 2746, 2747, 2748, 2749, 2750, 2751, 2752, 2753, 2754, 2755, 2756, 2757, 2758, 2759, 2760, 2761, 2762, 2763, 2764, 2765...
Plant nutrition and fertilization (D2200)
17, 121, 122, 123, 153, 176, 183, 224, 237, 311, 319, 333, 337, 338, 342, 444, 448, 450, 468, 472, 502, 506, 523, 524, 528, 539, 605, 609, 706, 716, 753, 778, 801, 1339, 1420, 1433, 1540, 2694, 2727, 2737, 2742, 2745, 2796, 2806, 2837, 2860, 2871, 2873, 2877, 2881, 5058, 5062, 5065, 5067, 5072, 5077, 5115, 5120, 5121, 5122, 5123, 5124, 5125, 5126, 5127, 5128, 5129, 5130, 5131, 5132, 5133, 5134, 5135, 5136, 5137, 5138, 5139, 5140, 5141, 5142, 5143, 5144, 5145, 5146, 5147, 5148, 5149, 5150, 5151, 5152, 5153, 5154, 5155, 5156, 5157, 5158, 5159, 5160, 5161, 5162, 5163, 5164, 5165, 5166, 5167, 5168, 5169, 5170, 5171, 5172, 5173, 5174, 5175, 5176, 5177,

5178, 5179, 5180, 5181, 5182, 5183, 5184, 5185, 5186, 5187...
Plant breeding (D2300)
2697, 2786, 2814, 5311, 6004, 6022, 6033, 6034, 6035, 6036, 6037, 6038, 6039, 6040, 6041, 6042, 6043, 6044, 6045, 6046, 6047, 6048, 6049, 6050, 6051, 6052, 6053, 6054, 6055, 6056, 6057, 6058, 6059, 6060, 6061, 6062, 6063, 6064, 6065, 6066, 6067, 6068, 6069, 6070, 6071, 6072, 6073, 6074, 6075, 6076, 6077, 6078, 6079, 6080, 6081, 6082, 6083, 6084, 6085, 6086, 6087, 6088, 6089, 6090, 6091, 6092, 6160, 8143, 9128, 9133, 20011
Plant protection (D2400)
108, 196, 397, 398, 1069, 2694, 2840, 5163, 7747, 7748, 7749, 7750, 7755, 7762, 7772, 7778, 7779, 7780, 7781, 7782, 7783, 7784, 7785, 7786, 7787, 7788, 7789, 7790, 7791, 7792, 7793, 7794, 7795, 7796, 7797, 7798, 7799, 7800, 7801, 7802, 7803, 7804, 7805, 7806, 7807, 7808, 7809, 7810, 7811, 7812, 7813, 7814, 7815, 7816, 7817, 7818, 7819, 7820, 7821, 7822, 7823, 7824, 7825, 7826, 7827, 7828, 9093
Pests of plants and pest control (D2410)
319, 794, 1040, 2243, 2291, 2791, 5174, 5319, 7958, 7961, 7973, 7988, 8035, 8036, 8037, 8038, 8039, 8040, 8041, 8042, 8043, 8044, 8045, 8046, 8047, 8048, 8049, 8050, 8051, 8052, 8053, 8054, 8055, 8056, 8057, 8058, 8059, 8060, 8061, 8062, 8063, 8064, 8065, 8066, 8067, 8068, 8069, 8070, 8071, 8072, 8073, 8074, 8075, 8076, 8077, 8078, 8079, 8080, 8081, 8082, 8083, 8084, 8085, 8086, 8087, 8088, 8089, 8090, 8091, 8092, 8093, 8094, 8095, 8096, 8097, 8098, 8099, 8100, 8101, 8102, 8103, 8104, 8105, 8106, 8107, 8108, 8109, 8110, 8111, 8112, 8113, 8114, 8115, 8116, 8117, 8118, 8119, 8120, 8121, 8122, 8123, 8124, 8125, 8126, 8127, 8128, 8129, 8130, 8131, 8132, 8133, 8134, 8135, 8136, 8137, 8138, 8139, 8140, 8141, 8142, 8143, 8144, 8145, 8146, 8147...
Plant diseases and disease control (D2420)
547, 892, 1067, 2798, 6033, 8038, 8054, 8055, 8086, 8098, 8106, 8107, 8108, 8261, 8284, 8287, 8378, 8382, 8391, 8951, 8973, 8975, 8976, 8977, 8978, 8987, 9008, 9009, 9010, 9011, 9012, 9013, 9014, 9015, 9016, 9017, 9018, 9019, 9020, 9021, 9022, 9023, 9024, 9025, 9026, 9027, 9028, 9029, 9030, 9031, 9032, 9033, 9034, 9035, 9036, 9037, 9038, 9039, 9040, 9041, 9042, 9043, 9044, 9045, 9046, 9047, 9048, 9049, 9050, 9051, 9052, 9053, 9054, 9055, 9056, 9057, 9058, 9059, 9060, 9061, 9062, 9063, 9064, 9065, 9066, 9067, 9068, 9069, 9070, 9071, 9072, 9073, 9074, 9075, 9076, 9077, 9078, 9079, 9080, 9081, 9082, 9083, 9084, 9085, 9086, 9087, 9088, 9089, 9090, 9091, 9092, 9093, 9094, 9095, 9096, 9097, 9098, 9099, 9100, 9101, 9102, 9103, 9104, 9105, 9106...
Weeds and weed control (D2430)
188, 322, 546, 627, 799, 904, 1067, 1543, 2761, 2762, 2763, 2877, 2881, 5319, 8043, 8378, 8404, 9060, 9199, 9220, 10058, 10065, 10085, 10086, 10087, 10088, 10089, 10090, 10091, 10092, 10093, 10094, 10095, 10096, 10097, 10098, 10099, 10100, 10101, 10102, 10103, 10104, 10105, 10106, 10107, 10108, 10109, 10110, 10111, 10112, 10113, 10114, 10115, 10116, 10117, 10118, 10119, 10120, 10121, 10122, 10123, 10124, 10125, 10126, 10127, 10128, 10129, 10130, 10131, 10132, 10133, 10134, 10135, 10136, 10137, 10138, 10139, 10140, 10141, 10142, 10143, 10144, 10145, 10146, 10147, 10148, 10149, 10150, 10151, 10152, 10153, 10154, 10155, 10156, 10157, 10158, 10159, 10160, 10161, 10162, 10163, 10164, 10165, 10166, 10167, 10168, 10169, 10170, 10171, 10172, 10173, 10174, 10175, 10176, 10177, 10178, 10179, 10180, 13550, 19814, 20009
Miscellaneous plant disorders (D2490)
5239, 10322, 10325, 10326, 10339, 10340, 10341, 10342, 10343, 10344, 10345, 10346, 10347, 10348, 10349, 10350,

10351, 10352, 10353, 10354, 10355, 10356, 10357, 13870

B3100 Cereals in general

76, 229, 341, 658, 749, 766, 838, 1377, 1408, 1753, 1767, 11621, 11708, 12156, 15144, 15190, 15199, 15203, 15204, 15205, 15206, 15207, 15222, 15254, 15334, 15363, 15364, 15401, 15402, 15479, 15702, 15743, 15765, 15772, 15791, 15859, 15907, 16032, 16034, 16035, 16036, 16037, 16038, 16039, 16040, 16043, 16044, 16049, 16059, 16060, 16289, 16316, 16317, 16318, 16439, 16476, 16497, 16498, 16499, 16502, 16506, 16508, 16509, 16512, 16514, 16517, 16518, 16519, 16522, 16523, 16524, 16525, 16526, 16527, 16536, 16539, 16543, 16544, 16545, 16550, 16551, 16552, 16594, 16600, 16604, 16607, 16616, 17307, 17394, 17440, 17441, 17447, 17753, 17773, 17799, 17847, 18045, 18104, 18111, 18199, 18982, 18983, 18984, 18986, 18987, 18988, 18989, 18990, 18991, 18992, 18993, 18996, 18997, 18999, 19000, 19001, 19002, 19003, 19004, 19006, 19010, 19012, 19015, 19016, 19018...

Plant production general and crop husbandry (D2100)

491, 493, 1439, 2600, 2823, 2896, 2897, 2898, 2899, 2900, 2901, 2902, 2903, 2904, 2905, 2906, 2907, 2908, 2909, 2910, 2911, 2912, 2913, 2914, 2915, 2916, 2917, 2918, 2919, 2920, 2921, 2922, 2923, 2924, 2925, 2926, 2927, 2928, 2929, 2930, 2931, 2932, 2933, 2934, 2935, 2936, 2937, 2938, 2939, 2940, 2941, 2942, 2943, 2944, 2945, 2946, 2947, 2948, 2949, 2950, 2951, 2952, 2953, 2954, 2955, 2956, 2957, 2958, 2959, 2960, 2961, 2962, 2963, 2964, 2965, 2966, 2967, 2968, 2969, 2970, 2971, 2972, 2973, 2974, 2975, 2976, 2977, 2978, 2979, 2980, 2981, 2982, 2983, 2984, 2985, 2986, 2987, 2988, 2989, 2990, 2991, 2992, 2993, 2994, 2995, 2996, 2997, 2998, 2999, 3000, 3001, 3002, 3003, 3004, 3005, 3006, 3007, 3282, 3390, 3926, 4362, 5386, 5545, 6100, 6110...

Plant nutrition and fertilization (D2200)

486, 2896, 2898, 2913, 5061, 5371, 5372, 5373, 5374, 5375, 5376, 5377, 5378, 5379, 5380, 5381, 5382, 5383, 5384, 5385, 5386, 5387, 5388, 5389, 5390, 5391, 5392, 5393, 5394, 5395, 5396, 5397, 5398, 5399, 5488, 5545, 6661, 9302

Plant breeding (D2300)

2897, 2913, 2935, 2957, 3001, 3002, 3003, 6041, 6093, 6094, 6095, 6096, 6097, 6098, 6099, 6100, 6101, 6102, 6103, 6104, 6105, 6106, 6107, 6108, 6109, 6110, 6111, 6112, 6113, 6114, 6115, 6116, 6117, 6118, 6119, 6120, 6121, 6122, 6123, 6124, 6125, 6126, 6127, 6128, 6129, 6130, 6131, 6132, 6133, 6134, 6135, 6136, 6137, 6138, 6139, 6140, 6141, 6142, 6143, 6144, 6145, 6146, 6147, 6148, 6149, 6150, 6151, 6152, 6153, 6154, 6155, 6156, 6570, 6661, 8454, 9313, 20069

Plant protection (D2400)

6144, 7744, 7752, 7754, 7770, 7771, 7829, 7830, 7831, 7832, 7833, 7834, 7835, 7836, 7837, 7838, 7839, 7840, 7841

Pests of plants and pest control (D2410)

7959, 7964, 7976, 8432, 8433, 8434, 8435, 8436, 8437, 8438, 8439, 8440, 8441, 8442, 8443, 8444, 8445, 8446, 8447, 8448, 8449, 8450, 8451, 8452, 8453, 8454, 8455, 8456, 8457, 8458, 8459, 8460, 8461, 8462, 8594, 9305, 9325, 9528, 15855

Plant diseases and disease control (D2420)

2898, 2996, 3007, 5394, 5396, 5398, 6099, 8440, 8452, 8956, 8958, 8961, 8965, 8969, 8970, 9266, 9267, 9268, 9269, 9270, 9271, 9272, 9273, 9274, 9275, 9276, 9277, 9278, 9279, 9280, 9281, 9282, 9283, 9284, 9285, 9286, 9287, 9288, 9289, 9290, 9291, 9292, 9293, 9294, 9295, 9296, 9297, 9298, 9299, 9300, 9301, 9302, 9303, 9304, 9305, 9306, 9307, 9308, 9309, 9310, 9311, 9312, 9313, 9314, 9315, 9316, 9317, 9318, 9319, 9320, 9321, 9322, 9323, 9324, 9325, 9326, 9327, 9328, 9329, 9330, 9331, 9332, 9333, 9334, 9335, 9336, 9337, 9338, 9339, 9340, 9341, 9445, 9528, 13519

Weeds and weed control (D2430)

1146, 8440, 9267, 10062, 10181, 10182, 10183, 10184, 10185, 10186, 10187, 10188, 10189, 10190, 10191, 10192, 10193, 10194, 10195, 10196, 10197

Miscellaneous plant disorders (D2490)

10358, 10359, 10360

B3110 Barley

11578, 11580, 11592, 12173, 12184, 12435, 12608, 12780, 12802, 15849, 15857, 15917, 16057, 16276, 16966, 16968, 16974, 16975, 16976, 16978, 16979, 16980, 17037, 17038, 17707, 17713, 17743, 17744, 17908, 19427, 19428, 19429, 19462, 19463, 19517

Plant production general and crop husbandry (D2100)

3008, 3009, 3010, 3011, 3012, 3013, 3014, 3015, 3016, 3017, 3018, 3019, 3020, 3021, 3022, 3023, 3024, 3025, 3026, 3027, 3028, 3029, 3030, 3031, 3032, 3033, 3034, 3035, 3056, 3079, 3096, 3666, 6159, 6428, 17446

Plant nutrition and fertilization (D2200)

5070, 5226, 5400, 5401, 5402, 5403, 5404, 5405, 5406, 5407, 5408, 5409, 5410, 5411, 5412, 5413, 5414, 5415, 5416, 5437, 5454

Plant breeding (D2300)

3008, 3019, 3020, 3034, 6010, 6094, 6157, 6158, 6159, 6160, 6161, 6162, 6163, 6164, 6165, 6166, 6167, 6168, 6169, 6170, 6171, 6172, 6173, 6174, 6175, 6176, 6177, 6178, 6179, 6180, 6181, 6182, 6183, 6184, 6185, 6186, 6187, 6188, 6189, 6190, 6191, 6192, 6193, 6194, 6195, 6196, 6197, 6198, 6199, 6200, 6201, 6202, 6203, 6204, 6205, 6206, 6207, 6208, 6209, 6210, 6211, 6212, 6213, 6214, 6215, 6216, 6217, 6218, 6219, 6220, 6221, 6222, 6223, 6224, 6225, 6323, 6339, 6395, 6409, 6428, 9353, 10198

Plant protection (D2400)

7842

Pests of plants and pest control (D2410)

8463, 15856

Plant diseases and disease control (D2420)

3008, 5414, 6169, 8950, 8985, 9342, 9343, 9344, 9345, 9346, 9347, 9348, 9349, 9350, 9351, 9352, 9353, 9354, 9355, 9356, 9357, 9358, 9359, 9360, 9361, 9362, 9363, 9364, 9365, 9366, 9367, 9368, 9369, 9370, 9371, 9408

Weeds and weed control (D2430)

3033, 10198

B3120 Maize

267, 485, 1405, 1505, 10401, 10765, 11549, 11576, 11950, 11990, 12191, 12381, 12802, 15367, 15814, 15822, 15921, 16033, 16070, 16272, 16329, 16491, 16520, 16596, 16602, 20702, 20812, 20891, 20930

Plant production general and crop husbandry (D2100)

494, 522, 1434, 3022, 3036, 3037, 3038, 3039, 3040, 3041, 3042, 3043, 3044, 3045, 3046, 3047, 3048, 3049, 3050, 3051, 3052, 3053, 3054, 3055, 3056, 3057, 3058, 3059, 3060, 3061, 3062, 3177, 6269, 16611

Plant nutrition and fertilization (D2200)

522, 682, 788, 1434, 3061, 5417, 5418, 5419, 5420, 5421, 5422, 5423, 5424, 5425, 5434, 5448, 5454, 6246, 7844

Plant breeding (D2300)

3052, 6226, 6227, 6228, 6229, 6230, 6231, 6232, 6233, 6234, 6235, 6236, 6237, 6238, 6239, 6240, 6241, 6242, 6243, 6244, 6245, 6246, 6247, 6248, 6249, 6250, 6251, 6252, 6253, 6254, 6255, 6256, 6257, 6258, 6259, 6260, 6261, 6262, 6263, 6264, 6265, 6266, 6267, 6268, 6269, 6270, 6271, 6347, 6348

Plant protection (D2400)

7844

Pests of plants and pest control (D2410)

8464, 8465, 8466, 8467, 8468, 8469, 8470, 8480, 9420

Plant diseases and disease control (D2420)

6236, 9372, 9373, 9374, 9375, 9376, 9377, 9378, 9379, 9380,

9381, 9382, 9383, 9412, 9414, 9416, 9419, 9420, 9579
Weeds and weed control (D2430)
6246, 9414, 9416, 9419, 9420, 10199, 10209, 10364
Miscellaneous plant disorders (D2490)
10361, 10364
B3130 Oats
11591, 12780, 16511
Plant production general and crop husbandry (D2100)
3063, 3064, 3065, 3066, 3067, 3068, 3069, 3070, 3079, 6273
Plant nutrition and fertilization (D2200)
5426, 5427
Plant breeding (D2300)
3067, 6094, 6209, 6272, 6273, 6274, 6275, 6276, 6277, 6278,
6279, 6280, 6281, 6282, 6283, 6284, 6285, 6286
Pests of plants and pest control (D2410)
7965, 8471, 8472
Plant diseases and disease control (D2420)
9345, 9384
B3140 Rice
16042, 16058, 16293, 16513, 17442, 17473, 17762, 18372,
20937
Plant production general and crop husbandry (D2100)
3071, 3072, 3073, 3074, 3075, 3076, 3077, 3078, 5432, 6305
Plant nutrition and fertilization (D2200)
3075, 5428, 5429, 5430, 5431, 5432, 5433, 5434, 5435
Plant breeding (D2300)
5435, 6287, 6288, 6289, 6290, 6291, 6292, 6293, 6294, 6295,
6296, 6297, 6298, 6299, 6300, 6301, 6302, 6303, 6304, 6305,
6306, 6307, 6308, 10202, 10203
Pests of plants and pest control (D2410)
8473, 8474
Plant diseases and disease control (D2420)
8474
Weeds and weed control (D2430)
10200, 10201, 10202, 10203
B3150 Rye
15466, 16505, 16507, 16528, 16546, 17892, 18460, 18461,
18462, 18994, 18998, 19965, 20252, 20618
Plant production general and crop husbandry (D2100)
3079, 3080, 3081, 3093
Plant breeding (D2300)
6094, 6309, 6310, 6311, 6312, 6313, 6314, 6315, 6316, 6321,
6330, 6338
B3160 Sorghum
1472, 18463
Plant production general and crop husbandry (D2100)
1434, 3057, 3082, 3083, 3084, 3085
Plant nutrition and fertilization (D2200)
1434
Plant breeding (D2300)
3083, 6270, 6317, 6318, 6319, 6320
Pests of plants and pest control (D2410)
8480, 9420
Plant diseases and disease control (D2420)
9381, 9414, 9419, 9420, 9579
Weeds and weed control (D2430)
3083, 9414, 9419, 9420
B3170 Wheat
499, 656, 12802, 15466, 16033, 16048, 16061, 16276, 16492,
16494, 16495, 16496, 16501, 16503, 16504, 16510, 16521,
16528, 16531, 16533, 16537, 16541, 16542, 16546, 16548,
16549, 16553, 16615, 16617, 18460, 18461, 18462, 18829,
18979, 18980, 18981, 18994, 18995, 19008, 19009, 19011,
19013, 19017, 19019, 19022, 19965, 20252, 20702, 20857,
20894, 20928
Plant production general and crop husbandry (D2100)

494, 526, 3015, 3022, 3086, 3087, 3088, 3089, 3090, 3091,
3092, 3093, 3094, 3095, 3096, 3097, 3098, 3099, 3100, 3101,
3102, 3103, 3104, 3105, 3106, 3107, 3108, 3109, 3110, 3111,
3112, 3113, 3114, 3115, 3116, 3117, 3118, 3119, 3120, 3121,
3122, 3123, 3124, 3125, 3126, 3127, 3128, 3129, 3130, 3131,
3177, 3873, 5449, 6342, 6391, 6428, 6429, 9411, 10204
Plant nutrition and fertilization (D2200)
788, 3090, 3129, 5436, 5437, 5438, 5439, 5440, 5441, 5442,
5443, 5444, 5445, 5446, 5447, 5448, 5449, 5450, 5451, 5452,
5453, 5454, 5455, 7844, 9411
Plant breeding (D2300)
3103, 3104, 6094, 6247, 6248, 6321, 6322, 6323, 6324, 6325,
6326, 6327, 6328, 6329, 6330, 6331, 6332, 6333, 6334, 6335,
6336, 6337, 6338, 6339, 6340, 6341, 6342, 6343, 6344, 6345,
6346, 6347, 6348, 6349, 6350, 6351, 6352, 6353, 6354, 6355,
6356, 6357, 6358, 6359, 6360, 6361, 6362, 6363, 6364, 6365,
6366, 6367, 6368, 6369, 6370, 6371, 6372, 6373, 6374, 6375,
6376, 6377, 6378, 6379, 6380, 6381, 6382, 6383, 6384, 6385,
6386, 6387, 6388, 6389, 6390, 6391, 6392, 6393, 6394, 6395,
6396, 6397, 6398, 6399, 6400, 6401, 6402, 6403, 6404, 6405,
6406, 6407, 6408, 6409, 6410, 6411, 6412, 6413, 6414, 6415,
6416, 6417, 6418, 6419, 6420, 6421, 6422, 6423, 6424, 6425,
6426, 6427, 6428, 6429, 6430, 6431, 6432, 6433, 6434, 6435,
6436, 6437, 6438, 6439, 6440...
Plant protection (D2400)
7843, 7844, 7845
Pests of plants and pest control (D2410)
2115, 8475, 8476, 8477, 8478, 8479, 8480, 8481, 9420
Plant diseases and disease control (D2420)
3127, 3129, 6340, 6341, 8950, 9345, 9348, 9385, 9386, 9387,
9388, 9389, 9390, 9391, 9392, 9393, 9394, 9395, 9396, 9397,
9398, 9399, 9400, 9401, 9402, 9403, 9404, 9405, 9406, 9407,
9408, 9409, 9410, 9411, 9412, 9413, 9414, 9415, 9416, 9417,
9418, 9419, 9420, 9421, 9422, 9423, 9424, 9425, 9426, 9427
Weeds and weed control (D2430)
9414, 9416, 9418, 9419, 9420, 10204, 10205, 10206, 10207,
10208, 10209
Miscellaneous plant disorders (D2490)
10362
B3190 Other cereals
11538, 18463
Plant production general and crop husbandry (D2100)
3083, 10210
Plant breeding (D2300)
3083, 6446, 6482
Weeds and weed control (D2430)
3083, 10210
B3200 Fibre plants and oil crops in general
16065, 16365, 16436, 16572, 16578, 16964, 17090, 17846,
17906, 19025, 20463, 20465, 20610
Plant production general and crop husbandry (D2100)
2999, 3132, 3133, 3134, 3135, 3324
Plant nutrition and fertilization (D2200)
3135
Plant protection (D2400)
7871
Plant diseases and disease control (D2420)
3135, 9341
Weeds and weed control (D2430)
10197
B3210 Flax
Plant production general and crop husbandry (D2100)
3136, 3137, 3138
Plant breeding (D2300)
3136, 3137, 6323, 6447, 6448, 6449, 6450, 6451, 6452, 6453
B3220 Olive

1480, 1481, 2152, 15377, 15793, 15794, 15795, 15818,
16064, 16075, 16118, 16143, 16144, 16145, 16564, 16565,
16566, 16567, 16568, 16569, 16570, 16571, 16573, 16574,
16575, 16576, 16577, 16657, 16658, 16659, 16661, 17590,
19051, 19052, 19053, 19054, 19105, 20369, 20897

Plant production general and crop husbandry (D2100)
3139, 3140, 3141, 3142, 3143, 3144, 3145, 3146, 3147, 3148,
3149, 3150, 3151, 3152, 3153, 3154, 3155, 3156, 3157, 3158,
3159, 3160, 4105, 4106, 4108, 4109, 4113, 4114, 4115, 4116,
4264, 4266, 4273, 4474, 4839, 6455, 6459, 6460, 6461, 6467,
6468, 6470, 15267

Plant nutrition and fertilization (D2200)
5456, 5457, 5458

Plant breeding (D2300)
3156, 6454, 6455, 6456, 6457, 6458, 6459, 6460, 6461, 6462,
6463, 6464, 6465, 6466, 6467, 6468, 6469, 6470

Pests of plants and pest control (D2410)
8482, 8483, 8484, 8485, 8486, 8487, 8488, 8489, 8490, 8491,
8492, 8493, 8494, 8495, 8496, 8497, 8498, 8499, 8500, 8501,
8502, 8503, 8504, 8690, 8777, 8778

Plant diseases and disease control (D2420)
9428, 9429, 9430

B3230 Rape
12020, 12658, 15845, 15892, 19487, 20217, 20689

Plant production general and crop husbandry (D2100)
494, 3006, 3161, 3162, 3163, 3164, 3165, 3166, 3167, 3168,
3169, 3170

Plant nutrition and fertilization (D2200)
3164, 5459, 5460

Plant breeding (D2300)
6471, 6472, 6473, 6474, 6475, 6476, 6477, 6478, 6479, 6480,
6481, 6482, 6509

Pests of plants and pest control (D2410)
7978, 7981, 8505, 8506, 8507, 8508, 8509, 9432, 9433

Plant diseases and disease control (D2420)
9431, 9432, 9433, 9434, 9435

Weeds and weed control (D2430)
9433, 10211, 10212

B3240 Soyabean
1023, 1026, 10401, 12657, 14377, 16553, 16580, 18133,
19093, 19416, 19500, 20200, 20798

Plant production general and crop husbandry (D2100)
3037, 3171, 3172, 3173, 3174, 3175, 3176, 3177, 3178, 3179,
3180, 3633, 3664, 3900, 3901, 5463, 6484, 6485

Plant nutrition and fertilization (D2200)
3633, 3901, 5461, 5462, 5463, 5464, 5465, 5466

Plant breeding (D2300)
6481, 6483, 6484, 6485, 6486, 6487, 6488, 6489

Weeds and weed control (D2430)
10209

B3250 Sunflower
1469

Plant production general and crop husbandry (D2100)
3061, 3181, 3182, 3183, 5468

Plant nutrition and fertilization (D2200)
3061, 5417, 5418, 5467, 5468, 5469

Plant breeding (D2300)
6262, 6306, 6490, 6491, 6492, 6493, 6494, 6495, 6496, 6497,
6498, 6499, 6500, 6501, 6502, 6503, 6504, 6505, 6506, 6507,
6508, 6509, 7146

Plant diseases and disease control (D2420)
9372, 9436, 9437

Weeds and weed control (D2430)
10213

B3290 Other fibre plants and oil crops
15396, 16560, 16563, 16579, 17935, 19046, 19416, 19988,

19989

Plant production general and crop husbandry (D2100)
3184, 3185, 3186, 3187, 3188, 3189, 3190, 5014

Plant nutrition and fertilization (D2200)
5470, 5471

Plant breeding (D2300)
6151, 6481, 6482, 6510, 6511, 6512, 6513, 6514, 6515

Pests of plants and pest control (D2410)
8510, 8511

B3300 Sugarbeets and starch producing plants in general
11543, 11768, 15728, 20389

Plant production general and crop husbandry (D2100)
2924, 3191, 3192, 3193

Pests of plants and pest control (D2410)
8512

Plant diseases and disease control (D2420)
9438, 9439

Weeds and weed control (D2430)
10197, 10209

B3310 Potatoes
341, 957, 960, 1090, 1377, 1632, 1766, 11959, 15208, 15209,
15210, 15211, 15212, 15213, 15214, 15215, 15216, 15217,
15219, 15249, 15389, 15394, 15773, 15774, 15775, 15800,
15836, 15848, 15858, 15940, 15942, 16071, 16076, 16077,
16078, 16079, 16080, 16081, 16082, 16083, 16084, 16111,
16154, 16592, 16595, 16598, 16599, 16600, 16604, 16605,
16606, 16607, 16608, 16610, 16612, 16618, 16619, 16620,
16621, 16622, 16623, 16624, 16625, 16626, 16627, 16628,
16629, 17050, 17113, 17124, 17125, 17191, 17308, 17453,
17455, 17465, 17799, 17848, 17900, 17901, 18809, 18990,
18993, 19062, 19064, 19065, 19070, 19071, 19084, 19124,
20041, 20163, 20280, 20282, 20480, 20483, 20719, 20720,
20729, 20858, 20860

Plant production general and crop husbandry (D2100)
2898, 3194, 3195, 3196, 3197, 3198, 3199, 3200, 3201, 3202,
3203, 3204, 3205, 3206, 3207, 3208, 3209, 3210, 3211, 3212,
3213, 3214, 3215, 3216, 3217, 3218, 3219, 3220, 3221, 3222,
3223, 3224, 3225, 3226, 3227, 3228, 3229, 3230, 3231, 3232,
3233, 3234, 3235, 3236, 3237, 3238, 3239, 3240, 3241, 3242,
3243, 3244, 3245, 3246, 3247, 3248, 3249, 3250, 3251, 3252,
3303, 3708, 3873, 3902, 15390

Plant nutrition and fertilization (D2200)
552, 976, 2898, 3240, 5439, 5454, 5472, 5473, 5474, 5475,
5476, 5477, 5478, 5479, 5480, 5481, 5482, 5733

Plant breeding (D2300)
3233, 6151, 6516, 6517, 6518, 6519, 6520, 6521, 6522, 6523,
6524, 6525, 6526, 6527, 6528, 6529, 6530, 6531, 6532, 6533,
6534, 6535, 6536, 6537, 6538, 6539, 6540, 6541, 6542, 6543,
6544, 6545, 6546, 6547, 6548, 6549, 6550, 6551, 6552, 6553,
6554, 6555, 6556, 6557, 6558, 6559, 6560, 6561, 6562, 6563,
6564, 6565, 6566, 6567, 6568, 6569, 6570, 6571, 6572, 6573,
6574, 6575, 6576, 6577, 7191, 8523, 8524, 9470, 9491, 9492

Plant protection (D2400)
3233, 7846, 7847, 7848, 7849, 7850, 7851, 7852, 7853, 7854

Pests of plants and pest control (D2410)
7979, 8446, 8513, 8514, 8515, 8516, 8517, 8518, 8519, 8520,
8521, 8522, 8523, 8524, 8525, 8526, 8527, 8528, 8529, 8530,
8531, 8532, 8533, 8534, 8535, 8536, 8537, 8538, 8539, 8540,
8541, 8542, 8543, 8544, 8545, 8546, 8547, 8548, 8549, 8550,
8551, 8552, 8553, 9500

Plant diseases and disease control (D2420)
2898, 8520, 8527, 8533, 8536, 8963, 8971, 8972, 8986, 9440,
9441, 9442, 9443, 9444, 9445, 9446, 9447, 9448, 9449, 9450,
9451, 9452, 9453, 9454, 9455, 9456, 9457, 9458, 9459, 9460,
9461, 9462, 9463, 9464, 9465, 9466, 9467, 9468, 9469, 9470,
9471, 9472, 9473, 9474, 9475, 9476, 9477, 9478, 9479, 9480,

9481, 9482, 9483, 9484, 9485, 9486, 9487, 9488, 9489, 9490, 9491, 9492, 9493, 9494, 9495, 9496, 9497, 9498, 9499, 9500, 9501, 9502, 9503, 9504, 9505, 9506, 9507, 9508, 9509, 9510, 9511, 9512, 9513, 9514, 9515, 9516, 9517, 9518, 9519, 9520, 9521, 9522, 9523, 9524, 9525, 9671, 9719, 9943, 16073, 20281

Weeds and weed control (D2430)
10061, 10214, 10215, 10216, 10275
Miscellaneous plant disorders (D2490)
10323, 10363

B3320 Sugarbeets and other sugar crops
961, 1452, 11670, 11698, 12023, 12208, 12349, 12874, 15134, 15339, 15362, 15375, 15392, 15397, 15797, 15828, 16066, 16068, 16074, 16581, 16589, 16590, 16591, 17444, 17457, 17467, 17586, 19697, 19966, 20343, 20365, 20515, 20516, 20517, 20518, 20801, 20830, 20896

Plant production general and crop husbandry (D2100)
684, 2600, 2997, 3087, 3246, 3252, 3253, 3254, 3255, 3256, 3257, 3258, 3259, 3260, 3261, 3262, 3263, 3264, 3265, 3266, 3267, 3268, 3269, 3270, 3271, 3272, 3273, 3274, 3275, 3276, 3277, 3278, 3279, 3280, 3281, 3282, 3283, 3284, 3285, 3286, 3287, 3288, 3289, 3290, 3291, 3292, 3293, 3294, 3295, 3296, 3297, 3298, 6595

Plant nutrition and fertilization (D2200)
1478, 3253, 3254, 3264, 3283, 5424, 5448, 5483, 5484, 5485, 5486, 5487, 5488, 5489, 5490, 5491, 5492, 5493, 5494, 5495, 5496, 5497, 5498, 5499, 5500, 5501, 5502, 5503, 5504, 5505, 5506, 5507, 10223

Plant breeding (D2300)
3254, 3270, 3279, 3285, 3295, 5494, 6219, 6570, 6578, 6579, 6580, 6581, 6582, 6583, 6584, 6585, 6586, 6587, 6588, 6589, 6590, 6591, 6592, 6593, 6594, 6595, 6596, 6597, 6598, 6599, 6600, 6601, 6602, 6603, 6604, 6605, 6606, 6607, 6608, 6609, 6610, 6611, 6612, 6613, 6614, 6615, 6616, 6617, 6618, 9531

Plant protection (D2400)
6585, 7855, 7856, 7857, 7858, 7859, 7860, 7861

Pests of plants and pest control (D2410)
5502, 7969, 8554, 8555, 8556, 8557, 8558, 8559, 8560, 8561, 8562, 8563, 8564, 8565, 8566, 8567, 8568, 8569, 8570, 8571, 8572, 8573, 8574, 8575, 8576, 8577, 8578, 8579, 8580, 8630, 9528

Plant diseases and disease control (D2420)
3263, 3286, 7855, 8555, 8560, 8567, 8570, 8573, 8957, 9526, 9527, 9528, 9529, 9530, 9531, 9532, 9533, 9534, 9535, 9536, 9537, 9567, 9840

Weeds and weed control (D2430)
7855, 8570, 10063, 10186, 10217, 10218, 10219, 10220, 10221, 10222, 10223, 10224, 10225, 10364

Miscellaneous plant disorders (D2490)
10364

B3390 Other starch producing plants
15404, 16114, 16553, 16595, 16610, 17124, 18045, 18370, 19064, 20280, 20719, 20720

Plant production general and crop husbandry (D2100)
3202, 3205, 3206, 3299, 3300, 3301, 3302, 3303, 3304, 3305
Plant nutrition and fertilization (D2200)
5474, 5508
Plant breeding (D2300)
6522, 6524, 6619, 6620, 6621, 17123
Plant diseases and disease control (D2420)
9441, 9444, 9446, 9448, 9449, 9450, 9451, 9456, 9508, 9538, 9539, 9540, 9541, 9542, 9718, 16073, 20281
Miscellaneous plant disorders (D2490)
10365

B3400 Grasses and forage crops in general
1406, 1421, 1473, 10561, 10720, 10764, 11545, 11546,

11563, 11571, 11581, 11585, 11659, 11661, 11678, 11724, 11762, 11771, 11774, 11802, 11811, 11821, 11865, 11868, 11933, 11934, 11985, 12016, 12150, 12199, 12216, 12219, 12227, 12244, 12283, 12294, 12315, 12393, 15380, 15385, 15475, 15707, 15723, 15728, 15750, 15765, 15767, 15770, 15781, 15782, 15783, 15784, 15821, 15840, 15842, 15847, 15853, 15866, 15899, 15903, 15918, 15925, 15928, 16268, 16269, 16277, 16280, 16289, 16292, 16317, 16322, 16337, 16338, 16339, 16341, 17201, 17482, 17501, 17504, 17547, 17588, 18130, 19312, 19551, 19940, 19941, 20159, 20317, 20931

Plant production general and crop husbandry (D2100)
1451, 1476, 1514, 2924, 3101, 3132, 3306, 3307, 3308, 3309, 3310, 3311, 3312, 3313, 3314, 3315, 3316, 3317, 3318, 3319, 3320, 3321, 3322, 3323, 3324, 3325, 3326, 3327, 3328, 3329, 3330, 3331, 3332, 3333, 3334, 3335, 3336, 3337, 3338, 3339, 3340, 3341, 3342, 3343, 3344, 3345, 3346, 3347, 3348, 3349, 3350, 3351, 3352, 3353, 3354, 3355, 3356, 3357, 3358, 3359, 3360, 3361, 3362, 3363, 3364, 3365, 3366, 3367, 3368, 3369, 3370, 3371, 3372, 3373, 3374, 3375, 5513, 10860, 10877, 11745, 12312, 13790, 15371, 15374, 15382, 15860, 15862

Plant nutrition and fertilization (D2200)
3309, 3331, 3344, 3364, 5509, 5510, 5511, 5512, 5513, 5514, 5515, 5516, 5517, 5518, 5519, 5520, 5521, 5522, 5523, 5524, 5525, 5526, 11973

Plant breeding (D2300)
3327, 3344, 3364, 6291, 6622, 6623, 6624, 6625, 6626, 6627, 6628, 6629, 6630, 6631, 6632, 6633, 6634, 6635, 6636, 6637, 6638

Plant protection (D2400)
7862

Pests of plants and pest control (D2410)
8581, 8582

Plant diseases and disease control (D2420)
9543

Weeds and weed control (D2430)
10197, 10226, 10227, 10228, 10229

B3410 Grasses
46, 1377, 1729, 10416, 10645, 11544, 11560, 11561, 11565, 11582, 11583, 11584, 11589, 11590, 11593, 11595, 11596, 11597, 11598, 11599, 11600, 11609, 11613, 11649, 11769, 11916, 11923, 12009, 12014, 12084, 12087, 12112, 12177, 12306, 12320, 12988, 15147, 15151, 15163, 15164, 15202, 15252, 15263, 15486, 15786, 15839, 15843, 15898, 15923, 16059, 16320, 16326, 19916, 20459, 20759, 20844, 20861, 20870

Plant production general and crop husbandry (D2100)
3134, 3376, 3377, 3378, 3379, 3380, 3381, 3382, 3383, 3384, 3385, 3386, 3387, 3388, 3389, 3390, 3391, 3392, 3393, 3394, 3395, 3396, 3397, 3398, 3399, 3400, 3401, 3402, 3403, 3404, 3404, 3405, 3406, 3407, 3408, 3409, 3410, 3411, 3412, 3413, 3414, 3415, 3416, 3417, 3418, 3419, 3420, 3421, 3422, 3423, 3424, 3425, 3426, 3427, 3428, 3429, 3430, 3431, 3432, 3433, 3434, 3435, 3436, 3437, 3438, 3439, 3440, 3441, 3442, 3443, 3444, 3445, 3446, 3447, 3448, 3449, 3450, 3451, 3452, 3453, 3454, 3455, 3456, 3457, 3458, 3459, 3460, 3461, 3462, 3463, 3464, 3465, 3466, 3467, 3468, 3469, 3470, 3471, 3472, 3473, 3474, 3475, 3476, 3477, 3478, 3479, 3480, 3481, 3482, 3483, 3484, 3485, 3486, 3487, 3488, 3489, 3490, 3491, 3492, 3493, 3494, 3495, 3496, 3497, 3498...

Plant nutrition and fertilization (D2200)
3489, 3490, 3530, 3542, 5061, 5374, 5379, 5527, 5528, 5529, 5530, 5531, 5532, 5533, 5534, 5535, 5536, 5537, 5538, 5539, 5540, 5541, 5542, 5543, 5544, 5545, 5546, 5547, 5548, 5549, 5550, 5551, 5552, 5553, 5554, 5555, 5556, 5557, 5558, 5559, 5560, 5561, 5562, 5563, 5571, 5578, 5613, 5787, 5789, 6661

B – SUBJECT AREAS

Plant breeding (D2300)
3376, 3428, 3429, 3430, 3431, 3436, 3448, 3492, 3493, 5531,
5538, 5539, 5553, 6151, 6639, 6640, 6641, 6642, 6643, 6644,
6645, 6646, 6647, 6648, 6649, 6650, 6651, 6652, 6653, 6654,
6655, 6656, 6657, 6658, 6659, 6660, 6661, 6662, 6663, 6664,
6665, 6666, 6667, 6668, 6669, 6670, 6671, 6672, 6673, 6674,
6675, 6676, 6677, 6678, 6679, 6680, 6681, 6682, 6683, 6684,
6685, 6686, 6687, 6688, 6689, 6690, 6691, 6692, 6693, 6694,
6695, 6696, 6697, 6698, 6699, 6700, 6701, 6702, 6703, 6704,
6705, 6706, 6707, 6708, 6709, 6710, 6711, 6712, 6713, 6714,
6715, 6716, 6717, 6718, 6719, 6720, 6721, 6722, 6723, 6724,
6725, 6726, 6727, 6728, 6729, 6730, 6783, 6796, 6808, 6818,
6819, 7626, 9555, 20149

Plant protection (D2400)
7863, 7864, 7865, 7866

Pests of plants and pest control (D2410)
7964, 8290, 8583, 8584, 8585, 8586, 8587, 8588, 8589, 8593

Plant diseases and disease control (D2420)
3452, 8965, 8969, 8970, 8983, 9280, 9286, 9337, 9339, 9445,
9544, 9545, 9546, 9547, 9548, 9549, 9550, 9551, 9552, 9553,
9554, 9555, 9556, 9557, 9558, 9559, 9564

Weeds and weed control (D2430)
3509, 5533, 10230, 10231, 10232, 10233, 10234

B3420 Pastures, grassland
254, 354, 369, 370, 381, 1032, 1051, 1147, 1160, 1161,
1164, 1170, 1205, 1557, 1563, 1575, 1576, 1612, 1613, 1623,
1628, 1640, 1660, 1685, 1737, 1773, 1777, 1785, 2010, 2103,
2108, 2134, 2181, 2182, 2220, 2232, 2233, 2234, 2241, 2297,
2329, 2392, 10402, 10659, 10709, 10786, 10847, 10862,
10891, 10907, 10924, 11550, 11618, 11619, 11620, 11709,
11723, 11748, 11772, 11825, 11844, 11871, 11910, 11937,
11961, 11962, 11999, 12003, 12045, 12047, 12057, 12086,
12096, 12158, 12159, 12180, 12181, 12206, 12212, 12215,
12231, 12232, 12239, 12241, 12248, 12278, 12303, 12318,
12350, 12365, 12369, 13936, 14103, 14355, 14398, 14402,
15133, 15265, 15403, 15422, 15463, 15479, 15480, 15785,
15788, 15852, 15854, 15863, 15898, 15924, 15927, 16313,
16321, 16324, 16340, 17104, 17206, 17208, 17496, 17508,
17520, 17528, 17544, 17550, 17551...

Plant production general and crop husbandry (D2100)
558, 1124, 1439, 2179, 3307, 3308, 3506, 3507, 3508, 3509,
3510, 3511, 3512, 3513, 3514, 3515, 3516, 3517, 3518, 3519,
3520, 3521, 3522, 3523, 3524, 3525, 3526, 3527, 3528, 3529,
3530, 3531, 3532, 3533, 3534, 3535, 3536, 3537, 3538, 3539,
3540, 3541, 3542, 3543, 3544, 3545, 3546, 3547, 3548, 3549,
3550, 3551, 3552, 3553, 3554, 3555, 3556, 3557, 3558, 3559,
3560, 3561, 3562, 3563, 3564, 3565, 3566, 3567, 3568, 3569,
3570, 3571, 3572, 3573, 3574, 3575, 3576, 3577, 3578, 3579,
3580, 3581, 3582, 3583, 3584, 3585, 3586, 3587, 3588, 3589,
3590, 3591, 3592, 3593, 3594, 3595, 3596, 3597, 3598, 3599,
3600, 3601, 3602, 3603, 3604, 3605, 3606, 3607, 3608, 3609,
3610, 3611, 3612, 3613, 3614, 3615, 3616, 3617, 3618, 3619,
3620, 3621, 3622, 3623, 3624...

Plant nutrition and fertilization (D2200)
11, 1574, 3506, 3508, 3513, 3530, 3537, 3542, 3543, 3544,
3590, 3592, 3601, 5060, 5074, 5080, 5536, 5564, 5565, 5566,
5567, 5568, 5569, 5570, 5571, 5572, 5573, 5574, 5575, 5576,
5577, 5578, 5579, 5580, 5581, 5582, 5583, 5584, 5585, 5586,
5587, 5588, 5589, 5590, 5591, 5592, 5593, 5594, 5595, 5596,
5597, 5598, 5599, 5600, 5601, 5602, 5603, 5604, 5605, 5606,
5607, 5608, 5609, 5610, 5611, 5612, 5613, 5614, 5615, 5616,
5617, 5618, 5619, 5620, 5621, 5622, 5623, 5624, 5625, 5626,
5627, 5628, 5629, 5630, 5631, 5632, 5633, 5634, 5635, 5636,
5637, 5638, 5639, 5640, 5641, 5642, 5643, 5644, 10247,
12002

Plant breeding (D2300)

3603, 3608, 6660, 6723, 6731, 6732, 6733, 6734, 6735, 6736

Plant protection (D2400)
7867

Pests of plants and pest control (D2410)
8590, 8591, 8592, 8593, 8594, 8595, 8596, 8597, 8598,
19555

Plant diseases and disease control (D2420)
9560, 9561, 9562, 9563, 9564

Weeds and weed control (D2430)
3509, 3547, 3620, 10235, 10236, 10237, 10238, 10239,
10240, 10241, 10242, 10243, 10244, 10245, 10246, 10247,
10248, 10249

B3430 Mangolds
12349, 12352, 12353, 12551, 12874

Plant production general and crop husbandry (D2100)
3295, 3627, 3628, 3629, 3818

Plant nutrition and fertilization (D2200)
5535, 5645, 5646

Plant breeding (D2300)
3295, 6606, 6608, 6611, 6612, 6613, 6614, 6737, 6738, 6739,
6740, 6741

Pests of plants and pest control (D2410)
7978, 8568, 8569

Plant diseases and disease control (D2420)
9565, 9566, 9567

Weeds and weed control (D2430)
10250

B3440 Legumes in general
1026, 11840, 12521, 12571, 12867, 17090

Plant production general and crop husbandry (D2100)
1439, 3487, 3630, 3631, 3632, 3633, 3634, 3635, 3636, 3637,
3638, 3899, 3900, 3901, 3911, 3920, 3926, 3931, 3933

Plant nutrition and fertilization (D2200)
3631, 3633, 3901, 5527, 5528, 5613, 5647, 5648, 5649, 5650,
5651, 5652, 5653, 5654, 5655, 5656, 5746, 5748, 5750

Plant breeding (D2300)
6639, 6640, 6645, 6740, 6742, 6743, 6744, 6745, 6746, 6747,
6748, 6817

Plant protection (D2400)
7744, 7868

Pests of plants and pest control (D2410)
7964

Plant diseases and disease control (D2420)
9286, 9336, 9546, 9568, 9569, 9578

B3441 Grassland legumes
1474, 1505, 1753, 10772, 11761, 12009, 12084, 12131,
12182, 15843, 16312, 16332, 20600, 20892

Plant production general and crop husbandry (D2100)
1475, 3288, 3420, 3486, 3609, 3639, 3640, 3641, 3642, 3643,
3644, 3645, 3646, 3647, 3648, 3649, 3650, 3651, 3652, 3653,
3654, 6773

Plant nutrition and fertilization (D2200)
3639, 3641, 5558, 5657, 5658, 5659, 5660, 5661, 5662, 5663,
5664, 5665, 5666, 5687

Plant breeding (D2300)
3654, 6096, 6749, 6750, 6751, 6752, 6753, 6754, 6755, 6756,
6757, 6758, 6759, 6760, 6761, 6762, 6763, 6764, 6765, 6766,
6767, 6768, 6769, 6770, 6771, 6772, 6773, 6774, 6775, 6776,
6777, 6778, 6779, 6780, 6781, 6782, 6783, 6784, 6818, 7869,
9573

Plant protection (D2400)
7869, 7870

Pests of plants and pest control (D2410)
8599, 8600, 8601, 8602, 8603, 8604, 8605

Plant diseases and disease control (D2420)
9086, 9564, 9570, 9571, 9572, 9573, 9574, 9575

Weeds and weed control (D2430)
10251

B3449 Other legumes
1408, 11710, 12185, 12201, 12210, 12584, 12597, 12648, 12781, 19414

Plant production general and crop husbandry (D2100)
3179, 3391, 3627, 3655, 3656, 3657, 3658, 3659, 3660, 3661, 3662, 3663, 3664, 3897, 4233

Plant nutrition and fertilization (D2200)
5435, 5667, 5668, 5742, 6794

Plant breeding (D2300)
5435, 6785, 6786, 6787, 6788, 6789, 6790, 6791, 6792, 6793, 6794, 6795, 6796, 6797, 6798, 6799, 6800, 6801, 6802, 7082, 7089

Plant diseases and disease control (D2420)
9576, 9577

Weeds and weed control (D2430)
3659

B3450 Cereals used for forage
229, 328, 1405, 1408, 1474, 1502, 11576, 11677, 11703, 11705, 11750, 11751, 11820, 11929, 11931, 11952, 12084, 12087, 12201, 12203, 12204, 12207, 12208, 12418, 12555, 12607, 12608, 12803, 15743, 15745, 15779, 15841, 15891, 15893, 15900, 15907, 15919, 15920, 15922, 15923, 16043, 16044, 16272, 16294, 16296, 16316, 16326, 16329, 16340, 17206, 17532, 17545, 17550, 19841

Plant production general and crop husbandry (D2100)
493, 1439, 1475, 3045, 3055, 3090, 3112, 3390, 3397, 3542, 3627, 3665, 3666, 3667, 3668, 3669, 3670, 3671, 3672, 3673, 3674, 3675, 3926, 4233, 5545, 6110, 6812, 6814, 6815

Plant nutrition and fertilization (D2200)
3090, 3542, 5420, 5422, 5542, 5543, 5544, 5545, 5669, 5670, 5671, 5672, 5673, 5674, 5675, 5676, 5677, 5678, 6661, 6814

Plant breeding (D2300)
6110, 6148, 6149, 6206, 6211, 6267, 6351, 6352, 6353, 6661, 6740, 6748, 6803, 6804, 6805, 6806, 6807, 6808, 6809, 6810, 6811, 6812, 6813, 6814, 6815, 6816, 6817, 6818, 6819, 6820, 6821

Pests of plants and pest control (D2410)
8594, 8606, 8607

Plant diseases and disease control (D2420)
9336, 9376, 9383, 9578, 9579

Weeds and weed control (D2430)
10199, 10252

B3460 Turnips
11842, 12349, 12354, 12868

Plant production general and crop husbandry (D2100)
3166, 3676, 3677

Plant breeding (D2300)
6578, 6740, 6822, 6823, 6824, 6825, 6826, 6827, 6828, 6829, 6830, 6831, 6832, 6833

Plant protection (D2400)
7871

Pests of plants and pest control (D2410)
8505

Plant diseases and disease control (D2420)
9580, 9581, 9641

Weeds and weed control (D2430)
10253, 10254

B3490 Other forage crops
11644, 12007, 12189, 15921, 16954

Plant production general and crop husbandry (D2100)
3639, 3678, 3679, 3680, 3681, 3682, 3683, 3684, 3685, 3686, 3687, 3688

Plant nutrition and fertilization (D2200)
3639, 3687, 5679

Plant breeding (D2300)
6740, 6826, 6827, 6828, 6829, 6830, 6831, 6832, 6833, 6834, 6835, 6836, 6837, 6838, 6839, 6840, 6841, 6842, 6843, 6844, 6845, 6846, 6847, 6848, 6849, 6850, 6851, 6852, 6853, 6854, 6855

Plant protection (D2400)
7872

Pests of plants and pest control (D2410)
8484

Plant diseases and disease control (D2420)
9582, 9641

B3500 Vegetables in general
194, 357, 435, 1216, 1217, 1527, 1538, 1570, 1571, 1573, 1592, 1593, 1616, 1617, 1618, 1638, 1642, 2454, 2464, 2536, 15131, 15157, 15298, 15304, 15356, 15358, 15395, 15578, 15595, 15596, 15601, 15607, 15628, 15699, 15719, 15798, 15838, 15867, 15944, 16087, 16088, 16089, 16091, 16095, 16107, 16111, 16128, 16134, 16135, 16140, 16141, 16142, 16152, 16156, 16159, 16160, 16162, 16164, 16167, 16168, 16170, 16171, 16173, 16176, 16273, 16630, 16650, 16664, 16665, 16668, 16671, 16673, 16674, 16675, 16677, 17100, 17103, 17126, 17129, 17130, 17131, 17132, 17133, 17134, 17135, 17136, 17137, 17167, 17228, 17231, 17245, 17246, 17302, 17565, 17566, 17567, 17568, 17578, 17579, 17583, 17597, 17609, 17611, 17612, 17616, 17618, 17620, 17626, 17630, 17710, 17737, 17741, 17742, 17781, 17877, 17879, 17883, 17885, 17893, 18029, 18042, 18155, 18271, 18444, 18477...

Plant production general and crop husbandry (D2100)
932, 1477, 2600, 2603, 3208, 3288, 3373, 3689, 3690, 3691, 3692, 3693, 3694, 3695, 3696, 3697, 3698, 3699, 3700, 3701, 3702, 3703, 3704, 3705, 3706, 3707, 3708, 3709, 3710, 3711, 3712, 3713, 3714, 3715, 3716, 3717, 3718, 3719, 3720, 3721, 3722, 3723, 3724, 3725, 3726, 3727, 3728, 3729, 3730, 3731, 3732, 3733, 3734, 3735, 3736, 3737, 3738, 3739, 3740, 3741, 3742, 3743, 3744, 3745, 3746, 3747, 3748, 3749, 3750, 3751, 3752, 3753, 3754, 3755, 3756, 3757, 3758, 3759, 3760, 3761, 3762, 3763, 3764, 3765, 3766, 3767, 3768, 3769, 3770, 3771, 3772, 3773, 3774, 3775, 3776, 3777, 3778, 3779, 3780, 3781, 3782, 3783, 3784, 3785, 3786, 3787, 3788, 3789, 4043, 4091, 4362, 4363, 4518, 6866, 15350, 15352, 17564

Plant nutrition and fertilization (D2200)
1, 3702, 3706, 3775, 3778, 5439, 5528, 5680, 5681, 5682, 5683, 5684, 5685, 5686, 5687, 5688, 5689, 5690, 5691, 5692, 5693, 5694, 5695, 5696, 5697, 5698, 5699, 5700, 5701, 5702, 5703, 5704, 5705, 5706, 5707, 5708, 5709, 5710, 5711, 5712, 5713, 5714, 5715, 5716, 5717, 5718, 5719, 5720, 5721, 5722, 5723, 5724, 5725, 5761, 5911, 16090

Plant breeding (D2300)
3752, 3785, 6005, 6856, 6857, 6858, 6859, 6860, 6861, 6862, 6863, 6864, 6865, 6866, 6867, 6868, 6869, 6870, 6871, 6872, 6873, 6874, 6875, 6876, 6877, 6878, 7169, 7495

Plant protection (D2400)
5692, 7753, 7768, 7873, 7874, 7875, 7876, 7877, 7878, 7879, 7880, 7881, 7882, 7883, 7884, 7885, 7886, 19130

Pests of plants and pest control (D2410)
7957, 7971, 7972, 8501, 8608, 8609, 8610, 8611, 8612, 8613, 8614, 8615, 8616, 8617, 8618, 8619, 8620, 8621, 8622, 8623, 8624, 8625, 8626, 8627, 9597, 9601

Plant diseases and disease control (D2420)
8613, 8954, 8960, 8962, 8974, 9583, 9584, 9585, 9586, 9587, 9588, 9589, 9590, 9591, 9592, 9593, 9594, 9595, 9596, 9597, 9598, 9599, 9600, 9601, 9719

Weeds and weed control (D2430)
10060, 10186, 10255, 10256, 10257, 10258, 10259, 10260, 10261, 10262, 10263, 10264, 10265, 10266, 10267, 10268,

10269, 10270

Miscellaneous plant disorders (D2490)
10366, 10367, 10368

B3510 Root, tuber and bulb vegetables
355, 15391, 15844, 15905, 16021, 16104, 16106, 16112, 16114, 16125, 16148, 16153, 16161, 16175, 17121, 18817, 18953, 19086, 19120, 20217, 20563, 20854, 20855

Plant production general and crop husbandry (D2100)
2823, 3166, 3202, 3300, 3790, 3791, 3792, 3793, 3794, 3795, 3796, 3797, 3798, 3799, 3800, 3801, 3802, 3803, 3804, 3805, 3806, 3807, 3808, 3809, 3810, 3811, 3812, 3813, 3814, 3815, 3816, 3817, 3818, 3819, 3820, 3821, 3822, 3835, 3838, 3839, 3878, 3966, 3977

Plant nutrition and fertilization (D2200)
3790, 5535, 5726, 5739

Plant breeding (D2300)
3792, 6829, 6879, 6880, 6881, 6882, 6883, 6884, 6885, 6886, 6887, 6888, 6889, 6890, 6891, 6892, 6893, 6894, 6895, 6896, 6897, 6898, 6899, 6900, 6901, 6902, 6903, 6904, 6905, 6906, 6907, 6908, 6909, 6910, 6911, 6912, 6913, 6914, 6915, 6916, 6917, 6918, 6919, 6920, 6921, 6922, 6923, 6924, 6925, 6926, 6927, 6928, 6929, 6930, 6931, 6932, 6946, 7191, 7302

Plant protection (D2400)
3790, 7887, 7888, 7889

Pests of plants and pest control (D2410)
7967, 8571, 8572, 8628, 8629, 8630, 8631, 8632, 8633, 8634, 8635, 8636, 8637, 8638, 8639, 8655

Plant diseases and disease control (D2420)
9438, 9439, 9591, 9602, 9603, 9604, 9605, 9606, 9607, 9608, 9609, 9610, 9611, 9612, 9613, 9614, 9615, 9616, 9617, 9618, 9641, 9699

Weeds and weed control (D2430)
10216, 10271, 10272, 10273, 10274, 10275, 10276, 10288

Miscellaneous plant disorders (D2490)
10369

B3520 Greens and leafy vegetables
365, 955, 1192, 1537, 15132, 15482, 15485, 15797, 15823, 15824, 15844, 16100, 16103, 16105, 16112, 16126, 16178, 16179, 16636, 16638, 16667, 16669, 17217, 17229, 17247, 17293, 17875, 17880, 19089, 19124, 19294, 20485, 20486

Plant production general and crop husbandry (D2100)
3790, 3822, 3823, 3824, 3825, 3826, 3827, 3828, 3829, 3830, 3831, 3832, 3833, 3834, 3835, 3836, 3837, 3838, 3839, 3840, 3841, 3842, 3843, 3844, 3845, 3846, 3847, 3848, 3849, 3850, 3851, 3852, 3853, 3854, 3855, 3856, 3857, 3858, 3859, 3860, 3861, 3862, 3863, 3864, 3865, 3866, 3867, 3868, 3869, 3870, 3871, 3872, 3873, 3874, 3875, 3876, 3877, 3878, 3879, 3880, 3881, 3882, 3883, 3884, 3963, 3972, 3977, 7140, 17233

Plant nutrition and fertilization (D2200)
1549, 3790, 3847, 3848, 5427, 5535, 5727, 5728, 5729, 5730, 5731, 5732, 5733, 5734, 5735, 5736

Plant breeding (D2300)
6306, 6417, 6826, 6827, 6829, 6887, 6933, 6934, 6935, 6936, 6937, 6938, 6939, 6940, 6941, 6942, 6943, 6944, 6945, 6946, 6947, 6948, 6949, 6950, 6951, 6952, 6953, 6954, 6955, 6956, 6957, 6958, 6959, 6960, 6961, 6962, 6963, 6964, 6965, 6966, 6967, 6968, 6969, 6970, 6971, 6972, 6973, 6974, 6975, 6976, 6977, 6978, 6979, 6980, 6981, 6982, 6983, 6984, 6985, 6986, 6987, 6988, 6989, 6990, 6991, 6992, 6993, 6994, 6995, 6996, 6997, 6998, 6999, 7000, 7001, 7002, 7003, 7004, 7005, 7006, 7007, 7008, 7009, 7010, 7011, 7012, 7013, 7014, 7015, 7016, 7017, 7018, 7019, 7020, 7021, 7022, 7023, 7024, 7025, 7026, 7027, 7028, 7029, 7030, 7031, 7032, 7033, 7034, 7140, 7191, 7302, 9633

Plant protection (D2400)
3790, 7746, 7871, 7887, 7890, 7891, 7892, 7893

Pests of plants and pest control (D2410)
8484, 8571, 8572, 8630, 8640, 8641, 8642, 8643, 8644, 8645, 8646, 8647, 8648, 8649, 8650, 8651, 8652, 8653, 8654, 9644

Plant diseases and disease control (D2420)
9445, 9603, 9608, 9619, 9620, 9621, 9622, 9623, 9624, 9625, 9626, 9627, 9628, 9629, 9630, 9631, 9632, 9633, 9634, 9635, 9636, 9637, 9638, 9639, 9640, 9641, 9642, 9643, 9644, 9645, 9646, 9647, 9648, 9649, 9889, 9926

Weeds and weed control (D2430)
10275, 10277, 10278, 10285, 10288

B3530 Vegetable fruits in general
Plant diseases and disease control (D2420)
9926

B3531 Leguminous vegetables
1026, 11594, 11602, 11607, 15375, 15393, 15822, 16059, 16060, 16113, 16114, 16126, 16637, 16662, 16663, 16954, 16959, 16961, 17090, 17854, 19023, 19086, 19087, 19102, 19414, 20353, 20700, 20856, 20881

Plant production general and crop husbandry (D2100)
2922, 3285, 3633, 3658, 3790, 3811, 3845, 3873, 3885, 3886, 3887, 3888, 3889, 3890, 3891, 3892, 3893, 3894, 3895, 3896, 3897, 3898, 3899, 3900, 3901, 3902, 3903, 3904, 3905, 3906, 3907, 3908, 3909, 3910, 3911, 3912, 3913, 3914, 3915, 3916, 3917, 3918, 3919, 3920, 3921, 3922, 3923, 3924, 3925, 3926, 3927, 3928, 3929, 3930, 3931, 3932, 3933, 3934, 3935, 3936, 3937, 3964, 3973, 3977, 7053, 13512, 13513

Plant nutrition and fertilization (D2200)
187, 516, 1549, 3633, 3790, 3898, 3901, 3903, 3907, 5435, 5653, 5654, 5655, 5656, 5733, 5737, 5738, 5739, 5740, 5741, 5742, 5743, 5744, 5745, 5746, 5747, 5748, 5749, 5750, 5751, 5752, 5753

Plant breeding (D2300)
3285, 3894, 3904, 3923, 3927, 5435, 6096, 6570, 6741, 6745, 6791, 6798, 6817, 6887, 7035, 7036, 7037, 7038, 7039, 7040, 7041, 7042, 7043, 7044, 7045, 7046, 7047, 7048, 7049, 7050, 7051, 7052, 7053, 7054, 7055, 7056, 7057, 7058, 7059, 7060, 7061, 7062, 7063, 7064, 7065, 7066, 7067, 7068, 7069, 7070, 7071, 7072, 7073, 7074, 7075, 7076, 7077, 7078, 7079, 7080, 7081, 7082, 7083, 7084, 7085, 7086, 7087, 7088, 7089, 7090, 7091, 7092, 7093, 7094, 7095, 7096, 7097, 7098, 7099, 7100, 7101, 7102, 7103, 7104, 7105, 7106, 7107, 7108, 7109, 7110, 7191, 9652, 9661, 9666

Plant protection (D2400)
3790, 7894

Pests of plants and pest control (D2410)
8644, 8655, 8656, 8657, 8658, 8659, 8660, 8661, 8662

Plant diseases and disease control (D2420)
8952, 9336, 9445, 9568, 9629, 9650, 9651, 9652, 9653, 9654, 9655, 9656, 9657, 9658, 9659, 9660, 9661, 9662, 9663, 9664, 9665, 9666, 9667, 9668, 9669, 9670, 9699

Weeds and weed control (D2430)
5747, 10275

Miscellaneous plant disorders (D2490)
10370

B3532 Tomatoes
198, 482, 1505, 15378, 15381, 15762, 15763, 15837, 16129, 16177, 16652, 16660, 17219, 17227, 17243, 17592, 17592, 19097

Plant production general and crop husbandry (D2100)
3242, 3285, 3843, 3844, 3845, 3846, 3848, 3872, 3932, 3938, 3939, 3940, 3941, 3942, 3943, 3944, 3945, 3946, 3947, 3948, 3949, 3950, 3951, 3952, 3953, 3954, 3955, 3956, 3957, 3958, 3959, 3960, 3961, 3962, 3963, 3964, 3965, 3966, 3967, 3968, 3969, 3970, 3971, 3972, 3973, 3974, 3975, 5756, 7124, 7140

Plant nutrition and fertilization (D2200)
1549, 3848, 5424, 5754, 5755, 5756, 5757, 5758, 5759, 5760,

5761, 5762, 5763, 5764

Plant breeding (D2300)
3285, 6417, 6887, 7041, 7111, 7112, 7113, 7114, 7115, 7116, 7117, 7118, 7119, 7120, 7121, 7122, 7123, 7124, 7125, 7126, 7127, 7128, 7129, 7130, 7131, 7132, 7133, 7134, 7135, 7136, 7137, 7138, 7139, 7140, 7141, 7142, 7143, 7144, 7145, 7146, 7147, 7148, 7149, 7150, 7151, 7152, 7153, 7154, 7155, 7156, 7157, 7158, 7159, 7160, 7161, 7162, 7163, 7164, 7165, 7191

Plant protection (D2400)
7895, 7896

Plant diseases and disease control (D2420)
9591, 9671, 9672, 9673, 9674, 9675, 9676, 9677, 9678, 9679, 9680, 9681, 9682, 9683, 9684, 9685, 9686, 9687, 9688, 9689, 9690

Weeds and weed control (D2430)
10279

B3533 Cucumbers
17600, 18817

Plant production general and crop husbandry (D2100)
3844, 3941, 3959, 3976, 3977, 3978, 3979, 3980, 3981, 3982

Plant nutrition and fertilization (D2200)
5765, 5766

Plant breeding (D2300)
6887, 7166, 7167, 7168, 7169, 7170, 7171, 7172, 7173, 7174, 7175, 7176

Plant diseases and disease control (D2420)
9591, 9642, 9671, 9685, 9688, 9691, 9692, 9693, 9694, 9695, 9699, 9889

B3539 Other vegetable fruits
15383, 16126, 17221, 17230, 17240, 17244, 17603, 17627

Plant production general and crop husbandry (D2100)
3873, 3932, 3959, 3967, 3972, 3983, 3984, 3985, 3986, 3987, 3988, 3989, 3990, 3991, 3992, 3993, 3994, 3995, 3996, 3997, 15388

Plant nutrition and fertilization (D2200)
1549, 5454, 5733

Plant breeding (D2300)
7169, 7174, 7177, 7178, 7179, 7180, 7181, 7182, 7183, 7184, 7185, 7186, 7187, 7188, 7189, 7190, 7191, 7192, 7193, 7194, 7195, 7196, 7197, 7198, 7199, 16651

Pests of plants and pest control (D2410)
8663, 8664, 8665

Plant diseases and disease control (D2420)
9642, 9643, 9671, 9672, 9679, 9680, 9694, 9695, 9696, 9697, 9698, 9699, 9700, 9701, 9702, 9889

Weeds and weed control (D2430)
10275

B3540 Mushrooms and other edible fungi
15269, 15306, 15632, 16122, 16284, 16285, 16646, 16670, 17166, 17220, 17852, 18813, 19023, 20794

Plant production general and crop husbandry (D2100)
3790, 3998, 3999, 4000, 4001, 4002, 4003, 4004, 4005, 4006, 4007, 4008, 4009, 4010, 4011, 4012, 4013, 4014, 4015, 4016, 4017, 4018, 4019, 4020, 4021, 4022, 4023, 4024, 4025, 4026, 4027, 4028, 4029, 4030, 4031, 4032, 4033, 4034, 4035, 4036, 4037, 4477, 5769

Plant nutrition and fertilization (D2200)
3790, 4000, 4008, 4013, 5767, 5768, 5769, 5770, 5771, 5772, 5773, 5774, 5775, 5776, 5777

Plant breeding (D2300)
3998, 4011, 4012, 7200, 7201, 7202, 7203, 7204, 7205, 7206, 7207, 7208

Plant protection (D2400)
3790

Pests of plants and pest control (D2410)
8666, 8667, 8668, 8669, 8670, 8671, 8672, 8673

Plant diseases and disease control (D2420)
3998, 8673, 8984, 9703, 9704, 9705, 9706, 9707, 9708, 9709, 9710, 9711, 9712, 9713, 9714, 9715

Miscellaneous plant disorders (D2490)
10371

B3600 Fruits in general
194, 354, 1249, 1371, 1416, 1450, 1519, 1520, 1527, 1532, 1538, 1556, 1577, 1642, 2454, 2464, 15139, 15167, 15300, 15355, 15699, 15719, 15801, 15802, 16096, 16099, 16102, 16111, 16116, 16131, 16134, 16135, 16138, 16141, 16152, 16156, 16160, 16162, 16165, 16167, 16168, 16170, 16263, 16273, 16630, 16639, 16640, 16656, 16664, 16666, 16668, 16671, 16672, 16674, 16675, 16677, 16972, 17100, 17126, 17127, 17129, 17130, 17131, 17132, 17133, 17134, 17135, 17136, 17137, 17225, 17565, 17566, 17567, 17568, 17578, 17579, 17597, 17601, 17607, 17612, 17615, 17617, 17621, 17631, 17632, 17633, 17635, 17710, 17737, 17741, 17742, 17747, 17781, 17877, 17879, 17883, 17885, 17893, 18042, 18155, 18271, 18477, 18950, 19076, 19079, 19083, 19094, 19098, 19099, 19100, 19103, 19104, 19110, 19111, 19112, 19113, 19114, 19118, 19119, 19121, 19123, 19127, 19435, 19977, 20168...

Plant production general and crop husbandry (D2100)
932, 1482, 3156, 3373, 3705, 3707, 3747, 3757, 3758, 3759, 3760, 3761, 3763, 3765, 3766, 3769, 3770, 3780, 3781, 3782, 3785, 3786, 3788, 3789, 4038, 4039, 4040, 4041, 4042, 4043, 4044, 4045, 4046, 4047, 4048, 4049, 4050, 4051, 4052, 4053, 4054, 4055, 4056, 4057, 4058, 4059, 4060, 4061, 4062, 4063, 4064, 4065, 4066, 4067, 4068, 4069, 4070, 4071, 4072, 4073, 4074, 4075, 4076, 4077, 4078, 4079, 4080, 4081, 4082, 4083, 4084, 4085, 4086, 4087, 4088, 4089, 4090, 4091, 4092, 4093, 4094, 4095, 4096, 4097, 4098, 4099, 4100, 4101, 4102, 4103, 4104, 4105, 4106, 4107, 4108, 4109, 4110, 4111, 4112, 4113, 4114, 4115, 4116, 4117, 4118, 4119, 4120, 4839

Plant nutrition and fertilization (D2200)
1, 4060, 4066, 4100, 5680, 5708, 5710, 5713, 5778, 5779, 5780, 5781, 5782, 5783, 5784, 5785, 5786, 5787, 5788, 5789, 5790

Plant breeding (D2300)
3156, 3785, 6857, 6858, 6872, 6873, 6875, 6877, 7209, 7210, 7211, 7212, 7213, 7214, 7215, 7216, 7217, 7218, 7219, 7220, 7221, 7222, 7223, 7495, 7654

Plant protection (D2400)
5782, 7209, 7751, 7753, 7878, 7879, 7880, 7881, 7897, 7898, 7899, 7900, 7901, 7902, 7903, 7904, 7905, 7906, 7941, 19130

Pests of plants and pest control (D2410)
8611, 8613, 8614, 8620, 8622, 8674, 8675, 8676, 8677, 8678, 8679, 8680, 8681, 8682, 8683, 8684, 8685, 8686, 8687, 8688, 8689, 8690, 8691, 8692, 8693, 8694, 8695, 8696, 8697, 9597, 9722

Plant diseases and disease control (D2420)
8613, 8684, 8685, 8688, 8697, 8953, 8968, 9583, 9597, 9716, 9717, 9718, 9719, 9720, 9721, 9722, 9723, 9724, 9725, 9726, 9727, 9728, 9729, 9730, 9731, 9732, 9733, 9734, 9735, 9736

Weeds and weed control (D2430)
4066, 10060, 10267, 10268, 10280, 10281, 10282

Miscellaneous plant disorders (D2490)
10366, 10372, 10373

B3610 Top fruit in general
29, 2106, 15266, 15365, 15384, 15799, 16631, 16633, 19077, 19126, 19843, 19900, 20417, 20845

Plant production general and crop husbandry (D2100)
2595, 2600, 4121, 4122, 4123, 4124, 4125, 4126, 4127, 4128, 4129, 4130, 4131, 4132, 4133, 4134, 4135, 4136, 4137, 4138, 4139, 4140, 4141, 4142, 4143, 4144, 4145, 4146, 4147, 4148,

9814, 9815, 9816, 9817, 9818, 9819, 9820, 9821, 9822, 9823, 9824, 9825, 9826

Weeds and weed control (D2430)
10281, 10287, 10288, 10289, 10290

Miscellaneous plant disorders (D2490)
10380

B3630 Citrus fruit
1065, 1483, 1494, 1500, 1515, 1516, 1530, 15268, 15376, 16150, 16303, 16632, 17035, 17867, 18802, 19106, 19107, 19125, 19461, 20909

Plant production general and crop husbandry (D2100)
1484, 4336, 4337, 4338, 4339, 4340, 4341, 4342, 4343, 4344, 4345, 4346, 4347, 4348, 4349, 4350, 4351, 4352, 4353, 4354, 4355, 4356, 4357, 4358, 4359, 4360

Plant nutrition and fertilization (D2200)
231, 4346, 5835, 5836, 5837, 5838, 5839, 5840

Plant breeding (D2300)
7330, 7331, 7332, 7333, 7334, 7335, 7336, 7337, 7338, 7339, 7340, 7341, 7342, 7343, 7344, 7345, 7346, 7347, 7348, 7349, 7350, 7351, 7352, 7353, 7354, 7355, 7356, 7357

Plant protection (D2400)
7922, 7923

Pests of plants and pest control (D2410)
8766, 8767, 8768, 8769, 8770, 8771, 8772, 8773, 8774, 8775, 8776, 8777, 8778

Plant diseases and disease control (D2420)
4358, 9827, 9828, 9829, 9830, 9831, 9832, 9833, 9834, 9835, 9836, 9837

Weeds and weed control (D2430)
10291

Miscellaneous plant disorders (D2490)
10381

B3640 Tropical and sub–tropical fruits
16136, 16151

Plant production general and crop husbandry (D2100)
2613, 3773, 3846, 4352, 4361, 4362, 4363, 4364, 4365, 4366

Plant nutrition and fertilization (D2200)
5721, 5841

Plant breeding (D2300)
6870, 6878, 7251, 7358, 7359, 7360, 7361, 7362, 7363, 7364, 7365

Plant diseases and disease control (D2420)
9838, 9839

Weeds and weed control (D2430)
10216

B3650 Grapes
164, 192, 216, 572, 638, 726, 759, 865, 1281, 1282, 1284, 1344, 1361, 1362, 1506, 1507, 1745, 1955, 2407, 2440, 15180, 15181, 15182, 15295, 15326, 15370, 15379, 15386, 15387, 15504, 15818, 15819, 15820, 16262, 16287, 16475, 16631, 16961, 16964, 16971, 16982, 16983, 16984, 16985, 16986, 16987, 16988, 16991, 16992, 16993, 16994, 16996, 16997, 16998, 16999, 17000, 17001, 17002, 17003, 17004, 17005, 17007, 17008, 17009, 17010, 17011, 17012, 17013, 17014, 17015, 17016, 17022, 17023, 17024, 17025, 17027, 17028, 17029, 17030, 17031, 17032, 17034, 17518, 17570, 17571, 17572, 17573, 17574, 17575, 17576, 17580, 17582, 17584, 17585, 17588, 17748, 17754, 17759, 17764, 17830, 17833, 17834, 17838, 17925, 17992, 17996, 18082, 18283, 18816, 18818, 19108, 19109, 19424, 19425, 19431, 19434, 19436, 19440, 19442, 19443, 19444, 19445, 19446, 19447, 19448...

Plant production general and crop husbandry (D2100)
2441, 2606, 2637, 3148, 3392, 4264, 4367, 4368, 4369, 4370, 4371, 4372, 4373, 4374, 4375, 4376, 4377, 4378, 4379, 4380, 4381, 4382, 4383, 4384, 4385, 4386, 4387, 4388, 4389, 4390,

4391, 4392, 4393, 4394, 4395, 4396, 4397, 4398, 4399, 4400, 4401, 4402, 4403, 4404, 4405, 4406, 4407, 4408, 4409, 4410, 4411, 4412, 4413, 4414, 4415, 4416, 4417, 4418, 4419, 4420, 4421, 4422, 4423, 4424, 4425, 4426, 4427, 4428, 4429, 4430, 4431, 4432, 4433, 4434, 4435, 4436, 4437, 4438, 4439, 4440, 4441, 4442, 4443, 4444, 4445, 4446, 4447, 4448, 4449, 4450, 4451, 4452, 4453, 4454, 4455, 4456, 4457, 4458, 4459, 4460, 4461, 4462, 4463, 4464, 4465, 4466, 4467, 4468, 4469, 4470, 4471, 4472, 4473, 4474, 4475, 7388, 7392, 15817, 17006

Plant nutrition and fertilization (D2200)
286, 1479, 4390, 4418, 4454, 5842, 5843, 5844, 5845, 5846, 5847, 5848, 5849, 5850, 5851, 5852, 5853, 5854, 5855, 5856, 5857, 5858, 5859, 5860, 5861, 5862, 5863, 5864, 5865, 5866, 5867, 5868, 8785, 8789, 9869

Plant breeding (D2300)
4382, 4416, 4418, 4423, 4443, 4448, 4449, 5868, 7366, 7367, 7368, 7369, 7370, 7371, 7372, 7373, 7374, 7375, 7376, 7377, 7378, 7379, 7380, 7381, 7382, 7383, 7384, 7385, 7386, 7387, 7388, 7389, 7390, 7391, 7392, 7393, 7394, 7395, 7396, 7397, 7398, 7399, 7400, 7401, 7402, 7403, 7404, 7405, 7406, 7407, 7408, 7409, 7410, 7411, 7412, 7413, 7414, 7415, 7416, 7417, 7418, 7419, 7420, 7421, 7422, 7423, 7424, 7425, 7426, 7427, 7428, 7429, 7430, 7431, 7432, 7433, 7434, 7435, 7436, 7437, 7438, 7439, 7440, 7441, 7442, 7443, 7444, 7445, 7446, 7447, 7448, 7449, 7450, 7451, 7452, 7453, 7454, 7455, 7456, 7457, 7458, 7459

Plant protection (D2400)
4425, 7430, 7924, 7925, 7926, 7927, 7928

Pests of plants and pest control (D2410)
8779, 8780, 8781, 8782, 8783, 8784, 8785, 8786, 8787, 8788, 8789, 8790, 8791, 8792, 8793, 8794, 8795, 8796, 8797, 8798, 8799, 8800, 8801, 8802, 8803, 8804, 8805, 8806, 8807, 8808, 8809, 8810, 8811, 8812, 8813, 8814, 8815, 9883, 15348

Plant diseases and disease control (D2420)
7417, 8798, 8800, 8806, 8814, 8815, 9835, 9840, 9841, 9842, 9843, 9844, 9845, 9846, 9847, 9848, 9849, 9850, 9851, 9852, 9853, 9854, 9855, 9856, 9857, 9858, 9859, 9860, 9861, 9862, 9863, 9864, 9865, 9866, 9867, 9868, 9869, 9870, 9871, 9872, 9873, 9874, 9875, 9876, 9877, 9878, 9879, 9880, 9881, 9882, 9883, 9884, 15348

Weeds and weed control (D2430)
193, 294, 10292, 10293

Miscellaneous plant disorders (D2490)
10382

B3660 Edible nut fruits
16149, 17587, 17673, 18820, 19081, 19082, 19565

Plant production general and crop husbandry (D2100)
3150, 4197, 4198, 4236, 4262, 4476, 4477, 4478, 4479, 4480, 4481, 4482, 4483, 4484, 4485, 4486, 5012, 5013, 7275, 7462, 7465, 7471, 7481, 10294

Plant nutrition and fertilization (D2200)
5869, 7473, 7478

Plant breeding (D2300)
4236, 7228, 7251, 7275, 7280, 7344, 7460, 7461, 7462, 7463, 7464, 7465, 7466, 7467, 7468, 7469, 7470, 7471, 7472, 7473, 7474, 7475, 7476, 7477, 7478, 7479, 7480, 7481, 7482, 7483

Plant protection (D2400)
4482

Pests of plants and pest control (D2410)
8816, 8817, 8818

Plant diseases and disease control (D2420)
5869, 9642, 9885, 9886, 9887, 9888, 9889, 9890, 9891

Weeds and weed control (D2430)
10294

B3690 Other fruits
Plant breeding (D2300)

7251, 7484

B3700 Ornamentals and ornamental products in general
357, 436, 1200, 1409, 1527, 1538, 1788, 2454, 2464, 13520, 15131, 15142, 15153, 15166, 15232, 15233, 15234, 15237, 15304, 15356, 15358, 15578, 15595, 15596, 15601, 15606, 15628, 15719, 17315, 17565, 17566, 17567, 17568, 17597, 17612, 17879, 17883, 18155, 18271, 20403, 20422, 20619, 20759, 20846, 20849

Plant production general and crop husbandry (D2100)
932, 3402, 3690, 3705, 3707, 3758, 3759, 3760, 3761, 3762, 3763, 3764, 3765, 3766, 3769, 3770, 3780, 3781, 3782, 3785, 3786, 3788, 3789, 4091, 4094, 4102, 4104, 4487, 4488, 4489, 4490, 4491, 4492, 4493, 4494, 4495, 4496, 4497, 4498, 4499, 4500, 4501, 4502, 4503, 4504, 4505, 4506, 4507, 4508, 4509, 4510, 4511, 4512, 4513, 4514, 4515, 4516, 4517, 4518, 4519, 4520, 4521

Plant nutrition and fertilization (D2200)
1, 5680, 5694, 5708, 5710, 5713, 5870, 5871, 5872, 5873

Plant breeding (D2300)
3785, 4517, 6817, 6856, 6858, 6859, 6872, 6873, 6875, 6877, 7485, 7486, 7487, 7488, 7489, 7490, 7491, 7492, 7493, 7494, 7495

Plant protection (D2400)
7878, 7879, 7880, 7881, 7929, 7930, 7931, 7932, 7933, 7934, 7935, 7936

Pests of plants and pest control (D2410)
7975, 7980, 8501, 8609, 8611, 8614, 8620, 8819, 8820, 8821, 8822, 8823, 8824, 8825, 8826, 8827, 8828, 8829, 8830, 8831, 8832, 8833, 9597

Plant diseases and disease control (D2420)
8959, 9583, 9592, 9597, 9892, 9893, 9894, 9895, 9896, 9897, 9898, 9899, 9900, 9901, 9902, 9903, 9904, 9905

Weeds and weed control (D2430)
10260, 10268

Miscellaneous plant disorders (D2490)
10383

Storage and conservation (D4420)
4493, 4501, 4502, 4503, 15838, 16092, 16167, 16168, 16170

Transport and handling (D4440)
16167, 17133, 17136, 17137

B3710 Bulbs
895, 1568, 1611, 17222, 17223, 17232, 17236, 17237, 17238, 17239, 17598, 17599, 17602, 17606, 17624, 17625, 17876, 17884, 18170, 18171, 18182, 20397, 20420, 20435, 20436, 20437, 20726, 20964, 20965

Plant production general and crop husbandry (D2100)
2897, 3872, 3928, 3966, 4522, 4523, 4524, 4525, 4526, 4527, 4528, 4529, 4530, 4531, 4532, 4533, 4534, 4535, 4536, 4537, 4538, 4539, 4540, 4541, 4542, 4543, 4544, 4545, 4546, 4547, 4548, 4549, 4550, 4551, 4552, 4553, 4554, 4555, 4556, 4557, 4558, 4559, 4560, 4561, 4562, 4563, 4564, 4565, 4566, 4567, 4568, 4569, 4570, 4571, 4572, 4573, 4574, 4575, 4608, 4660, 9919, 15390

Plant nutrition and fertilization (D2200)
4608, 5874, 5875, 5876, 5877, 5878, 5879

Plant breeding (D2300)
2897, 4528, 4560, 4565, 4566, 4570, 7496, 7497, 7498, 7499, 7500, 7501, 7502, 7503, 7504, 7505, 7506, 7507, 7508, 7509, 7510, 7511, 7512, 7513, 7514, 7515, 7516, 7517, 7518, 7519, 7520, 7521, 7522, 7523, 7602

Plant protection (D2400)
4560, 4565, 4566, 4567, 4568, 4570, 4572, 4573, 7937, 7938, 7939

Pests of plants and pest control (D2410)
8834, 8835, 8836

Plant diseases and disease control (D2420)

9525, 9906, 9907, 9908, 9909, 9910, 9911, 9912, 9913, 9914, 9915, 9916, 9917, 9918, 9919, 9920, 9921, 9922, 9923, 9924, 9925

Weeds and weed control (D2430)
10295, 10296, 10297

Miscellaneous plant disorders (D2490)
10384, 10385

Storage and conservation (D4420)
4559, 4560, 4565, 4566, 4567, 4568, 4570, 4572, 4574, 15869, 15870, 15871, 15872, 15873, 15874, 15878, 15879, 15880, 15881, 15882, 15883, 15941

Transport and handling (D4440)
15870, 17107, 17108, 17109

B3720 Flowers and pot plants
194, 1570, 2536, 15270, 15307, 15357, 15359, 15399, 15478, 15484, 17161, 17226, 17234, 17242, 17596, 17604, 17605, 17609, 17611, 17613, 17618, 17619, 17622, 17623, 17626, 17628, 17735, 17779, 17884, 17890, 17896, 18183, 19968, 20218, 20420, 20434, 20438, 20803, 20964, 20965

Plant production general and crop husbandry (D2100)
2599, 3772, 3774, 3775, 3778, 3779, 3843, 3844, 3907, 3941, 3942, 3960, 3965, 3989, 4575, 4576, 4577, 4578, 4579, 4580, 4581, 4582, 4583, 4584, 4585, 4586, 4587, 4588, 4589, 4590, 4591, 4592, 4593, 4594, 4595, 4596, 4597, 4598, 4599, 4600, 4601, 4602, 4603, 4604, 4605, 4606, 4607, 4608, 4609, 4610, 4611, 4612, 4613, 4614, 4615, 4616, 4617, 4618, 4619, 4620, 4621, 4622, 4623, 4624, 4625, 4626, 4627, 4628, 4629, 4630, 4631, 4632, 4633, 4634, 4635, 4636, 4637, 4638, 4639, 4640, 4641, 4642, 4643, 4644, 4645, 4646, 4647, 4648, 4649, 4650, 4651, 4652, 4653, 4654, 4655, 4656, 4657, 4658, 4659, 4660, 4661, 4662, 4663, 4664, 4665, 4666, 4667, 4668, 4669, 4670, 4671, 4672, 4673, 4674, 4675, 4686, 5888, 5889, 5890, 5905, 5907, 5910, 7575, 15352

Plant nutrition and fertilization (D2200)
3775, 3778, 3907, 4606, 4607, 4608, 4619, 4655, 4658, 5712, 5715, 5717, 5718, 5719, 5720, 5721, 5722, 5723, 5724, 5880, 5881, 5882, 5883, 5884, 5885, 5886, 5887, 5888, 5889, 5890, 5891, 5892, 5893, 5894, 5895, 5896, 5897, 5898, 5899, 5900, 5901, 5902, 5903, 5904, 5905, 5906, 5907, 5908, 5909, 5910, 5911, 5912, 5913, 5914, 5915, 5916, 5917, 5918, 5919, 5920, 5921, 5922

Plant breeding (D2300)
4578, 4666, 4675, 5887, 6011, 7502, 7524, 7525, 7526, 7527, 7528, 7529, 7530, 7531, 7532, 7533, 7534, 7535, 7536, 7537, 7538, 7539, 7540, 7541, 7542, 7543, 7544, 7545, 7546, 7547, 7548, 7549, 7550, 7551, 7552, 7553, 7554, 7555, 7556, 7557, 7558, 7559, 7560, 7561, 7562, 7563, 7564, 7565, 7566, 7567, 7568, 7569, 7570, 7571, 7572, 7573, 7574, 7575, 7576, 7577, 7578, 7579, 7580, 7581, 7582, 7583, 7584, 7585, 7586, 7587, 7588, 7589, 7590, 7591, 7592, 7593, 7594, 7595, 7596, 7597, 7598, 7599, 7600, 7601, 7622

Plant protection (D2400)
7552, 7883, 7884, 7940

Pests of plants and pest control (D2410)
280, 7968, 8610, 8627, 8837, 8838, 8839, 8840, 8841, 8842, 8843, 8844, 8845, 8846, 8847, 8848, 8849, 8850, 8851, 8852, 8853, 8854, 8855, 8856, 9601

Plant diseases and disease control (D2420)
5884, 8955, 9599, 9600, 9601, 9679, 9685, 9688, 9907, 9926, 9927, 9928, 9929, 9930, 9931, 9932, 9933, 9934, 9935, 9936, 9937, 9938, 9939, 9940, 9941, 9942, 9943, 9944, 9945, 9946, 9947, 9948, 9949, 9950, 9951, 9952, 9953, 9954, 9955, 9956, 9957, 9958, 9959, 9960, 9961, 9962, 9963, 9964, 9965, 9966, 9971

Weeds and weed control (D2430)
10269, 10298

1730, 16356, 16368, 16405, 17667, 18104, 20359, 20607, 20885, 20904

Plant production general and crop husbandry (D2100)
4879, 4880, 4881, 4882, 4883, 4884, 4885, 4886, 4887, 4888, 4889, 4890, 4891, 4892, 4893, 4894, 4895, 4896, 4897, 4898, 4899, 4900, 4901, 4902, 4903, 4904, 4905, 4906, 4907, 4908, 4909, 4910, 4911, 4916, 4973, 4975

Plant nutrition and fertilization (D2200)
5975, 5976, 5977, 5978, 5979, 5980

Plant breeding (D2300)
7662, 7663, 7664, 7665, 7666, 7667, 7668, 7669, 7670, 7671, 7672, 10010

Pests of plants and pest control (D2410)
8917, 8918, 8919, 8920, 8921, 8922, 8923, 8925

Plant diseases and disease control (D2420)
10001, 10002, 10003, 10004, 10005, 10006, 10007, 10008, 10009, 10010, 10011

Weeds and weed control (D2430)
5978

Miscellaneous plant disorders (D2490)
10394

B3812 Spruce and fir forests
345, 1242, 11568, 12875, 15366, 15735, 15747, 15967, 16356, 16368, 16387, 16437, 17251, 20208, 20318, 20754, 20901

Plant production general and crop husbandry (D2100)
1396, 4879, 4884, 4887, 4894, 4910, 4912, 4913, 4914, 4915, 4916, 4917, 4918, 4919, 4920, 4921, 4922, 4923, 4924, 4925, 4926, 4927, 4928, 4929, 4930, 4931, 4932, 4933, 4934, 4935, 4936, 4937, 4938, 4939, 4940, 4941, 4942, 4943, 4944, 4945, 4946, 4947, 4948, 4949, 4950, 4951, 4952, 7677

Plant nutrition and fertilization (D2200)
4927, 5976, 5977, 5979, 5981, 5982, 5983, 5984, 5985, 5986, 5987, 5988, 5989, 5990, 7948

Plant breeding (D2300)
7611, 7673, 7674, 7675, 7676, 7677, 7678, 7679, 7680, 7681, 7682, 7683, 7684, 7685, 7686, 10018, 10023

Plant protection (D2400)
7947, 7948, 7949

Pests of plants and pest control (D2410)
8034, 8924, 8925, 8926, 8927, 8928, 8929

Plant diseases and disease control (D2420)
7678, 10012, 10013, 10014, 10015, 10016, 10017, 10018, 10019, 10020, 10021, 10022, 10023

Weeds and weed control (D2430)
10314

Miscellaneous plant disorders (D2490)
8927, 10395, 10396, 10397

B3813 Larch forests
15742, 16267, 16377

Plant production general and crop husbandry (D2100)
4902, 4937, 4938, 4953, 4954, 4955, 4956, 7677

Plant nutrition and fertilization (D2200)
4955

Plant breeding (D2300)
7677, 7687, 7688

Pests of plants and pest control (D2410)
8930

Plant diseases and disease control (D2420)
9840

B3819 Other pine forests
Plant production general and crop husbandry (D2100)
4957, 4975

Pests of plants and pest control (D2410)
8931, 8932

B3820 Leafwoods in general

2084, 11567, 17264, 20324, 20333

Plant production general and crop husbandry (D2100)
4002, 4865, 4958, 4959, 4960, 4961, 4962, 4963, 4964, 4965, 4966

Plant nutrition and fertilization (D2200)
4960, 4966, 5695, 5991

Plant breeding (D2300)
7656, 7689, 7690

Plant protection (D2400)
4966

Plant diseases and disease control (D2420)
10024

Miscellaneous plant disorders (D2490)
10393

B3821 Oak tree stands
13, 782, 2070, 2078, 2122, 11568, 15753, 20206, 20603, 20796

Plant production general and crop husbandry (D2100)
1926, 4879, 4916, 4967, 4968, 4969, 4970, 4971, 4972, 4973, 4974, 4975, 4976, 4977, 4978, 4979, 4980, 4981, 4982, 4983, 4984, 4985

Plant nutrition and fertilization (D2200)
4976, 4977, 5976, 5979

Plant breeding (D2300)
7691, 7692, 7693

Plant protection (D2400)
7950

Pests of plants and pest control (D2410)
8925, 8933, 8934, 8935, 8936, 8937

Plant diseases and disease control (D2420)
9998, 10025, 10026, 10027, 10028

Miscellaneous plant disorders (D2490)
10398

B3822 Ash tree stands
Plant production general and crop husbandry (D2100)
4984, 4985, 5992

Plant nutrition and fertilization (D2200)
5992

Pests of plants and pest control (D2410)
8927

Plant diseases and disease control (D2420)
10029

Miscellaneous plant disorders (D2490)
8927

B3823 Poplar tree stands
1463, 1464, 17824, 20884

Plant production general and crop husbandry (D2100)
382, 4733, 4918, 4986, 4987, 4988, 4989, 4990, 4991, 4992, 4993, 4994, 4995, 4996, 4997, 4998, 4999, 5000, 5001, 5002, 5003, 5004, 5005, 5006, 5007, 5993, 7697, 7700, 10032

Plant nutrition and fertilization (D2200)
5993, 5994, 5995

Plant breeding (D2300)
4995, 7673, 7692, 7694, 7695, 7696, 7697, 7698, 7699, 7700, 7701, 7702, 7703, 7704, 7705, 7706, 7707, 7708, 7709, 7710, 7711, 7712

Plant diseases and disease control (D2420)
10030, 10031, 10032, 10033, 10034, 10035, 10036

B3829 Other leafwoods
571, 20900

Plant production general and crop husbandry (D2100)
4989, 5008, 5009, 5010, 5011. 5012, 5013, 5992

Plant nutrition and fertilization (D2200)
5992

Plant breeding (D2300)
7711, 7713, 7714

Pests of plants and pest control (D2410)
8938, 8939
Plant diseases and disease control (D2420)
10030, 10037, 10038
B3890 Other forests
2035, 20898, 20899
Plant production general and crop husbandry (D2100)
1909, 5014, 5015, 5016, 5017, 5018, 5019, 5020, 5021, 5022, 5023
Plant nutrition and fertilization (D2200)
5996
Plant breeding (D2300)
7715, 7716, 7717, 7718
Pests of plants and pest control (D2410)
8940
B3910 Stimulant crops
16556, 17036, 17928, 19420, 19421, 19422, 19432, 19438, 19439, 20677
Plant production general and crop husbandry (D2100)
4278, 5024, 5025, 5026, 5044
Plant nutrition and fertilization (D2200)
5997
Plant breeding (D2300)
6872, 7719, 7720
Plant protection (D2400)
7951, 7952, 7953
Pests of plants and pest control (D2410)
8941, 8942, 8943, 8944, 8945, 10039
Plant diseases and disease control (D2420)
10039, 10040, 10041, 10042, 10043
Weeds and weed control (D2430)
10136, 10315
B3920 Spice and seasoning plants of warm climates
15830, 16952, 19031, 19407, 19408, 20677
Plant production general and crop husbandry (D2100)
5027, 5028, 5029, 5030, 5035
Plant nutrition and fertilization (D2200)
5028
Plant breeding (D2300)
7721, 7722, 7723
B3930 Spice and seasoning plants of temperate climates
15825, 15830, 16952, 16968, 17927, 19028, 19124, 19437, 19679, 20288, 20662, 20677
Plant production general and crop husbandry (D2100)
5027, 5031, 5032, 5033, 5034, 5035, 5036, 5037, 5038
Plant nutrition and fertilization (D2200)
5998, 5999, 6000, 7724
Plant breeding (D2300)
5032, 7724, 7725, 7726
Plant protection (D2400)
7724, 7954
Pests of plants and pest control (D2410)
8946
Plant diseases and disease control (D2420)
5032, 10044, 10045, 10046, 10047, 10048, 10049, 10050, 10051
Weeds and weed control (D2430)
10273, 10285
B3940 Perfume plants
17044, 19029, 19030, 20677, 20678
Plant production general and crop husbandry (D2100)
4555, 5039, 5040, 5041, 5042, 5043
Plant nutrition and fertilization (D2200)
5042
Plant breeding (D2300)
5042, 7727

B3950 Rubber, gum, wax and resin plants
428
Plant production general and crop husbandry (D2100)
5044
B3970 Drugs and medicine plants
15368, 15373, 16304, 17017, 17018, 17019, 17020, 17591, 20677, 20803
Plant production general and crop husbandry (D2100)
3791, 4063, 5029, 5045, 5046, 5047, 5048, 5049, 5050, 6002, 7728
Plant nutrition and fertilization (D2200)
1545, 5048, 5049, 5424, 6001, 6002, 6003
Plant breeding (D2300)
7572, 7728, 7729, 7730, 7731, 7732, 7733, 7734, 7735, 7736, 7737, 7738, 7739, 7740, 7741
Plant protection (D2400)
6003, 7955
Pests of plants and pest control (D2410)
8501, 8947, 8948
Plant diseases and disease control (D2420)
10052, 10053, 10054, 10055
Weeds and weed control (D2430)
10316, 17021
B3990 Other crops
12774, 18647
Plant production general and crop husbandry (D2100)
5051, 5052, 5053, 5054
Plant breeding (D2300)
7742, 7743
Pests of plants and pest control (D2410)
8949
Plant diseases and disease control (D2420)
10056, 10057
Weeds and weed control (D2430)
10317
B4000 Domestic animals in general
1873, 2103, 2219, 2221, 2614, 2618, 2620, 2622, 2791, 3482, 3661, 6664, 7996, 7998, 7999, 8002, 8007, 8008, 8009, 8010, 8011, 8012, 8013, 8014, 8015, 8305, 15143, 15150, 15238, 15260, 15261, 15420, 15421, 15536, 15541, 15545, 15548, 15549, 15638, 15641, 15650, 15668, 15684, 15755, 15764, 15807, 15889, 16014, 16182, 16184, 16197, 16295, 16692, 16693, 16699, 16700, 16701, 16711, 16712, 16715, 16955, 16957, 17070, 17091, 17096, 17140, 17158, 17316, 17496, 17511, 17559, 17711, 17731, 17734, 17739, 17755, 17767, 17777, 17784, 17785, 17819, 17857, 17860, 17864, 17887, 17939, 17979, 18058, 18194, 18199, 18807, 18946, 18947, 18975, 19033, 19036, 19065, 19132, 19133, 19134, 19135, 19136, 19141, 19143, 19145, 19147, 19151, 19152, 19158, 19160, 19161, 19174, 19176, 19178, 19179, 19180, 19188, 19189, 19193, 19207, 19213, 19220, 19222, 19223, 19224...
Animal management general and animal husbandry (D3100)
2170, 2393, 2751, 2848, 3536, 3683, 10439, 10440, 10441, 10442, 10443, 10444, 10445, 10446, 10447, 10448, 10449, 10450, 10451, 10452, 10453, 10454, 10455, 10456, 10457, 10458, 10459, 10460, 10461, 10462, 10463, 10464, 10465, 10466, 10467, 10468, 10469, 10470, 10471, 10472, 10473, 10474, 10475, 10476, 10477, 10478, 10479, 10480, 10481, 10482, 10483, 10484, 10485, 10486, 10487, 10488, 10489, 10490, 10491, 10492, 10493, 10494, 10495, 10496, 10497, 10498, 10499, 10500, 10501, 10502, 10503, 10504, 10505, 10506, 10507, 10508, 10509, 10510, 10511, 10512, 10513, 10514, 10515, 10516, 10517, 10518, 10519, 10520, 10521, 10522, 10523, 10524, 10525, 10526, 10527, 10528, 10529, 10530, 10531, 10532, 10533, 10534, 10535, 10536, 10537.

16697, 16706, 16707, 16714, 16745, 16747, 16749, 16750, 16751, 16752, 16753, 16754, 16757, 16759, 16760, 16761, 16762, 16764...

Animal management general and animal husbandry (D3100)

2076, 2085, 3367, 3613, 3665, 5590, 10399, 10400, 10404, 10620, 10621, 10622, 10623, 10624, 10625, 10626, 10627, 10628, 10629, 10630, 10631, 10632, 10633, 10634, 10635, 10636, 10637, 10638, 10639, 10640, 10641, 10642, 10643, 10644, 10645, 10646, 10647, 10648, 10649, 10650, 10651, 10652, 10653, 10654, 10655, 10656, 10657, 10658, 10659, 10660, 10661, 10662, 10663, 10664, 10665, 10666, 10667, 10668, 10669, 10670, 10671, 10672, 10673, 10674, 10675, 10676, 10677, 10678, 10679, 10680, 10681, 10682, 10683, 10684, 10685, 10686, 10687, 10688, 10689, 10690, 10691, 10692, 10693, 10694, 10695, 10696, 10697, 10698, 10699, 10700, 10701, 10702, 10703, 10704, 10705, 10706, 10707, 10708, 10709, 10710, 10711, 10712, 10713, 10714, 10715, 10716, 10717, 10718, 10719, 10720, 10721, 10722, 10723, 10724, 10725, 10726, 10727, 10728, 10729, 10730, 10731, 10732, 10733, 10734, 10735...

Animal nutrition (D3200)

3084, 3344, 3373, 3505, 3522, 3595, 3613, 3621, 5612, 10658, 10666, 10688, 10698, 10720, 10725, 10736, 10738, 10741, 10742, 10746, 10757, 10761, 10785, 10886, 10887, 10895, 10899, 10909, 11361, 11539, 11540, 11544, 11546, 11547, 11549, 11551, 11552, 11561, 11564, 11565, 11572, 11573, 11574, 11620, 11800, 11944, 11945, 11946, 11947, 11948, 11949, 11950, 11951, 11952, 11953, 11954, 11955, 11956, 11957, 11958, 11959, 11960, 11961, 11962, 11963, 11964, 11965, 11966, 11967, 11968, 11969, 11970, 11971, 11972, 11973, 11974, 11975, 11976, 11977, 11978, 11979, 11980, 11981, 11982, 11983, 11984, 11985, 11986, 11987, 11988, 11989, 11990, 11991, 11992, 11993, 11994, 11995, 11996, 11997, 11998, 11999, 12000, 12001, 12002, 12003, 12004, 12005, 12006, 12007, 12008, 12009, 12010, 12011, 12012, 12013, 12014, 12015, 12016, 12017, 12018, 12019, 12020, 12021, 12022, 12023...

Animal breeding (D3300)

10644, 10645, 10716, 10739, 10741, 10744, 10759, 10770, 10821, 10822, 10867, 10890, 10896, 10919, 12025, 12153, 12221, 12870, 12871, 12872, 12967, 12968, 12969, 12970, 12971, 12972, 12973, 12974, 12975, 12976, 12977, 12978, 12979, 12980, 12981, 12982, 12983, 12984, 12985, 12986, 12987, 12988, 12989, 12990, 12991, 12992, 12993, 12994, 12995, 12996, 12997, 12998, 12999, 13000, 13001, 13002, 13003, 13004, 13005, 13006, 13007, 13008, 13009, 13010, 13011, 13012, 13013, 13014, 13015, 13016, 13017, 13018, 13019, 13020, 13021, 13022, 13023, 13024, 13025, 13026, 13027, 13028, 13029, 13030, 13031, 13032, 13033, 13034, 13035, 13036, 13037, 13038, 13039, 13040, 13041, 13042, 13043, 13044, 13045, 13046, 13047, 13048, 13049, 13050, 13051, 13052, 13053, 13054, 13055, 13056, 13057, 13058, 13059, 13060, 13061, 13062, 13063, 13064, 13065, 13066, 13067, 13068, 13069, 13070, 13071...

Animal diseases, veterinary medicine (D3400)

5083, 10691, 10711, 10732, 10848, 10850, 10854, 10889, 10911, 11035, 12017, 12040, 12077, 12151, 12201, 12253, 13032, 13037, 13091, 13097, 13473, 13475, 13476, 13484, 13493, 13494, 13495, 13504, 13506, 13507, 14066, 14067, 14068, 14069, 14070, 14071, 14072, 14073, 14074, 14075, 14076, 14077, 14078, 14079, 14080, 14081, 14082, 14083, 14084, 14085, 14086, 14087, 14088, 14089, 14090, 14091, 14092, 14093, 14094, 14095, 14096, 14097, 14098, 14099, 14100, 14101, 14102, 14103, 14104, 14105, 14106, 14107, 14108, 14109, 14110, 14111, 14112, 14113, 14114, 14115, 14116, 14117, 14118, 14119, 14120, 14121, 14122, 14123, 14124, 14125, 14126, 14127, 14128, 14129, 14130, 14131, 14132, 14133, 14134, 14135, 14136, 14137, 14138, 14139, 14140, 14141, 14142, 14143, 14144, 14145, 14146, 14147, 14148, 14149, 14150, 14151, 14152, 14153, 14154, 14155, 14156, 14157, 14158, 14159, 14160...

B4220 Sheep

95, 1010, 1981, 2090, 2634, 3520, 3598, 6015, 15189, 15434, 15436, 15439, 15440, 15542, 15673, 15785, 15938, 15939, 16200, 16707, 16857, 16858, 16874, 16875, 16881, 17306, 17495, 17502, 17509, 17533, 17546, 17554, 17831, 17832, 19209, 19214, 19215, 19231, 19374, 19383, 19667, 19942, 20177, 20442, 20862

Animal management general and animal husbandry (D3100)

3370, 3613, 4829, 10402, 10678, 10701, 10702, 10704, 10735, 10742, 10747, 10750, 10872, 10880, 10884, 10911, 10912, 10925, 10926, 10927, 10928, 10929, 10930, 10931, 10932, 10933, 10934, 10935, 10936, 10937, 10938, 10939, 10940, 10941, 10942, 10943, 10944, 10945, 10946, 10947, 10948, 10949, 10950, 10951, 10952, 10953, 10954, 10955, 10956, 10957, 10958, 10959, 10960, 10961, 10962, 10963, 10964, 10965, 10966, 10967, 10968, 10969, 10970, 10971, 10972, 10973, 10974, 10975, 10976, 10977, 10978, 10979, 10980, 10981, 10982, 10983, 10984, 10985, 10986, 10987, 10988, 10989, 10990, 10991, 10992, 10993, 10994, 10995, 10996, 10997, 10998, 10999, 11000, 11001, 11002, 11003, 11004, 11005, 11006, 11007, 11008, 11009, 11010, 11011, 11012, 11013, 11014, 11015, 11016, 11017, 11018, 11019, 11020, 11021, 11022, 11023, 11024, 11025, 11026, 11027, 11028, 11029, 11030, 11031, 11032...

Animal nutrition (D3200)

3544, 3599, 3613, 10742, 10925, 10948, 10973, 10978, 11048, 11569, 11947, 12067, 12071, 12073, 12181, 12206, 12261, 12262, 12267, 12270, 12280, 12281, 12282, 12283, 12284, 12285, 12286, 12287, 12288, 12289, 12290, 12291, 12292, 12293, 12294, 12295, 12296, 12297, 12298, 12299, 12300, 12301, 12302, 12303, 12304, 12305, 12306, 12307, 12308, 12309, 12310, 12311, 12312, 12313, 12314, 12315, 12316, 12317, 12318, 12319, 12320, 12321, 12322, 12323, 12324, 12325, 12326, 12327, 12328, 12329, 12330, 12331, 12332, 12333, 12334, 12335, 12336, 12337, 12338, 12339, 12340, 12341, 12342, 12343, 12344, 12345, 12346, 12347, 12348, 12349, 12350, 12351, 12352, 12353, 12354, 12355, 12356, 12357, 12358, 12359, 12360, 12361, 12362, 12363, 12364, 12365, 12366, 12367, 13215, 14454, 14574, 15669, 15670

Animal breeding (D3300)

10972, 10976, 11038, 11050, 11051, 11062, 12317, 12327, 12328, 12345, 13065, 13066, 13173, 13174, 13175, 13176, 13177, 13178, 13179, 13180, 13181, 13182, 13183, 13184, 13185, 13186, 13187, 13188, 13189, 13190, 13191, 13192, 13193, 13194, 13195, 13196, 13197, 13198, 13199, 13200, 13201, 13202, 13203, 13204, 13205, 13206, 13207, 13208, 13209, 13210, 13211, 13212, 13213, 13214, 13215, 13216, 13217, 13218, 13219, 13220, 13221, 13222, 13223, 13224, 13225, 13226, 13227, 13228, 13229, 13230, 13231, 13232, 13233, 13234, 13235, 13236, 13237, 13238, 13239, 13240, 13241, 13242, 13243, 13244, 13245, 13246, 13247, 13248, 13249, 13250, 13251, 13252, 13253, 13254, 13255, 14440, 14462, 14478

Animal diseases, veterinary medicine (D3400)

10911, 10939, 11034, 11035, 11036, 11047, 12315, 13216, 14099, 14104, 14161, 14196, 14395, 14396, 14398, 14402, 14403, 14404, 14440, 14441, 14442, 14443, 14444, 14445, 14446, 14447, 14448, 14449, 14450, 14451, 14452, 14453,

14454, 14455, 14456, 14457, 14458, 14459, 14460, 14461, 14462, 14463, 14464, 14465, 14466, 14467, 14468, 14469, 14470, 14471, 14472, 14473, 14474, 14475, 14476, 14477, 14478, 14479, 14480, 14481, 14482, 14483, 14484, 14485, 14486, 14487, 14488, 14489, 14490, 14491, 14492, 14493, 14494, 14495, 14496, 14497, 14498, 14499, 14500, 14501, 14502, 14503, 14504, 14505, 14506, 14507, 14508, 14509, 14510, 14511, 14512, 14513, 14514, 14515, 14516, 14517, 14518, 14519, 14520, 14521, 14522, 14523, 14524, 14525, 14526, 14527, 14528, 14529, 14530, 14531, 14532, 14533, 14534, 14535, 14536, 14537, 14538, 14539, 14540, 14541, 14542, 14543, 14544, 14545, 14546...

B4290 Goats, farm deer and other ruminants
6015, 6628, 15785, 16707, 16714, 16858, 16867, 16868, 16869, 16870, 16881, 17106, 18827, 19208, 19380, 19667

Animal management general and animal husbandry (D3100)
3370, 10625, 10678, 10742, 10872, 10884, 11052, 11056, 11058, 11066, 11067, 11068, 11069, 11070, 11071, 11072, 11073, 11074, 11075, 11076, 11077, 11078, 11079, 11080, 11081, 11082, 11083, 11084, 11085, 11086, 11087, 11088, 11519, 12357, 14587, 15385

Animal nutrition (D3200)
1825, 1861, 10742, 12357, 12368, 12369, 12370, 12371, 12372, 12373, 12374, 12375, 12957

Animal breeding (D3300)
12957, 13139, 13140, 13238, 13256, 13257, 13258, 13259, 13260, 13261, 14587

Animal diseases, veterinary medicine (D3400)
1888, 14455, 14587, 14588, 14589, 14590, 14591, 14592, 14593, 14594, 14595, 14596, 14597, 14598, 14599, 14600, 14601, 14670, 14671

B4300 Pigs
292, 523, 720, 1047, 2461, 3566, 5300, 5306, 5524, 5560, 15256, 15439, 15535, 15537, 15642, 15643, 15645, 15651, 15657, 15658, 15687, 15808, 15831, 15861, 16183, 16194, 16204, 16209, 16211, 16307, 16707, 16708, 16710, 16713, 17071, 17142, 17143, 17144, 17162, 17195, 17297, 17476, 17477, 17478, 17494, 17503, 17514, 17521, 17526, 17534, 17786, 17789, 17798, 17843, 17881, 18036, 18141, 18172, 18408, 18826, 19032, 19138, 19140, 19148, 19150, 19165, 19166, 19168, 19170, 19171, 19175, 19206, 19211, 19219, 19221, 19225, 19226, 19229, 19232, 19242, 19247, 19250, 19252, 19544, 19576, 19628, 19684, 19696, 19700, 19715, 19815, 19827, 19967, 20076, 20106, 20133, 20177, 20222, 20410, 20488, 20753, 20877, 21072, 21074

Animal management general and animal husbandry (D3100)
10589, 10590, 10628, 10689, 10701, 10702, 10704, 10866, 10871, 10911, 10918, 11054, 11089, 11090, 11091, 11092, 11093, 11094, 11095, 11096, 11097, 11098, 11099, 11100, 11101, 11102, 11103, 11104, 11105, 11106, 11107, 11108, 11109, 11110, 11111, 11112, 11113, 11114, 11115, 11116, 11117, 11118, 11119, 11120, 11121, 11122, 11123, 11124, 11125, 11126, 11127, 11128, 11129, 11130, 11131, 11132, 11133, 11134, 11135, 11136, 11137, 11138, 11139, 11140, 11141, 11142, 11143, 11144, 11145, 11146, 11147, 11148, 11149, 11150, 11151, 11152, 11153, 11154, 11155, 11156, 11157, 11158, 11159, 11160, 11161, 11162, 11163, 11164, 11165, 11166, 11167, 11168, 11169, 11170, 11171, 11172, 11173, 11174, 11175, 11176, 11177, 11178, 11179, 11180, 11181, 11182, 11183, 11184, 11185, 11186, 11187, 11188, 11189, 11190, 11191, 11192, 11193, 11194, 11195, 11196, 11197, 11198, 11199, 11200, 11201...

Animal nutrition (D3200)
1860, 11095, 11123, 11138, 11164, 11187, 11191, 11213,

11241, 11242, 11254, 11566, 11575, 11712, 11748, 11799, 11803, 11808, 11820, 11986, 12033, 12034, 12218, 12223, 12224, 12261, 12262, 12264, 12267, 12269, 12270, 12271, 12273, 12274, 12275, 12277, 12297, 12376, 12377, 12378, 12379, 12380, 12381, 12382, 12383, 12384, 12385, 12386, 12387, 12388, 12389, 12390, 12391, 12392, 12393, 12394, 12395, 12396, 12397, 12398, 12399, 12400, 12401, 12402, 12403, 12404, 12405, 12406, 12407, 12408, 12409, 12410, 12411, 12412, 12413, 12414, 12415, 12416, 12417, 12418, 12419, 12420, 12421, 12422, 12423, 12424, 12425, 12426, 12427, 12428, 12429, 12430, 12431, 12432, 12433, 12434, 12435, 12436, 12437, 12438, 12439, 12440, 12441, 12442, 12443, 12444, 12445, 12446, 12447, 12448, 12449, 12450, 12451, 12452, 12453, 12454, 12455, 12456, 12457, 12458, 12459, 12460, 12461, 12462, 12463...

Animal breeding (D3300)
11170, 11190, 11210, 11211, 11214, 11232, 11242, 11253, 11264, 12556, 12559, 12902, 12992, 13001, 13009, 13133, 13136, 13154, 13181, 13188, 13262, 13263, 13264, 13265, 13266, 13267, 13268, 13269, 13270, 13271, 13272, 13273, 13274, 13275, 13276, 13277, 13278, 13279, 13280, 13281, 13282, 13283, 13284, 13285, 13286, 13287, 13288, 13289, 13290, 13291, 13292, 13293, 13294, 13295, 13296, 13297, 13298, 13299, 13300, 13301, 13302, 13303, 13304, 13305, 13306, 13307, 13308, 13309, 13310, 13311, 13312, 13313, 13314, 13315, 13316, 13317, 13318, 13319, 13320, 13321, 13322, 13323, 13324, 13325, 13326, 13327, 13328, 13329, 13330, 13331, 13332, 13333, 13334, 13335, 13336, 13337, 13338, 13339, 13340, 13341, 13342, 13343, 13344, 13345, 13346, 13347, 13348, 13349, 13350, 13351, 13352, 13353, 13354, 13355, 14647, 14755, 14777, 19256

Animal diseases, veterinary medicine (D3400)
10911, 11107, 11214, 11264, 12427, 12494, 12562, 12861, 13263, 13352, 13477, 13485, 13490, 13493, 14063, 14106, 14135, 14156, 14173, 14382, 14406, 14409, 14411, 14413, 14443, 14602, 14603, 14604, 14605, 14606, 14607, 14608, 14609, 14610, 14611, 14612, 14613, 14614, 14615, 14616, 14617, 14618, 14619, 14620, 14621, 14622, 14623, 14624, 14625, 14626, 14627, 14628, 14629, 14630, 14631, 14632, 14633, 14634, 14635, 14636, 14637, 14638, 14639, 14640, 14641, 14642, 14643, 14644, 14645, 14646, 14647, 14648, 14649, 14650, 14651, 14652, 14653, 14654, 14655, 14656, 14657, 14658, 14659, 14660, 14661, 14662, 14663, 14664, 14665, 14666, 14667, 14668, 14669, 14670, 14671, 14672, 14673, 14674, 14675, 14676, 14677, 14678, 14679, 14680, 14681, 14682, 14683, 14684, 14685, 14686, 14687, 14688, 14689, 14690, 14691, 14692, 14693, 14694, 14695, 14696, 14697, 14698, 14699, 14700, 14701...

B4400 Poultry and domestic birds in general
15236, 15239, 15481, 15483, 15539, 15644, 15769, 15868, 15943, 16187, 16215, 16696, 16709, 16948, 16949, 17114, 17138, 17141, 17205, 17523, 17524, 17525, 17796, 18036, 18122, 18141, 18188, 18815, 18822, 19056, 19177, 19186, 19216, 19217, 19402, 19415, 19654, 19655, 19684, 19704, 19715, 19761, 19892, 19893, 20164, 20165, 20409, 20467, 20666, 20679, 20877

Animal management general and animal husbandry (D3100)
11269, 11270, 11271, 11272, 11273, 11274, 11275, 11276, 11277, 11278, 11279, 11280, 11281, 11282, 11283, 11284, 11285, 11286, 11287, 11288, 11289, 11290, 11291, 11292, 11293, 11294, 11295, 11296, 11297, 11298, 11299, 11300, 11301, 11302, 11303, 11304, 11305, 11306, 11307, 11308, 11309, 11310, 11311, 11312, 11313, 17155, 19169

Animal nutrition (D3200)
11308, 11575, 12264, 12271, 12273, 12274, 12275, 12566,

12568, 12571, 12573, 12576, 12577, 12578, 12579, 12580,
12581, 12582, 12583, 12584, 12585, 12586, 12587, 12588,
12589, 12590, 12591, 12592, 12593, 12594, 12595, 12596,
12597, 12598, 12599, 12600, 12601, 12602, 12603, 12604,
12605, 12606, 12607, 12608, 12609, 12610, 12611, 12612,
12613, 12614, 12615, 12616, 12617, 12618, 12619, 12620,
12621, 12622, 12623, 12624, 12625, 12626, 12627, 12628,
12629, 12630, 12631, 12632, 12633, 12634, 12635, 12636,
12804, 13358, 13359, 13361, 13364, 14301, 14828, 19403

Animal breeding (D3300)

11296, 13356, 13357, 13358, 13359, 13360, 13361, 13362,
13363, 13364, 13365, 13366, 13367, 13368, 13369, 13370,
13371, 13372, 13373, 13374, 13375, 13376, 13377, 13378,
13379, 13380, 13381, 14858

Animal diseases, veterinary medicine (D3400)

2185, 14061, 14135, 14301, 14782, 14783, 14784, 14785,
14786, 14787, 14788, 14789, 14790, 14791, 14792, 14793,
14794, 14795, 14796, 14797, 14798, 14799, 14800, 14801,
14802, 14803, 14804, 14805, 14806, 14807, 14808, 14809,
14810, 14811, 14812, 14813, 14814, 14815, 14816, 14817,
14818, 14819, 14820, 14821, 14822, 14823, 14824, 14825,
14826, 14827, 14828, 14829, 14830, 14831, 14832, 14833,
14834, 14835, 14836, 14837, 14838, 14839, 14840, 14841,
14842, 14843, 14844, 14845, 14846, 14847, 14848, 14849,
14850, 14851, 14852, 14853, 14854, 14855, 14856, 14857,
14858, 14859, 14860, 14861, 14862, 14863, 14864, 14865,
14866, 14867, 14868, 14869, 14870, 14871, 14872

B4410 Chickens

15691, 15698, 16201, 16214, 16681, 17537, 17751, 17871,
18454, 19139, 19212, 19231, 19232, 19252, 19405, 19576,
19713, 19964, 20065, 20905, 21071

Animal management general and animal husbandry (D3100)

10633, 11314, 11315, 11316, 11317, 11318, 11319, 11320,
11321, 11322, 11323, 11324, 11325, 11326, 11327, 11328,
11329, 11330, 11331, 11332, 11333, 11334, 11335, 11336,
11337, 11338, 11339, 11340, 11341, 11342, 11343, 11344,
11345, 11346, 11347, 11348, 11349, 11350, 11351, 11352,
11353, 11354, 11355, 11356, 11357, 11358, 11359, 17281

Animal nutrition (D3200)

11340, 11341, 11342, 11343, 11555, 11556, 11557, 11558,
11559, 11986, 12574, 12637, 12638, 12639, 12640, 12641,
12642, 12643, 12644, 12645, 12646, 12647, 12648, 12649,
12650, 12651, 12652, 12653, 12654, 12655, 12656, 12657,
12658, 12659, 12660, 12661, 12662, 12663, 12664, 12665,
12666, 12667, 12668, 12669, 12670, 12671, 12672, 12673,
12674, 12675, 12676, 12677, 12678, 12679, 12680, 12681,
12682, 12683, 12684, 12685, 12686, 12687, 12688, 12689,
12690, 12691, 12692, 12693, 12694, 12695, 12696, 12697,
12698, 12699, 12700, 12701, 12702, 12703, 12704, 12705,
12706, 12707, 12708, 12709, 12710, 12711, 12712, 12713,
12714, 12845, 12846, 12854, 13404, 14887

Animal breeding (D3300)

11346, 13382, 13383, 13384, 13385, 13386, 13387, 13388,
13389, 13390, 13391, 13392, 13393, 13394, 13395, 13396,
13397, 13398, 13399, 13400, 13401, 13402, 13403, 13404,
13405, 13406, 13407, 13408, 13409, 13410, 13411, 13412,
13413, 13414, 13415

Animal diseases, veterinary medicine (D3400)

11327, 12704, 13391, 13472, 14873, 14874, 14875, 14876,
14877, 14878, 14879, 14880, 14881, 14882, 14883, 14884,
14885, 14886, 14887, 14888, 14889, 14890, 14891, 14892,
14893, 14894, 14895, 14896, 14897, 14898, 14899, 14900,
14901, 14902, 14903, 14904, 14905, 14906, 14907, 14908,
14909, 14910, 14911, 14912, 14913, 14914, 14915, 14916,
14917, 14918, 14919, 14920, 14921, 14922, 14923, 14924,

14925, 14926, 14927, 14928, 14929, 14930, 14931, 14932,
14933, 14934, 14935, 14936, 14937, 14938, 14939, 14940,
14941, 14942

B4490 Geese, turkeys and other domestic birds

16707, 17530, 19210, 20916

Animal management general and animal husbandry (D3100)

10421, 11360, 11361, 11362, 11363, 11364, 11365, 12716

Animal nutrition (D3200)

11361, 11362, 12704, 12715, 12716, 12717, 12718, 12719,
12720, 12721, 12722, 12723, 12724, 12725

Animal breeding (D3300)

11363, 11365, 12715, 13189, 13245, 13416, 13417, 13418

Animal diseases, veterinary medicine (D3400)

12704, 13496, 14943, 14944, 14945, 14946, 14947, 14948,
14949, 14950

B4500 Fishes, crustacea, shell fish and frogs in general

1360, 1932, 1933, 2029, 2041, 2060, 15405, 15438, 15864,
16197, 17059, 17278, 17279, 17696, 18149, 18812, 18825,
19263, 19264, 19266, 19267, 19270, 20074, 21069

Animal management general and animal husbandry (D3100)

2043, 2056, 10422, 11366, 11367, 11368, 11369, 11370,
11371, 11372, 11373, 11374, 11375, 11376, 11377, 11378,
11379, 11380, 11381, 11382, 11383, 11384, 11385, 11386,
11387, 11388, 11389, 11390, 11391, 11392, 11393, 11394,
11395

Animal nutrition (D3200)

12440, 12726, 12727, 12728, 12729, 12730, 12731, 12732,
12733, 12734

Animal breeding (D3300)

12873, 13419

Animal diseases, veterinary medicine (D3400)

1934, 2042, 2062, 2165, 2285, 12732, 14951, 14952, 14953,
14954, 14955, 14956, 14957, 14958, 14959, 14960, 14961,
14962

B4510 Carp

1663, 2334, 10070, 21070

Animal management general and animal husbandry (D3100)

2055, 11396, 11397, 11398, 11399, 11400

Animal nutrition (D3200)

2333, 11396, 12735, 12736, 12737, 12738, 12739, 12740

Animal diseases, veterinary medicine (D3400)

14963, 14964, 14965, 14966, 14967, 14968

B4520 Salmon

Animal management general and animal husbandry (D3100)

2055, 2057, 11401, 11402, 11403, 11404, 11405, 11406,
11407, 11408, 11409, 11410, 11411

Animal nutrition (D3200)

12750

Animal breeding (D3300)

13420, 13421, 13425

Animal diseases, veterinary medicine (D3400)

14964, 14969, 14970, 14971, 14972

B4530 Trout

Animal management general and animal husbandry (D3100)

2055, 11402, 11403, 11404, 11405, 11409, 11412, 11413,
11414, 11415, 11416, 11417, 11418, 11419, 11420, 11421,
11422, 11423, 11424, 11425, 11426, 11427, 11428, 11429,
11430, 11431, 11432

Animal nutrition (D3200)

1407, 11425, 11429, 12737, 12741, 12742, 12743, 12744,
12745, 12746, 12747, 12748, 12749, 12750, 12751, 12752,

12753, 12754, 12755, 12756, 12757, 12758
Animal breeding (D3300)
13421, 13422, 13423, 13424, 13425, 13426
Animal diseases, veterinary medicine (D3400)
14135, 14964, 14970, 14973, 14974, 14975, 14976, 14977, 14978
B4540 Mullet
Animal nutrition (D3200)
12759
B4550 Eel
16731
Animal management general and animal husbandry (D3100)
2262, 11396, 11433, 11434, 11435, 11436
Animal nutrition (D3200)
11396, 11433, 12760, 12761, 12762, 12763
Animal diseases, veterinary medicine (D3400)
14979, 14980, 14981
B4560 Crustacea, shell fish, frogs
2148, 17146
Animal management general and animal husbandry (D3100)
11269, 11437, 11438, 11439, 11440, 11441, 11442, 11443, 11444, 11445, 11446, 11447, 11448, 11449, 11450
Animal nutrition (D3200)
11444, 12764, 12765
Animal breeding (D3300)
13427
Animal diseases, veterinary medicine (D3400)
14982, 14983
B4590 Other fishes
16730, 16731, 17862, 21067, 21068
Animal management general and animal husbandry (D3100)
11396, 11451, 11452, 11453, 11454, 11455, 11456, 11457
Animal nutrition (D3200)
11396, 12766, 12767, 12768
Animal breeding (D3300)
13422
B4600 Invertebrates (bees, silk–worm)
2099, 2717, 2803, 3166, 3757, 4212, 7960, 8003, 8263, 8264, 8265, 8266, 8267, 8268, 8269, 8270, 8271, 8272, 8273, 8277, 8281, 8780, 8794, 16438, 18800, 19066, 20067, 20276, 20675
Animal management general and animal husbandry (D3100)
7987, 8896, 11458, 11459, 11460, 11461, 11462, 11463, 11464, 11465, 11466, 11467, 11468, 11469, 11470, 11471, 11472, 11473, 11474, 11475, 11476, 11477, 11478, 11479, 11480, 11481, 11482, 11483, 11484, 11485, 11486, 11487, 11488, 11489, 11490, 11491, 11492
Animal nutrition (D3200)
2832, 12769, 12770, 12771, 12772, 12773, 12774, 12775, 12776
Animal breeding (D3300)
13428, 13429, 13430, 13431, 13432, 13433, 13434, 13435
Animal diseases, veterinary medicine (D3400)
2050, 11487, 11488, 14984, 14985, 14986, 14987, 14988, 14989, 14990, 14991, 14992, 14993, 14994, 14995
B4910 Rabbits
16707, 18827
Animal management general and animal husbandry (D3100)
10589, 10590, 10668, 10871, 10949, 11107, 11115, 11308, 11309, 11362, 11493, 11494, 11495, 11496, 11497, 11498, 11499, 11500, 11501, 11502, 11503, 11504, 11505, 11506,

11507, 11508, 11509, 11510, 11530, 12794, 12795, 13439
Animal nutrition (D3200)
11308, 11362, 11503, 11530, 11553, 11554, 11799, 12460, 12777, 12778, 12779, 12780, 12781, 12782, 12783, 12784, 12785, 12786, 12787, 12788, 12789, 12790, 12791, 12792, 12793, 12794, 12795, 12796, 12797, 12798, 12799, 12800, 12801, 12802, 12803, 12804, 12805, 12806, 12807, 12808, 12809, 12810, 12811, 12812, 12813, 12814, 12815, 12816, 12817, 12818, 12819, 12820, 12821, 12822, 12823, 12824, 16853
Animal breeding (D3300)
11508, 13189, 13436, 13437, 13438, 13439, 13440, 13441, 13442, 13443, 13444, 13445, 13446, 13447, 13448
Animal diseases, veterinary medicine (D3400)
11107, 14614, 14657, 14996, 14997, 14998, 14999, 15000, 15001, 15002, 15003, 15004, 15005, 15006, 15007, 15008, 15125
B4920 Domestic pets and zoo animals
3687, 15931, 20523, 21073, 21079
Animal management general and animal husbandry (D3100)
11511, 11512, 11513, 11514, 11515, 11516, 11517, 11518, 11519, 11520, 11521, 11522, 11523, 11524, 11525, 11526, 11527, 11528
Animal nutrition (D3200)
12825, 12826
Animal breeding (D3300)
13449, 13450, 13451, 13452, 13453, 13454, 13455, 13456, 13457, 13458, 13459, 13460, 15010, 15056, 15066
Animal diseases, veterinary medicine (D3400)
1978, 1979, 13457, 13682, 13970, 14422, 14433, 14621, 15009, 15010, 15011, 15012, 15013, 15014, 15015, 15016, 15017, 15018, 15019, 15020, 15021, 15022, 15023, 15024, 15025, 15026, 15027, 15028, 15029, 15030, 15031, 15032, 15033, 15034, 15035, 15036, 15037, 15038, 15039, 15040, 15041, 15042, 15043, 15044, 15045, 15046, 15047, 15048, 15049, 15050, 15051, 15052, 15053, 15054, 15055, 15056, 15057, 15058, 15059, 15060, 15061, 15062, 15063, 15064, 15065, 15066, 15067, 15099
B4930 Fur animals
20322
Animal management general and animal husbandry (D3100)
11529, 11530
Animal nutrition (D3200)
11530, 12827, 12828, 12829, 12830, 12831, 12832, 12833, 12834, 12835
Animal breeding (D3300)
15073, 15075
Animal diseases, veterinary medicine (D3400)
14443, 14621, 15068, 15069, 15070, 15071, 15072, 15073, 15074, 15075, 15076, 15077, 15078, 15079, 15080
B4990 Other domestic animals
18827
Animal nutrition (D3200)
12376
Animal breeding (D3300)
15081
Animal diseases, veterinary medicine (D3400)
15081, 15082
B5100 Viruses (including mycoplasms)
1083, 1978, 2067, 2187, 2200, 2654, 3196, 3205, 3234, 4358, 4560, 4567, 4572, 5791, 6012, 6099, 6288, 6296, 6301, 6302, 6340, 6405, 6518, 6528, 6558, 6606, 6621, 6649, 6728, 6877, 6924, 6939, 6940, 6998, 7009, 7040, 7103, 7106, 7114, 7118, 7119, 7134, 7160, 7171, 7174, 7177, 7180, 7185, 7190, 7230,

7243, 7277, 7330, 7363, 7364, 7372, 7415, 7417, 7476, 7612, 7721, 7733, 7749, 7750, 7751, 7752, 7756, 7790, 7818, 7840, 7854, 7860, 7869, 7890, 7945, 8045, 8074, 8080, 8086, 8164, 8172, 8173, 8174, 8176, 8177, 8178, 8197, 8198, 8210, 8319, 8379, 8418, 8423, 8424, 8474, 8560, 8579, 8665, 8707, 8717, 8739, 8758, 8788, 8806, 8815, 8951, 8973, 8975, 8976, 8977, 8983, 8998, 9000, 9001, 9002, 9003, 9004, 9007, 9010, 9011, 9012, 9014, 9017, 9019, 9030...

B5200 Fungi, sea weed, algae, lichen, yeast

477, 544, 601, 1000, 1022, 1042, 1043, 1048, 1065, 1068, 1071, 1076, 1077, 1078, 1079, 1080, 1081, 1082, 1083, 1084, 1086, 1089, 1090, 1097, 1100, 1129, 1131, 1133, 1135, 1137, 1138, 1139, 1142, 1144, 1446, 1676, 1687, 1865, 1874, 1875, 1941, 1951, 2138, 2254, 2255, 2275, 2298, 2323, 2327, 2329, 2331, 2350, 2596, 2605, 2654, 2849, 3003, 3007, 3127, 3135, 3187, 3205, 3286, 3998, 4348, 4369, 4380, 4381, 4464, 4560, 4565, 4566, 4567, 4570, 4572, 4759, 4809, 4835, 4836, 4837, 4838, 4895, 4896, 4945, 5000, 5051, 5151, 5233, 5394, 5395, 5396, 5398, 5399, 5401, 5406, 5414, 5435, 5441, 5746, 5806, 5869, 5884, 6010, 6017, 6021, 6042, 6099, 6146, 6155, 6161, 6162, 6165, 6169, 6177, 6182, 6185, 6210, 6212, 6213, 6214, 6217, 6220, 6221, 6222, 6240...

B5300 Bacteria

97, 141, 469, 549, 586, 601, 732, 743, 744, 763, 803, 804, 814, 816, 890, 982, 993, 994, 995, 996, 1005, 1006, 1012, 1013, 1014, 1015, 1023, 1026, 1030, 1042, 1043, 1052, 1063, 1065, 1066, 1072, 1073, 1075, 1083, 1085, 1088, 1098, 1100, 1101, 1102, 1103, 1105, 1106, 1347, 1349, 1687, 1696, 1979, 2051, 2176, 2185, 2186, 2201, 2231, 2252, 2289, 2598, 2654, 2659, 2849, 2996, 3123, 3205, 3631, 3641, 3653, 3892, 3899, 3922, 3931, 4030, 4126, 4560, 4567, 5010, 5066, 5139, 5162, 5174, 5189, 5281, 5313, 5314, 5374, 5430, 5435, 5462, 5466, 5505, 5647, 5648, 5649, 5652, 5653, 5655, 5656, 5660, 5743, 5751, 5752, 5753, 5768, 5769, 5774, 5869, 5990, 5995, 6009, 6019, 6099, 6405, 6489, 6801, 6802, 6822, 6877, 7120, 7195, 7415, 7513...

B5400 Arthropods

477, 1007, 1011, 1013, 1034, 1093, 1099, 1132, 1718, 1826, 1833, 1859, 1880, 1956, 1957, 1959, 1961, 2016, 2017, 2025, 2034, 2069, 2070, 2071, 2095, 2101, 2106, 2123, 2127, 2140, 2147, 2195, 2223, 2234, 2251, 2253, 2256, 2270, 2271, 2276, 2284, 2285, 2291, 2292, 2294, 2295, 2296, 2297, 2319, 2340, 2357, 2363, 2607, 2621, 2717, 2734, 3150, 3208, 3556, 3600, 3640, 3677, 4213, 4565, 4570, 4572, 5083, 5319, 5502, 5616, 5806, 5817, 6162, 6218, 6237, 6238, 6242, 6259, 6296, 6437, 6458, 6478, 6888, 6910, 6920, 6922, 7003, 7033, 7161, 7172, 7174, 7374, 7745, 7746, 7754, 7756, 7757, 7768, 7827, 7831, 7854, 7858, 7869, 7873, 7889, 7908, 7913, 7944, 7956, 7957, 7959, 7964, 7966, 7967, 7969, 7970, 7971, 7972, 7974, 7975, 7976, 7977, 7981, 7982, 7984...

B5500 Molluscs and worms

23, 477, 945, 1012, 1021, 1032, 1033, 1035, 1036, 1037, 1038, 1039, 1040, 1041, 1042, 1043, 1044, 1045, 1049, 1093, 1095, 1096, 1106, 1742, 1768, 1942, 1943, 1944, 1945, 1947, 1977, 2016, 2017, 2025, 2034, 2101, 2147, 2150, 2186, 2195, 2201, 2235, 2236, 2251, 2253, 2256, 2259, 2276, 2284, 2285, 2290, 2363, 2882, 3233, 4572, 5079, 5616, 6068, 6069, 6144, 6212, 6437, 6560, 6614, 6618, 6621, 6762, 6832, 6833, 6883, 6908, 6909, 7092, 7741, 7756, 7765, 7766, 7767, 7818, 7823, 7827, 7850, 7851, 7852, 7888, 7919, 7941, 7944, 7958, 7961, 7965, 7966, 7968, 7969, 7973, 7974, 7978, 7979, 7980, 7983, 8031, 8036, 8038, 8040, 8041, 8052, 8053, 8064, 8065, 8066, 8073, 8083, 8087, 8088, 8089, 8090, 8091, 8092, 8093, 8097, 8105, 8107, 8142, 8143, 8144...

B5600 Fishes in general

1153, 1553, 1663, 1833, 1834, 1921, 1932, 1933, 1934, 1935,

1936, 1937, 1948, 1949, 2011, 2012, 2013, 2014, 2015, 2016, 2017, 2018, 2019, 2020, 2021, 2022, 2023, 2024, 2025, 2026, 2027, 2029, 2041, 2042, 2043, 2055, 2056, 2058, 2062, 2119, 2141, 2144, 2149, 2165, 2226, 2251, 2253, 2262, 2263, 2264, 2265, 2266, 2268, 2269, 2271, 2272, 2273, 2274, 2276, 2277, 2278, 2279, 2280, 2281, 2282, 2283, 2284, 2285, 2286, 2287, 2288, 2316, 2320, 2321, 2334, 2341, 2342, 2344, 8356, 10403, 11375, 11376, 11377, 11378, 11379, 11380, 11381, 11382, 11383, 11384, 11385, 11386, 11388, 11393, 11399, 12727, 12729, 12730, 12733, 12830, 12832, 13498, 13502, 13890, 14951, 14952, 14953, 14956, 14957, 14959, 14961, 15405, 15406, 15424, 15425, 15426, 15427, 15428, 15429, 15430, 15453, 15454, 15455, 15521, 15522...

B5700 Birds

1845, 1882, 1889, 1916, 1923, 1958, 1966, 1978, 1979, 1990, 2040, 2062, 2063, 2064, 2065, 2066, 2094, 2098, 2120, 2122, 2125, 2154, 2155, 2160, 2162, 2163, 2165, 2170, 2172, 2185, 2186, 2187, 2191, 2192, 2198, 2199, 2202, 2204, 2205, 2207, 2208, 2210, 2212, 2213, 2214, 2217, 2227, 2228, 2229, 2230, 2232, 2233, 2238, 2252, 2253, 2261, 2293, 2307, 2310, 2316, 2317, 2348, 2349, 2351, 2353, 2354, 2355, 2356, 2358, 2361, 2363, 2364, 2408, 2505, 2506, 2507, 6008, 7777, 7913, 8104, 8305, 8318, 8325, 8326, 8329, 8331, 8335, 8341, 8345, 8349, 8355, 8356, 8358, 8400, 8449, 8558, 8594, 8774, 8893, 10421, 10451, 11269, 12885, 13256, 13479, 13487, 13488, 13491, 13695, 13854, 13867, 13890, 14802, 14803, 14804, 14805, 15014, 15015, 19132, 20166, 20193, 20274, 20544

B5800 Mammals

412, 1070, 1181, 1827, 1840, 1860, 1861, 1869, 1883, 1884, 1885, 1886, 1887, 1888, 1914, 1915, 1916, 1917, 1918, 1919, 1920, 1922, 1924, 1925, 1945, 1947, 1958, 1959, 1960, 1965, 1967, 1968, 1969, 1970, 1972, 1973, 1974, 1977, 1988, 1990, 2033, 2067, 2072, 2098, 2126, 2128, 2139, 2153, 2156, 2157, 2158, 2159, 2165, 2170, 2193, 2195, 2196, 2200, 2201, 2202, 2206, 2210, 2214, 2219, 2222, 2227, 2243, 2253, 2256, 2305, 2310, 2314, 2349, 2357, 2364, 2408, 7962, 7963, 8214, 8318, 8325, 8326, 8329, 8331, 8335, 8339, 8341, 8345, 8349, 8355, 8356, 8358, 8429, 8449, 8558, 8592, 8774, 8865, 8888, 8890, 8891, 8894, 8897, 10314, 10408, 10409, 10451, 11104, 11105, 11117, 11521, 11522, 11523, 11524, 11529, 12885, 13489, 13511, 13695, 13867, 13887, 13890, 15000, 15014, 15017...

B5910 Other non–cultivated plants

188, 231, 294, 297, 322, 546, 567, 627, 698, 699, 904, 933, 1283, 1297, 1543, 1553, 1654, 1663, 1717, 1718, 1848, 1852, 1854, 1890, 1901, 1902, 2001, 2029, 2031, 2055, 2057, 2058, 2145, 2179, 2180, 2203, 2215, 2224, 2226, 2241, 2255, 2324, 2325, 2346, 2349, 2355, 2356, 2646, 2650, 2695, 2762, 2763, 3547, 3659, 3884, 4601, 4983, 5145, 5627, 5692, 6149, 6246, 6274, 6310, 6345, 7756, 7776, 8378, 8404, 8570, 8589, 9060, 9199, 9414, 9416, 9418, 9419, 9420, 9515, 9860, 10032, 10059, 10064, 10066, 10067, 10073, 10079, 10081, 10086, 10089, 10090, 10091, 10095, 10096, 10097, 10098, 10099, 10100, 10101, 10102, 10105, 10109, 10110, 10111, 10113, 10114, 10117, 10123, 10124, 10125, 10126, 10127, 10131, 10133, 10134, 10135, 10137, 10148, 10149, 10151, 10152, 10153, 10154, 10156, 10157...

B5920 Other non–domesticated animals

412, 1012, 1031, 1046, 1048, 1130, 1136, 1212, 1825, 1941, 1943, 1944, 1945, 1946, 1948, 1949, 1991, 2093, 2098, 2141, 2161, 2194, 2195, 2253, 2310, 2345, 2349, 5079, 7754, 7944, 8326, 8558, 9075, 11326, 11374, 11518, 11690, 11826, 11831, 12209, 12249, 13389, 13482, 13483, 13489, 13536, 13538, 13539, 13543, 13549, 13553, 13582, 13596, 13603, 13604, 13605, 13606, 13627, 13628, 13629, 13632, 13633, 13634, 13635, 13636, 13637, 13639, 13640, 13641, 13867,

13890, 13900, 13903, 13904, 13927, 13947, 14080, 14081,
14144, 14145, 14414, 14423, 14434, 14443, 14449, 14581,
14589, 14596, 14598, 14599, 14600, 14654, 14655, 14656,
14817, 14818, 14884, 14885, 14935, 14938, 14988, 15014,
15029, 15034, 15035, 15050, 15085, 15099, 15100, 15106,
15107, 15111, 16221, 16629, 17066, 17067, 19071, 19164,
19180, 19619, 19631, 20193, 20522, 20675, 20804...

B6000 Man–made resources in general
15274, 15549, 17804, 20043

B6100 Implements, tools and machinery in general
951, 952, 956, 972, 2299, 2883, 3137, 4061, 10344, 15768,
17115, 17163, 17175, 17184, 17192, 17238, 17258, 17261,
17282, 17423, 17695, 17699, 17700, 17703, 17794, 18056,
18058, 18139, 18167, 18282, 18333, 18418, 19649, 20147,
20386, 20391, 20397, 20398, 20401, 20402, 20403, 20404,
20405, 20406, 20408, 20437, 20492, 20493

Engineering – equipments (D4100)
965, 2811, 2835, 15271, 15272, 15273, 15274, 15275, 15276,
15277, 15278, 15279, 15280, 15281, 15282, 15283, 15284,
15285, 15286, 15287, 15288, 15289, 15290, 15291, 15292,
15293, 15294, 15295, 15296, 15297, 15298, 15299, 15300,
15301, 15302, 15303, 15304, 15305, 15306, 15307, 15308,
15309, 15310, 15311, 15312, 15313, 15314, 15315, 15316,
15317, 15725, 17427, 19644, 19653

B6110 Soil working, tilling and fertilization equipment
9, 189, 884, 893, 919, 931, 943, 948, 949, 955, 959, 963,
973, 1574, 1776, 5031, 5325, 5350, 5434, 5507, 5635, 5786,
10266, 17190, 17272, 18281, 20387, 20388

Engineering – equipments (D4100)
558, 15270, 15318, 15319, 15320, 15321, 15322, 15323,
15324, 15325, 15326, 15327, 15328, 15329, 15330, 15331,
15332, 15333, 15334, 15462

B6120 Equipment for sowing, planting and setting
189, 864, 959, 963, 973, 2612, 2885, 2991, 3602, 4627,
4799, 5956, 17272, 20388, 20517

Engineering – equipments (D4100)
558, 2834, 4767, 15133, 15134, 15334, 15335, 15336

B6130 Equipment and implements for crop husbandry and crop protection
189, 304, 963, 973, 977, 2316, 2613, 2677, 2827, 2872,
2885, 3294, 3703, 3749, 3784, 3792, 3826, 3828, 3842, 3907,
4110, 4239, 4256, 4394, 4399, 4452, 4554, 4615, 4677, 4716,
4725, 4752, 4790, 4874, 4926, 4954, 5199, 5242, 5243, 5346,
5754, 5939, 5940, 7776, 7779, 7825, 7883, 7924, 10173,
10175, 10252, 10369, 15617, 15625, 15634, 15635, 17249,
17255, 17293, 17613, 17675, 19708, 20278, 20279, 20348,
20395, 20407, 20620

Engineering – equipments (D4100)
2753, 3868, 4767, 7390, 7792, 7857, 7954, 8098, 9877,
15131, 15270, 15334, 15337, 15338, 15339, 15340, 15341,
15342, 15343, 15344, 15345, 15346, 15347, 15348, 15349,
15350, 15351, 15352, 15353, 15354, 15355, 15356, 15357,
15358, 15359, 15360, 15590

B6140 Implements and machinery for harvesting crop products
189, 3139, 3144, 3151, 3153, 3235, 3285, 3289, 3293, 3303,
3361, 3367, 3370, 3514, 3703, 3755, 3882, 3963, 3967, 3973,
3990, 4330, 4338, 4350, 4447, 4450, 4453, 4473, 4474, 4475,
5047, 6002, 6569, 7053, 7140, 10860, 10877, 12874, 15749,
15773, 15777, 15778, 15781, 15790, 15791, 15792, 15795,
15802, 17006, 17225, 17229, 17233, 17247, 17254, 17262,
17268, 17271, 17290, 17291, 17293, 17586, 17590, 17591,
20389, 20394

Engineering – equipments (D4100)
3282, 3964, 4356, 4461, 10185, 15132, 15265, 15266, 15268,
15269, 15360, 15361, 15362, 15363, 15364, 15365, 15366,
15367, 15368, 15369, 15370, 15371, 15372, 15373, 15374,

15375, 15376, 15377, 15378, 15379, 15380, 15381, 15382,
15383, 15384, 15385, 15386, 15387, 15388, 15389, 15390,
15391, 15392, 15393, 15394, 15395, 15396, 15397, 15398,
15399, 15400, 15401, 15402, 15403, 15404, 17251

B6150 Equipment for harvesting animal products and raising of animals
3309, 10422, 10675, 10888, 10893, 10896, 10898, 10914,
10920, 11084, 11138, 11192, 11228, 11230, 11231, 11392,
11453, 11502, 11665, 12234, 12238, 12240, 13336, 13397,
14399, 14683, 14755, 15651, 15812, 16763, 17278, 17279,
17299, 17556, 19170, 19217, 19572, 19573, 19594, 19704,
19714, 19757, 20392, 20468, 20470

Engineering – equipments (D4100)
10894, 11102, 11391, 11622, 12447, 12448, 12542, 14151,
14390, 14391, 14407, 15374, 15385, 15405, 15406, 15407,
15408, 15409, 15410, 15411, 15412, 15413, 15414, 15415,
15416, 15417, 15418, 15419, 15420, 15421, 15422, 15423,
15424, 15425, 15426, 15427, 15428, 15429, 15430, 15431,
15432, 15433, 15434, 15435, 15436, 15437, 15438, 15439,
15440, 15441, 15442, 15443, 15444, 15445, 15446, 15447,
15448, 15449, 15450, 15451, 15452, 15453, 15454, 15455,
15654, 17280, 19720, 20074

B6160 Machines and equipment for processing of products
2549, 6569, 15695, 15704, 15741, 15780, 15812, 15847,
15870, 15871, 15884, 15885, 15925, 15980, 15993, 16000,
16001, 16030, 16031, 16035, 16059, 16068, 16084, 16182,
16191, 16252, 16253, 16255, 16284, 16285, 16295, 16317,
16337, 16338, 16339, 16340, 16341, 16372, 16373, 16439,
16442, 16455, 16459, 16460, 16471, 16489, 16515, 16554,
16568, 16622, 16662, 16673, 16674, 16680, 16689, 16696,
16706, 16745, 16775, 16776, 16777, 16781, 16805, 16806,
16814, 16837, 16902, 16915, 16917, 16919, 16922, 16929,
16938, 16943, 16944, 16945, 16949, 16967, 16972, 17008,
17014, 17016, 17018, 17019, 17020, 17064, 17094, 17107,
17110, 17117, 17120, 17131, 17132, 17133, 17136, 17137,
17138, 17139, 17140, 17144, 17145, 17148, 17152, 17153,
17154, 17157, 17191, 17206, 17222, 17236, 17256, 17257,
17268, 17274, 17290, 17291, 17494, 17814, 17856, 18920,
18970, 18978, 19123, 19143, 19620...

Engineering – equipments (D4100)
15135, 15368, 15369, 15397, 15400, 15456, 15457, 15458,
15459, 15460, 15461, 15462, 15463, 15464, 15465, 15466,
15467, 15468, 15469, 15470, 15471, 15472, 15473, 15474,
15475, 15476, 15477, 15478, 15479, 15480, 15481, 15482,
15483, 15484, 15485, 15486, 15487, 15698, 15703, 15709,
15745, 15985, 16028, 16268, 16289, 16474, 16578, 16620,
16621, 17081, 17113, 17114, 17125, 17151, 17280, 18930,
19632, 19659

B6170 Transport equipment, – installations and facilities
2485, 2522, 4565, 8400, 8792, 10459, 10930, 11121, 11350,
13613, 15730, 15790, 15812, 16022, 16030, 16031, 16167,
16247, 17107, 17130, 17135, 17136, 17137, 17142, 17145,
17146, 17147, 17158, 17178, 17231, 17245, 17268, 17276,
17283, 17290, 17291, 17669, 17885, 19550, 20390, 20393,
20469, 20703

Engineering – equipments (D4100)
15406, 15429, 15453, 15454, 15455, 15475, 15478, 15484,
15488, 15489, 15490, 15491, 15492, 15493, 15494, 15495,
15496, 15497, 15498, 15499, 15500, 15501, 15502, 15503,
15504, 15505, 15506, 15507, 15508, 15509, 15510, 15511,
15512, 15513, 15514, 15515, 15516, 15517, 15518, 15519,
15520, 15521, 15522, 16158, 17129, 17151, 17280, 20249

B6180 Household equipment
18433, 18434, 18436, 18446, 18976

Engineering – equipments (D4100)
15523

B6190 Other machinery and equipment
889, 1141, 15596, 19825, 20399
Engineering – equipments (D4100)
15524, 15525, 15526, 15527, 15528, 15529, 15530

B6200 Structures and other permanent installations in general
2197, 2421, 2457, 2458, 2496, 2501, 5888, 15725, 16366, 17163, 17428, 17810, 18058, 19744, 20066, 20386, 20408, 20620
Engineering – buildings (D4200)
15281, 15301, 15302, 15308, 15552, 15553, 15554, 15555, 15556, 15557, 15558, 15559, 15560, 15561, 15562, 15563, 15564, 15565, 15566, 15567, 15568, 15569, 15570, 15571, 15572, 15573, 15574, 15575, 15576, 15577, 15578, 15579, 15580, 15581, 15582, 15583, 15584, 15585, 15586, 15587, 15588

B6210 Glasshouses, nurseries
758, 894, 905, 1409, 2536, 2613, 2826, 3694, 3707, 3759, 3767, 3768, 3769, 3770, 3780, 3781, 3782, 3843, 3844, 3845, 3846, 3907, 3940, 3941, 3942, 3944, 3972, 4355, 4388, 4408, 4497, 4624, 4625, 4667, 4726, 4939, 5708, 6004, 6859, 7192, 7557, 7559, 7755, 7975, 7980, 8075, 8116, 8610, 8611, 8614, 8823, 8860, 8974, 8978, 9214, 10116, 10271, 11182, 15131, 15298, 15307, 15341, 15838, 16159, 17593, 17611, 19098, 20081, 20405
Engineering – buildings (D4200)
2753, 15531, 15532, 15589, 15590, 15591, 15592, 15593, 15594, 15595, 15596, 15597, 15598, 15599, 15600, 15601, 15602, 15603, 15604, 15605, 15606, 15607, 15608, 15609, 15610, 15611, 15612, 15613, 15614, 15615, 15616, 15617, 15618, 15619, 15620, 15621, 15622, 15623, 15624, 15625, 15626, 15627, 15628, 15629, 15630, 15631, 15632, 15633, 15634, 15635, 15636, 15660

B6220 Stables
1360, 1445, 1492, 5895, 10357, 10478, 10629, 10646, 10672, 10699, 10705, 10709, 10721, 10723, 10727, 10736, 10868, 10892, 10895, 10897, 10899, 10901, 10902, 10906, 10935, 10939, 11056, 11103, 11128, 11129, 11130, 11131, 11133, 11135, 11138, 11139, 11143, 11146, 11163, 11173, 11179, 11184, 11189, 11192, 11245, 11249, 11250, 11251, 11259, 11262, 11272, 11274, 11275, 11277, 11278, 11282, 11310, 11311, 11312, 11317, 11322, 11324, 11327, 11328, 11418, 11432, 11444, 11493, 11496, 11575, 12001, 12213, 12233, 12238, 12407, 13611, 13612, 13646, 13651, 14136, 14425, 14647, 15136, 15441, 15442, 15444, 15446, 15449, 15846, 17069, 17193, 17194, 17198, 17211, 17212, 17281, 18823, 19217, 19576, 19617, 19618, 19628, 19654, 19656, 19712, 19714, 19715, 19716, 19717, 19718, 19740, 20053, 20084, 20396, 20401
Engineering – buildings (D4200)
10404, 10450, 10698, 10714, 10718, 10726, 10840, 10856, 10857, 10903, 10905, 11063, 11100, 11101, 11167, 11174, 11176, 11177, 11178, 11235, 11248, 11252, 11283, 11309, 11397, 11417, 11462, 14397, 15380, 15439, 15450, 15451, 15532, 15533, 15535, 15536, 15537, 15538, 15552, 15637, 15638, 15639, 15640, 15641, 15642, 15643, 15644, 15645, 15646, 15647, 15648, 15649, 15650, 15651, 15652, 15653, 15654, 15655, 15656, 15657, 15658, 15659, 15660, 15661, 15662, 15663, 15664, 15665, 15666, 15667, 15668, 15669, 15670, 15671, 15672, 15673, 15674, 15675, 15676, 15677, 15678, 15679, 15680, 15681, 15682, 15683, 15684, 15685, 15686, 15687, 15688, 15689, 15690, 15691, 15692, 15693, 15694, 15695, 15696, 15697, 15698, 16352, 17299, 19746

B6230 Accomodation for storing and processing of products
3235, 3883, 8071, 11345, 12227, 13877, 15407, 15475, 15476, 15787, 15835, 15836, 15848, 15849, 15871, 15872, 15885, 15886, 15897, 15902, 15905, 15920, 15923, 15925,

15991, 15996, 16020, 16022, 16030, 16031, 16039, 16044, 16056, 16059, 16060, 16066, 16068, 16072, 16077, 16081, 16083, 16097, 16111, 16138, 16160, 16167, 16169, 16170, 16171, 16178, 16179, 16224, 16233, 16241, 16247, 16252, 16260, 16262, 16298, 16340, 16622, 16790, 16791, 16830, 17016, 17107, 17120, 17133, 17144, 17146, 17196, 17274, 17288, 17289, 17550, 17711, 18435, 19098, 19598, 19620, 19672, 19715, 19742, 19743, 19747, 20417, 20419, 20421, 20422, 20469, 20703
Engineering – buildings (D4200)
15461, 15480, 15676, 15680, 15688, 15695, 15699, 15700, 15701, 15702, 15703, 15704, 15705, 15706, 15707, 15708, 15709, 15710, 15723, 16076, 16096

B6240 Houses and furniture
1856, 2486, 2515, 18296, 18393, 18433, 18451, 18455, 20219
Engineering – buildings (D4200)
15711

B6250 Domestic and community water supply facilities and systems
1394, 1590, 1674, 1688, 1693, 1724, 2288, 19686
Engineering – buildings (D4200)
1335

B6260 Drainage and irrigation facilities and systems
176, 279, 816, 830, 1162, 1163, 1166, 1167, 1169, 1171, 1172, 1174, 1188, 1192, 1193, 1194, 1195, 1200, 1202, 1206, 1216, 1217, 1221, 1223, 1249, 1250, 1252, 1256, 1257, 1259, 1261, 1264, 1268, 1273, 1285, 1287, 1298, 1308, 1315, 1317, 1318, 1340, 1344, 1345, 1361, 1362, 1367, 1374, 1382, 1383, 1384, 1385, 1389, 1390, 1412, 1417, 1437, 1461, 1462, 1463, 1464, 1468, 1470, 1471, 1480, 1482, 1485, 1486, 1487, 1489, 1491, 1494, 1497, 1499, 1500, 1501, 1502, 1506, 1512, 1513, 1515, 1519, 1520, 1521, 1522, 1524, 1525, 1526, 1527, 1528, 1529, 1530, 1531, 1535, 1536, 1538, 1540, 1550, 1554, 1556, 1574, 1575, 1576, 1577, 1585, 1586, 1593, 1594, 1600, 1602, 1609, 1613, 1617, 1618, 1623, 1625, 1627, 1641, 1645, 1646, 1670, 1671, 1674, 1678, 1680...
Engineering – buildings (D4200)
1395, 1441, 2699, 5181, 5186, 15624, 15712, 15713, 15714, 15715, 15716, 15717, 15718, 15719

B6270 Sewage and waste disposal facilities and systems
17, 559, 560, 1134, 1140, 1632, 1648, 1724, 1725, 15764, 16622, 16626, 16664, 17052, 17059, 17081, 17094, 19570, 19582, 19583, 19590, 19591, 19595, 19610, 19627, 19638, 19694, 19696, 19705, 19709, 19723, 19730, 19731, 19732, 19734, 19736, 19737, 19738, 19745, 19752, 19759
Engineering – buildings (D4200)
1356, 11374, 11397, 15534, 15680, 15708, 15710, 15720, 15721, 15722, 17159

B6280 Roads, farm yards, fences
852, 1204, 1586, 1590, 1635, 1852, 1971, 2031, 2182, 2183, 2238, 2300, 2355, 2376, 2377, 2397, 2423, 2426, 2438, 2439, 2464, 2465, 2468, 2475, 2485, 2500, 2508, 2520, 2526, 2553, 2554, 2585, 2610, 2648, 3906, 4085, 4776, 4779, 4980, 8981, 10308, 10330, 10367, 12255, 15528, 15529, 15729, 15730, 15731, 17686, 18012, 19613
Engineering – buildings (D4200)
5047, 15375, 15551, 15723

B6290 Other structures and facilities
2032, 2132, 2281, 2437, 2581, 4755, 11435, 15530, 19663
Engineering – buildings (D4200)
15686, 15724

B6300 Feeding stuffs and drinking water for animals in general
313, 1974, 3063, 3688, 5509, 10319, 10471, 10561, 10588, 10658, 10691, 10736, 10738, 10741, 10746, 10870, 10883, 10887, 10895, 11095, 11102, 11241, 11361, 11433, 11503,

11538, 11540, 11541, 11543, 11545, 11555, 11558, 11559, 11563, 11603, 11605, 11606, 11628, 11633, 11635, 11638, 11639, 11640, 11644, 11645, 11646, 11647, 11648, 11650, 11651, 11655, 11665, 11666, 11667, 11668, 11669, 11671, 11674, 11678, 11693, 11694, 11695, 11703, 11752, 11753, 11755, 11759, 11762, 11765, 11766, 11774, 11775, 11776, 11779, 11780, 11782, 11783, 11785, 11786, 11788, 11790, 11793, 11794, 11796, 11805, 11806, 11809, 11818, 11821, 11824, 11832, 11834, 11836, 11837, 11839, 11925, 11932, 11935, 11947, 11949, 11950, 11951, 11966, 11972, 11977, 11982, 11989, 11992, 11996, 11997, 12004, 12025, 12028, 12029, 12041, 12043, 12046, 12048, 12050, 12051, 12055, 12058, 12061, 12065, 12068...

Storage and conservation (D4420)
2702, 3318, 15576, 15887, 15888, 15889, 15890, 15990, 15997, 16026

Processing (D4430)
12037, 12056, 15472, 15888, 16289, 16290, 16291, 16292, 16293, 16406, 16457, 16815, 16892, 17100

Transport and handling (D4440)
17100, 17111

B6310 Concentrates
720, 2663, 3637, 3897, 6211, 6438, 6821, 10499, 10698, 10706, 10720, 11164, 11542, 11547, 11553, 11554, 11556, 11557, 11564, 11577, 11621, 11629, 11637, 11643, 11654, 11664, 11679, 11680, 11697, 11698, 11702, 11705, 11706, 11707, 11756, 11757, 11760, 11767, 11770, 11773, 11784, 11787, 11789, 11791, 11800, 11803, 11807, 11813, 11819, 11822, 11823, 11827, 11833, 11860, 11928, 11930, 11933, 11934, 11939, 11940, 11941, 11943, 11953, 11958, 11959, 11960, 11964, 11969, 11975, 11979, 11983, 11984, 12000, 12001, 12003, 12022, 12027, 12030, 12031, 12032, 12035, 12052, 12053, 12075, 12076, 12078, 12083, 12164, 12165, 12175, 12178, 12179, 12180, 12188, 12190, 12191, 12192, 12193, 12194, 12198, 12199, 12200, 12205, 12207, 12208, 12210, 12211, 12213, 12214, 12216, 12217, 12234, 12235, 12237, 12246, 12257, 12258, 12259, 12260, 12261, 12262, 12263, 12264, 12265, 12266...

Storage and conservation (D4420)
15891, 15892, 15893, 16043, 16323

Processing (D4430)
11663, 11711, 12815, 16043, 16274, 16294, 16295, 16296, 16297, 16298, 16299, 16300, 16301, 16302, 16303, 16304, 16305, 16306, 16307, 16308, 16309, 16310, 16311, 16312, 16313, 16314, 16315, 16316, 16317, 16318, 16323, 16488, 16538, 16540, 16853, 16955, 16958, 16960, 16961, 16963, 16964, 17089, 17098, 19269

Transport and handling (D4440)
17120

B6320 Roughage
2625, 2626, 2627, 2628, 2629, 2632, 2633, 3307, 3314, 3316, 3343, 3346, 3361, 3373, 3375, 3397, 3402, 3405, 3406, 3484, 3504, 3505, 3510, 3519, 3522, 3523, 3545, 3598, 3665, 5067, 5257, 5523, 5560, 5576, 5580, 5594, 5597, 5621, 5623, 5626, 5627, 5630, 5635, 5639, 5642, 5657, 5658, 5670, 5671, 5676, 5677, 6202, 6624, 6627, 6628, 6734, 6788, 6805, 6806, 6820, 10659, 10720, 10725, 10860, 10877, 10899, 10928, 11546, 11548, 11549, 11550, 11560, 11566, 11570, 11572, 11573, 11574, 11610, 11611, 11612, 11614, 11615, 11616, 11619, 11620, 11622, 11656, 11659, 11661, 11670, 11696, 11697, 11699, 11700, 11705, 11708, 11709, 11750, 11760, 11772, 11797, 11802, 11811, 11812, 11820, 11825, 11827, 11840, 11841, 11842, 11864, 11885, 11924, 11931, 11933, 11937, 11938, 11942, 11952, 11954, 11956, 11957, 11985, 11990, 11998...

Harvesting (D4410)

3280, 3367, 3370, 3529, 11751, 15385, 15778, 15779, 15780, 15781, 15782, 15783, 15784, 15785, 15786, 15787, 15788, 15814

Storage and conservation (D4420)
3280, 3367, 3391, 3661, 11571, 11701, 11723, 11771, 12084, 14200, 15786, 15787, 15821, 15822, 15839, 15840, 15841, 15842, 15843, 15854, 15894, 15895, 15896, 15897, 15898, 15899, 15900, 15901, 15902, 15903, 15904, 15905, 15906, 15907, 15908, 15909, 15910, 15911, 15912, 15913, 15914, 15915, 15916, 15917, 15918, 15919, 15920, 15921, 15922, 15923, 15924, 15925, 15926, 15927, 15966, 16044, 16323, 16340

Processing (D4430)
3084, 3319, 3621, 3657, 11701, 11929, 12222, 12236, 12279, 15907, 15927, 16297, 16313, 16319, 16320, 16321, 16322, 16323, 16324, 16325, 16326, 16327, 16328, 16329, 16330, 16331, 16332, 16333, 16334, 16335, 16336, 16337, 16338, 16339, 16340, 16341, 16433

Transport and handling (D4440)
15966, 17104

B6330 Drinking water for animals
1625, 10925, 10928, 11153, 11791, 11967, 12553, 14132, 19630

B6390 Other feeding stuffs
11554, 11608, 11781, 11955, 11987, 11988, 12256, 12282, 12554, 12705, 12760, 12804, 12829, 13874, 14394, 14431, 14454, 15174, 15253, 17872, 19902, 19966

Storage and conservation (D4420)
16344, 16345

Processing (D4430)
16342, 16343, 16344, 16345, 16346, 16347, 16348, 16853, 16962, 17062

B6400 Fertilizers and water for plants in general
1, 18, 19, 122, 165, 169, 190, 207, 224, 231, 298, 313, 319, 338, 448, 450, 470, 491, 497, 499, 502, 503, 521, 524, 820, 825, 848, 878, 929, 988, 1194, 1235, 1295, 1339, 1343, 1377, 1382, 1418, 1571, 1574, 1608, 1615, 1638, 1752, 1754, 2046, 2694, 2898, 3094, 3240, 3344, 3506, 3537, 3542, 3544, 3898, 3901, 4008, 4013, 4060, 4162, 4491, 4606, 4608, 4615, 4782, 5056, 5060, 5073, 5077, 5081, 5088, 5118, 5119, 5132, 5147, 5149, 5163, 5166, 5173, 5177, 5191, 5209, 5215, 5218, 5219, 5221, 5232, 5240, 5243, 5244, 5251, 5256, 5261, 5262, 5264, 5265, 5281, 5298, 5312, 5326, 5328, 5331, 5338, 5349, 5355, 5365, 5379, 5380, 5410, 5426, 5428, 5431, 5442, 5443, 5445, 5447, 5448, 5456, 5463, 5465, 5468, 5474, 5495, 5503...

Storage and conservation (D4420)
15862, 15928

B6410 Organic fertilizers
11, 23, 161, 174, 187, 193, 328, 342, 441, 444, 468, 522, 523, 528, 534, 556, 560, 609, 612, 720, 727, 734, 746, 753, 788, 789, 802, 804, 807, 808, 947, 999, 1044, 1087, 1341, 1647, 1660, 1705, 1728, 1734, 1749, 1756, 2809, 2837, 2877, 2881, 2924, 3075, 3569, 3590, 3673, 3847, 4035, 4036, 4658, 5062, 5071, 5082, 5086, 5093, 5140, 5152, 5159, 5161, 5162, 5168, 5169, 5172, 5175, 5196, 5202, 5204, 5205, 5214, 5229, 5235, 5239, 5275, 5295, 5299, 5300, 5303, 5305, 5319, 5321, 5323, 5324, 5332, 5334, 5339, 5350, 5356, 5357, 5358, 5367, 5432, 5433, 5435, 5440, 5502, 5509, 5510, 5522, 5523, 5533, 5567, 5573, 5574, 5575, 5582, 5593, 5600, 5609, 5614, 5616, 5617, 5621, 5629, 5630, 5631, 5634, 5635, 5643, 5647, 5648...

Storage and conservation (D4420)
15709, 15929, 16264

Processing (D4430)
1012, 15709, 15929, 16349, 16350, 16351, 16352, 16353,

16354, 16355, 16356, 17061, 17065, 17078, 17088, 17093
Transport and handling (D4440)
17112

B6420 Mineral fertilizers
17, 23, 93, 121, 123, 133, 134, 138, 153, 161, 175, 176, 183,
227, 237, 285, 311, 438, 449, 464, 467, 468, 473, 500, 509,
510, 516, 520, 522, 528, 571, 605, 609, 610, 612, 618, 634,
635, 651, 662, 664, 670, 674, 675, 680, 689, 766, 780, 781,
844, 1202, 1373, 1414, 1434, 1478, 1660, 1735, 1743, 1761,
1984, 2030, 2680, 2837, 2996, 3054, 3090, 3113, 3128, 3164,
3262, 3264, 3268, 3355, 3374, 3480, 3486, 3510, 3530, 3595,
3637, 3687, 3699, 3706, 3847, 4153, 4234, 4383, 4607, 4618,
4655, 5048, 5049, 5057, 5059, 5061, 5063, 5064, 5065, 5072,
5091, 5120, 5121, 5125, 5133, 5137, 5138, 5141, 5143, 5156,
5157, 5161, 5168, 5171, 5174, 5183, 5187, 5198, 5199, 5201,
5203, 5205, 5207, 5210, 5213, 5214...
Storage and conservation (D4420)
3568, 3837, 4131
Processing (D4430)
16581, 17053, 17084, 17092

B6430 Pouring, sprinkling, and irrigation water
285, 293, 310, 492, 578, 932, 1206, 1216, 1257, 1348, 1349,
1350, 1352, 1353, 1358, 1376, 1378, 1380, 1381, 1397, 1405,
1406, 1408, 1409, 1414, 1415, 1416, 1419, 1423, 1424, 1425,
1426, 1427, 1433, 1434, 1435, 1437, 1441, 1450, 1463, 1464,
1467, 1469, 1470, 1471, 1472, 1473, 1475, 1476, 1477, 1478,
1479, 1481, 1482, 1483, 1484, 1486, 1489, 1490, 1492, 1494,
1495, 1496, 1500, 1501, 1502, 1503, 1504, 1507, 1513, 1516,
1520, 1522, 1525, 1526, 1529, 1531, 1532, 1535, 1536, 1537,
1538, 1540, 1542, 1544, 1545, 1547, 1549, 1550, 1552, 1556,
1567, 1567, 1577, 1592, 1593, 1617, 1618, 1625, 1629, 1680,
1738, 2727, 2742, 2860, 2873, 3061, 3138, 3257, 3331, 3693,
3907, 4193, 4866, 5070, 5104, 5107, 5167, 5178, 5179, 5180,
5181, 5182, 5194, 5253...
Processing (D4430)
1449, 16357

B6440 Amendments
485, 610, 613, 614, 680, 897, 940, 1014, 1161, 1300, 1640,
1732, 1740, 1743, 1747, 1748, 1758, 1762, 1769, 1802, 2823,
5140, 5216, 5258, 5339, 5356, 5358, 5409, 5458, 5586, 5612,
5652, 5695, 5890, 5968, 15324, 20325, 20383
Processing (D4430)
16349, 17086

B6490 Other fertilizers
3754, 5840, 5941

B6500 Propagation materials in general
1838, 1839, 2602, 3145, 4167, 4471, 4719, 6226, 7237, 9061,
14174, 18194, 20245
Storage and conservation (D4420)
15930

B6510 Seed for sowing
558, 2694, 2698, 2739, 2740, 2741, 2749, 2834, 2838, 2841,
2855, 2856, 2857, 2860, 2862, 2885, 2899, 2907, 2917, 2922,
3015, 3061, 3066, 3092, 3135, 3160, 3181, 3236, 3237, 3238,
3241, 3288, 3291, 3302, 3347, 3393, 3491, 3494, 3499, 3500,
3501, 3502, 3503, 3515, 3516, 3528, 3602, 3620, 3629, 3701,
3705, 3751, 3786, 3793, 3817, 3825, 3896, 3929, 3962, 3963,
4047, 4102, 4109, 4420, 4463, 4586, 4630, 4688, 4714, 4731,
4741, 4774, 4843, 4844, 4850, 4920, 4933, 4981, 5043, 5561,
5562, 5563, 6005, 6007, 6037, 6097, 6342, 6448, 6602, 6637,
6642, 6656, 6728, 6730, 6867, 6873, 6919, 6949, 7013, 7047,
7136, 7164, 7175, 7583, 7631, 7638, 7650, 7671, 7672, 7684,
7685, 7688, 7696, 7817, 7866, 8378, 8539, 8599, 8878, 8952,
9201, 9220, 9230, 9275, 9336...
Storage and conservation (D4420)
2697, 2858, 3095, 4598, 7812, 15931, 15932, 15933, 15934,

15935, 15936, 15937, 16059, 16060
Transport and handling (D4440)
16060

B6520 Sperm
10433, 10435, 10444, 10455, 10474, 10716, 10732, 10844,
10951, 10954, 10956, 10959, 10960, 10963, 10964, 11029,
11032, 11033, 11078, 11115, 11194, 11196, 11346, 11372,
11379, 11380, 11382, 11506, 12870, 12938, 12970, 12972,
13292, 13293, 13297, 13362, 13485, 14175, 14180, 14588,
16288
Storage and conservation (D4420)
10668, 11061, 11258, 15938, 15939

B6590 Other propagation materials
2660, 2661, 2664, 2698, 2721, 2722, 2723, 2829, 2863, 2865,
2866, 2867, 2868, 3157, 3159, 3234, 3245, 3247, 3249, 3707,
3893, 3943, 4046, 4086, 4089, 4094, 4103, 4104, 4105, 4107,
4108, 4114, 4119, 4129, 4197, 4198, 4232, 4236, 4237, 4238,
4253, 4254, 4272, 4285, 4332, 4339, 4340, 4344, 4349, 4357,
4358, 4359, 4416, 4417, 4435, 4436, 4438, 4439, 4440, 4441,
4457, 4463, 4571, 4602, 4629, 4690, 4721, 4722, 4723, 4724,
4766, 4811, 4839, 4844, 4876, 4877, 4900, 4913, 4958, 4965,
5020, 5780, 5809, 5941, 6012, 6022, 6037, 6072, 6455, 6460,
6468, 6470, 6558, 6560, 6569, 7216, 7218, 7219, 7228, 7230,
7249, 7258, 7270, 7271, 7274, 7275, 7296, 7336, 7344, 7353,
7376, 7397, 7417, 7418, 7429, 7433, 7447, 7457, 7461, 7462,
7465, 7471, 7481, 7483, 7484...
Storage and conservation (D4420)
2697, 4560, 4565, 4566, 4567, 4570, 4574, 4613, 5893,
10563, 10598, 15875, 15940, 15941, 15942, 15943, 15944,
16076, 16077, 16079, 16080, 16130
Transport and handling (D4440)
15698, 16079, 17113, 17114, 17125

B6600 Biocides, detergents, food additives and growth regulators in general
437, 727, 1109, 1843, 1864, 1865, 2113, 2114, 2159, 4068,
4491, 5389, 6159, 6273, 7762, 7858, 8250, 8251, 9597, 9769,
10142, 10204, 10210, 10536, 10627, 11551, 11552, 12632,
13472, 13672, 13891, 13896, 15991, 16559, 18464, 18475,
18950, 18975, 19229, 19291, 19551, 19749, 19773, 20008,
20015, 20461, 20462, 20463, 20464, 20465, 20466, 20467,
20920, 20941

B6610 Pesticides
2, 3, 14, 20, 100, 142, 157, 188, 195, 196, 208, 231, 267,
280, 297, 317, 319, 320, 321, 322, 397, 398, 474, 477, 478,
545, 546, 547, 548, 625, 627, 698, 699, 725, 726, 794, 799,
904, 933, 945, 985, 988, 1001, 1002, 1003, 1005, 1007,
1019, 1020, 1025, 1028, 1029, 1030, 1037, 1067, 1072, 1073,
1082, 1084, 1086, 1089, 1090, 1091, 1092, 1094, 1107, 1114,
1116, 1146, 1219, 1283, 1284, 1297, 1321, 1373, 1543, 1555,
1578, 1716, 1717, 1718, 1831, 1875, 1932, 1976, 1986, 2026,
2031, 2050, 2051, 2052, 2054, 2069, 2072, 2106, 2115, 2163,
2165, 2202, 2207, 2208, 2210, 2217, 2226, 2228, 2259, 2316,
2319, 2342, 2364, 2646, 2694, 2761, 2791, 2792, 2840, 2877,
2881, 2882, 2898, 3007, 3008, 3033, 3098, 3113...

B6620 Growth regulators
669, 2597, 2599, 2600, 2634, 2660, 2689, 2703, 2724, 2836,
2839, 2840, 2844, 2847, 2849, 2854, 2861, 2901, 2905, 2938,
3007, 3008, 3014, 3016, 3021, 3035, 3079, 3086, 3096, 3106,
3108, 3123, 3128, 3136, 3139, 3143, 3149, 3155, 3200, 3242,
3245, 3247, 3248, 3250, 3286, 3287, 3296, 3393, 3509, 3760,
3763, 3764, 3776, 3785, 3872, 3900, 3923, 3935, 3946, 3961,
3966, 3968, 3969, 3970, 3971, 3989, 3995, 4052, 4055, 4056,
4062, 4093, 4095, 4102, 4111, 4112, 4113, 4115, 4120, 4153,
4157, 4190, 4191, 4194, 4195, 4196, 4205, 4207, 4221, 4240,
4261, 4264, 4265, 4266, 4269, 4273, 4280, 4338, 4351, 4402,
4432, 4446, 4466, 4470, 4520, 4521, 4522, 4550, 4559, 4560,

4566, 4567, 4570, 4573, 4599, 4609, 4623, 4624, 4659, 4660, 4663, 4675, 4718, 4720, 4722...

B6630 Animal medicines

5162, 9446, 10647, 10662, 10839, 10852, 10858, 11028, 11029, 11030, 11031, 11034, 11036, 11047, 11237, 11416, 11667, 11773, 12596, 12621, 12852, 12871, 12872, 13477, 13478, 13486, 13500, 13526, 13528, 13529, 13535, 13544, 13545, 13549, 13559, 13560, 13561, 13562, 13563, 13568, 13570, 13582, 13584, 13601, 13618, 13621, 13623, 13625, 13626, 13633, 13634, 13652, 13655, 13656, 13661, 13671, 13686, 13700, 13710, 13713, 13715, 13716, 13717, 13718, 13720, 13721, 13876, 13880, 13882, 13883, 13884, 13892, 13901, 13919, 13920, 13929, 13932, 13933, 13935, 13942, 13943, 13959, 13974, 13975, 13984, 13986, 14063, 14070, 14073, 14078, 14079, 14082, 14084, 14085, 14090, 14091, 14113, 14125, 14139, 14140, 14141, 14142, 14151, 14152, 14157, 14159, 14173, 14177, 14200, 14201, 14366, 14368, 14370, 14374, 14393, 14394, 14395, 14396, 14398, 14402, 14403, 14405, 14406, 14411, 14412...

B6640 Disinfectants

11345, 11359, 13589, 13590, 13665, 14108, 14392, 14396, 14399, 14789, 14869, 14872, 16929, 16938, 16939, 19572, 19594, 19629, 19630, 19666, 19714, 19720, 19742, 19743, 19757, 19761, 20468, 20470, 20489, 21089

B6650 Detergents, food additives and other preservations

720, 1311, 2053, 2345, 3318, 9082, 10335, 10387, 10889, 11092, 11539, 11548, 11664, 11711, 11940, 11941, 12168, 12222, 12236, 12279, 12415, 12493, 12548, 12558, 12566, 12568, 12653, 15471, 15585, 15821, 15828, 15834, 15861, 15870, 15873, 15878, 15882, 15884, 15891, 15895, 15908, 15922, 15936, 15937, 15942, 15947, 15948, 15958, 15963, 15964, 15972, 15977, 15978, 15984, 15989, 16014, 16030, 16031, 16062, 16066, 16077, 16078, 16083, 16084, 16117, 16139, 16140, 16142, 16150, 16155, 16160, 16161, 16166, 16172, 16190, 16194, 16195, 16207, 16209, 16212, 16238, 16244, 16250, 16251, 16260, 16262, 16286, 16297, 16310, 16316, 16323, 16325, 16326, 16328, 16337, 16368, 16445, 16446, 16458, 16473, 16475, 16482, 16483, 16493, 16494, 16495, 16496, 16500, 16522, 16525, 16536, 16543, 16544, 16545, 16547, 16548, 16549, 16550, 16551, 16573, 16624, 16628, 16656, 16657, 16661...

B6700 Fibre materials and wood products

648, 816, 1745, 2311, 2609, 2617, 2624, 4828, 4846, 4858, 4932, 4953, 4991, 5073, 5855, 5953, 5990, 7947, 7948, 10927, 10928, 11043, 11530, 11670, 11841, 11942, 12842, 13177, 13515, 15135, 15170, 15400, 15458, 15459, 15460, 15554, 15555, 15556, 15560, 15561, 15562, 15563, 15564, 15565, 15566, 15567, 15568, 15569, 15734, 15735, 15736, 15737, 15738, 15739, 15740, 15744, 15746, 15747, 15749, 15751, 15752, 15753, 15756, 15757, 17164, 17260, 17268, 17274, 17287, 17294, 17637, 17639, 17662, 17708, 17778, 17807, 17808, 17809, 17810, 17823, 17824, 17825, 17826, 17827, 17861, 17874, 17882, 17886, 18111, 18112, 18501, 19007, 19658, 19698, 19857, 20206, 20207, 20227, 20228, 20235, 20236, 20305, 20329, 20548, 20549, 20560, 20607, 20648, 20649, 20650, 20651, 20652, 20653, 20654, 20655, 20656, 20680, 20681, 20682, 20698, 20699, 20712, 20972

Harvesting (D4410)

2423, 4756, 4819, 15789, 15790, 17115, 17264, 17658

Storage and conservation (D4420)

15826, 15901, 15945, 15946, 15947, 15948, 15949, 15950, 15951, 15952, 15953, 15954, 15955, 15956, 15957, 15958, 15959, 15960, 15961, 15962, 15963, 15964, 15965, 15966, 15967, 15968, 16026

Processing (D4430)

16267, 16319, 16358, 16359, 16360, 16361, 16362, 16363,

16364, 16365, 16366, 16367, 16368, 16369, 16370, 16371, 16372, 16373, 16374, 16375, 16376, 16377, 16378, 16379, 16380, 16381, 16382, 16383, 16384, 16385, 16386, 16387, 16388, 16389, 16390, 16391, 16392, 16393, 16394, 16395, 16396, 16397, 16398, 16399, 16400, 16401, 16402, 16403, 16404, 16405, 16406, 16407, 16408, 16409, 16410, 16411, 16412, 16413, 16414, 16415, 16416, 16417, 16418, 16419, 16420, 16421, 16422, 16423, 16424, 16425, 16426, 16427, 16428, 16429, 16430, 16431, 16432, 16433, 16434, 16435, 16436, 16437, 16438, 16439, 16452, 17089

Transport and handling (D4440)

15966, 15967, 17115, 17116

B6800 Food and table luxuries in general

2623, 3797, 5072, 5344, 7758, 8019, 11586, 11640, 11746, 13682, 15523, 15741, 15758, 17722, 17723, 17726, 17727, 17734, 17739, 17757, 17761, 17763, 17766, 17767, 17774, 17776, 17782, 17787, 17793, 17800, 17802, 17835, 17837, 17845, 17858, 17860, 17888, 17891, 17903, 17938, 17940, 17945, 17959, 17970, 18011, 18059, 18084, 18090, 18100, 18112, 18128, 18140, 18159, 18164, 18214, 18233, 18234, 18263, 18341, 18349, 18390, 18391, 18392, 18432, 18435, 18437, 18445, 18446, 18447, 18448, 18454, 18456, 18457, 18458, 18473, 18474, 18476, 18481, 18484, 18486, 18488, 18489, 18490, 18491, 18492, 18493, 18496, 18502, 18510, 18511, 18543, 18556, 18583, 18586, 18597, 18598, 18618, 18620, 18622, 18629, 18637, 18639, 18641, 18642, 18643, 18644, 18645, 18646, 18648, 18654, 18655, 18656, 18657, 18658, 18659, 18660, 18661, 18662, 18663, 18664, 18666, 18667, 18668, 18669, 18670...

Storage and conservation (D4420)

2702, 8325, 8351, 15703, 15969, 15970, 15971, 15972, 15973, 15974, 15975, 15976, 15977, 15978, 15979, 15980, 15981, 15982, 15983, 15984, 15985, 15986, 15987, 15988, 15989, 15990, 15991, 15992, 15993, 15994, 15995, 15996, 15997, 15998, 15999, 16000, 16001, 16002, 16003, 16004, 16005, 16006, 16007, 16008, 16009, 16010, 16011, 16012, 16013, 16014, 16015, 16016, 16017, 16018, 16019, 16020, 16021, 16022, 16023, 16024, 16025, 16026, 16027, 16028, 16029, 16030, 16031, 16471, 16695

Processing (D4430)

2828, 15971, 15976, 15978, 16014, 16016, 16027, 16030, 16031, 16265, 16271, 16314, 16440, 16441, 16442, 16443, 16444, 16445, 16446, 16447, 16448, 16449, 16450, 16451, 16452, 16453, 16454, 16455, 16456, 16457, 16458, 16459, 16460, 16461, 16462, 16463, 16464, 16465, 16466, 16467, 16468, 16469, 16470, 16471, 16472, 16473, 16474, 16475, 16476, 16477, 16478, 16479, 16479, 16480, 16481, 16482, 16483, 16484, 16485, 16486, 16487, 16488, 16489, 16490, 16695, 17096, 17100, 17179, 18841, 18883, 18892, 18922, 18938

Transport and handling (D4440)

15518, 16008, 16030, 16031, 17100, 17117, 17118, 17119, 17120, 17121

Food composition (D7210)

313, 5183, 9006, 11541, 11542, 15984, 16443, 16462, 16463, 18814, 18833, 18834, 18835, 18836, 18837, 18838, 18839, 18840, 18841, 18842, 18843, 18844, 18845, 18846, 18847, 18848, 18849, 18850, 18851, 18852, 18853, 18854, 18855, 18856, 18857, 18858, 18859, 18860, 18861, 18862, 18863, 18864, 18865, 18866, 18867, 18868, 18869, 18870, 18871, 18872, 18873, 18874, 18875, 18876, 18877, 18878, 18879, 18880, 18881, 18882, 18883, 18884, 18885, 18886, 18887, 18888, 18889, 18890, 18891, 18892, 18893, 18894, 18895, 18896, 18897, 18898, 18899, 18900, 18901, 18902, 18903, 18904, 18905, 18906, 18907, 18908, 18909, 18910, 18911, 18912, 18913, 18914, 18915, 18916, 18917, 18918, 18919,

18920, 18921, 18922, 18923, 18924, 18925, 18926, 18927,
18928, 18929, 18930, 18931, 18932, 18933, 18934, 18935,
18936, 18937, 18938, 18939, 18940, 18941, 18942, 18943,
18944, 18945, 18946, 18947...

B6810 Mill and bakery products

2998, 3119, 5455, 5583, 6192, 6432, 6438, 8017, 9367,
15743, 15765, 17753, 17773, 17799, 17892, 18460, 18463,
18595, 18613, 18634, 18711, 18738, 18739, 18740, 18741,
18742, 18743, 18744, 18745, 18746, 18747, 18748, 18749,
18750, 18751, 18752, 18753, 18754, 18755, 18756, 18808,
19542, 19665, 20197, 20252, 20514

Harvesting (D4410)

15791, 15792

Storage and conservation (D4420)

16032, 16033, 16034, 16035, 16036, 16037, 16038, 16039,
16040, 16041, 16042, 16043, 16044, 16045, 16046, 16047,
16048, 16049, 16050, 16051, 16052, 16053, 16054, 16055,
16056, 16057, 16058, 16059, 16060, 16061

Processing (D4430)

2995, 6365, 16036, 16040, 16043, 16491, 16492, 16493,
16494, 16495, 16496, 16497, 16498, 16499, 16500, 16501,
16502, 16503, 16504, 16505, 16506, 16507, 16508, 16509,
16510, 16511, 16512, 16513, 16514, 16515, 16516, 16517,
16518, 16519, 16520, 16521, 16522, 16523, 16524, 16525,
16526, 16527, 16528, 16529, 16530, 16531, 16532, 16533,
16534, 16535, 16536, 16537, 16538, 16539, 16540, 16541,
16542, 16543, 16544, 16545, 16546, 16547, 16548, 16549,
16550, 16551, 16552, 16553, 16554, 17733, 19001, 19017

Transport and handling (D4440)

16060

Food composition (D7210)

6337, 16055, 16506, 16537, 16539, 18829, 18979, 18980,
18981, 18982, 18983, 18984, 18985, 18986, 18987, 18988,
18989, 18990, 18991, 18992, 18993, 18994, 18995, 18996,
18997, 18998, 18999, 19000, 19001, 19002, 19003, 19004,
19005, 19006, 19007, 19008, 19009, 19010, 19011, 19012,
19013, 19014, 19015, 19016, 19017, 19018, 19019, 19020,
19021, 19022, 19099, 19664

B6820 Oils, fats and related products

2152, 3155, 5468, 6477, 12534, 17750, 17846, 18785, 19485,
19486, 19487, 19488, 19987, 19988, 19989, 20200, 20265,
20266, 20267, 20268, 20269, 20463, 20465, 20466, 20572,
20638

Harvesting (D4410)

15793, 15794, 15795, 15818

Storage and conservation (D4420)

15794, 16062, 16063, 16064, 16065, 16066, 16067, 16068,
16572

Processing (D4430)

16064, 16067, 16271, 16493, 16532, 16555, 16556, 16557,
16558, 16559, 16560, 16561, 16562, 16563, 16564, 16565,
16566, 16567, 16568, 16569, 16570, 16571, 16572, 16573,
16574, 16575, 16576, 16577, 16578, 16579, 16580

Transport and handling (D4440)

17122

Food composition (D7210)

16064, 16576, 16579, 18908, 19023, 19024, 19025, 19026,
19027, 19028, 19029, 19030, 19031, 19032, 19033, 19034,
19035, 19036, 19037, 19038, 19039, 19040, 19041, 19042,
19043, 19044, 19045, 19046, 19047, 19048, 19049, 19050,
19051, 19052, 19053, 19054, 19055, 19056, 19057, 19058,
19059, 19060, 19061, 19358, 19668

B6830 Sugar and starch products

239, 1632, 2832, 3209, 3236, 3240, 3248, 3250, 3285, 3293,
5439, 5474, 6524, 6563, 6609, 6619, 8986, 9449, 9450, 9456,
10323, 10365, 12772, 13428, 15375, 17455, 17465, 17586,

17795, 17799, 17849, 17900, 17901, 17963, 18786, 18800,
19481, 20041, 20202, 20281, 20515, 20547, 20586, 20719,
20720, 20801

Harvesting (D4410)

3289, 3303, 12874, 15796, 15797

Storage and conservation (D4420)

3205, 15828, 16069, 16070, 16071, 16072, 16073, 16074,
16075, 16076, 16077, 16078, 16079, 16080, 16081, 16082,
16083, 16084, 16085, 16111, 16614

Processing (D4430)

3276, 15796, 16070, 16071, 16079, 16493, 16501, 16581,
16582, 16583, 16584, 16585, 16586, 16587, 16588, 16589,
16590, 16591, 16592, 16593, 16594, 16595, 16596, 16597,
16598, 16599, 16600, 16601, 16602, 16603, 16604, 16605,
16606, 16607, 16608, 16609, 16610, 16611, 16612, 16613,
16614, 16615, 16616, 16617, 16618, 16619, 16620, 16621,
16622, 16623, 16624, 16625, 16626, 16627, 16628, 16629,
17050, 17087, 19064, 19067

Transport and handling (D4440)

16079, 16082, 17123, 17124, 17125

Food composition (D7210)

9451, 16077, 16594, 16612, 16624, 18809, 18990, 18993,
19062, 19063, 19064, 19065, 19066, 19067, 19068, 19069,
19070, 19071, 19084, 19423

B6840 Fruit and vegetable products

239, 1642, 2615, 2637, 3144, 3373, 3973, 6463, 6464, 6465,
6466, 6467, 6468, 6470, 6894, 7087, 7088, 7090, 7139, 7144,
7278, 7418, 7422, 9507, 15172, 15377, 15699, 15761, 15762,
15763, 17577, 17578, 17579, 17620, 17710, 17725, 17747,
17781, 17867, 17868, 17877, 17879, 17880, 17883, 17885,
17893, 18042, 18478, 18480, 18482, 18483, 18487, 18494,
18495, 18497, 18498, 18499, 18500, 18504, 18505, 18509,
18534, 18535, 18536, 18537, 18538, 18539, 18540, 18541,
18542, 18544, 18545, 18546, 18547, 18548, 18549, 18550,
18551, 18593, 18594, 18596, 18612, 18614, 18627, 18632,
18649, 18650, 18651, 18652, 18653, 18679, 18680, 18704,
18705, 18706, 18707, 18708, 18709, 18710, 18712, 18713,
18714, 18715, 18720, 18722, 18724, 18737, 19977, 20168,
20195, 20196, 20216, 20353, 20417, 20418

Harvesting (D4410)

3703, 3815, 4199, 15796, 15797, 15798, 15799, 15800,
15801, 15802, 16096

Storage and conservation (D4420)

3837, 4068, 4127, 4131, 4152, 4617, 4627, 5728, 7085, 7289,
7321, 9630, 15823, 15824, 15867, 16059, 16060, 16064,
16085, 16086, 16087, 16088, 16089, 16090, 16091, 16092,
16093, 16094, 16095, 16096, 16097, 16098, 16099, 16100,
16101, 16102, 16103, 16104, 16105, 16106, 16107, 16108,
16109, 16110, 16111, 16112, 16113, 16114, 16115, 16116,
16117, 16118, 16119, 16120, 16121, 16122, 16123, 16124,
16125, 16126, 16127, 16128, 16129, 16130, 16131, 16132,
16133, 16134, 16135, 16136, 16137, 16138, 16139, 16140,
16141, 16142, 16143, 16144, 16145, 16146, 16147, 16148,
16149, 16150, 16151, 16152, 16153, 16154, 16155, 16156,
16157, 16158, 16159, 16160, 16161, 16162, 16163, 16164,
16165, 16166, 16167, 16168, 16169, 16170, 16171, 16172,
16173, 16174, 16175, 16176, 16177, 16178, 16179, 16661

Processing (D4430)

3707, 3747, 3925, 4225, 4263, 6073, 6074, 7045, 15796,
15867, 16064, 16303, 16593, 16630, 16631, 16632, 16633,
16634, 16635, 16636, 16637, 16638, 16639, 16640, 16641,
16642, 16643, 16644, 16645, 16646, 16647, 16648, 16649,
16650, 16651, 16652, 16653, 16654, 16655, 16656, 16657,
16658, 16659, 16660, 16661, 16662, 16663, 16664, 16665,
16666, 16667, 16668, 16669, 16670, 16671, 16672, 16673,
16674, 16675, 16676, 16677, 16983, 17027, 18444, 19101,

B – SUBJECT AREAS

Food composition (D7210)
19132, 19133, 19134, 19282, 19283, 19284, 19285

B6861 Milk products
3373, 3522, 3530, 3596, 5613, 6014, 10442, 10497, 10600, 10616, 10626, 10652, 10653, 10660, 10664, 10665, 10666, 10674, 10675, 10692, 10695, 10696, 10698, 10699, 10706, 10707, 10708, 10714, 10730, 10731, 10733, 10735, 10737, 10750, 10753, 10773, 10836, 10841, 10844, 10845, 10849, 10852, 10853, 10856, 10864, 10870, 10872, 10881, 10886, 10887, 10893, 10894, 10896, 10898, 10902, 10904, 10905, 10908, 10909, 10910, 10914, 10920, 10943, 11042, 11044, 11049, 11052, 11066, 11085, 11547, 11770, 11775, 11794, 11835, 11843, 11949, 11964, 11965, 11966, 11968, 11969, 11972, 11977, 11979, 11980, 11985, 11989, 11994, 11998, 12001, 12002, 12004, 12028, 12029, 12035, 12036, 12043, 12044, 12046, 12047, 12048, 12057, 12059, 12060, 12061, 12062, 12063, 12066, 12085, 12086, 12087, 12176, 12177, 12178, 12179, 12180, 12184, 12186, 12187, 12188, 12189, 12196, 12199, 12200, 12211...

Harvesting (D4410)
3367, 11084, 15385, 15434, 15781, 15787, 15815, 15816

Storage and conservation (D4420)
3367, 15695, 15787, 15850, 15924, 15927, 16228, 16229, 16230, 16231, 16232, 16233, 16234, 16235, 16236, 16237, 16238, 16239, 16240, 16241, 16242, 16243, 16244, 16245, 16246, 16247, 16248, 16249, 16250, 16251, 16252, 16253, 16254, 16255, 16256, 16257, 16258, 16259, 16260, 16840, 16849, 16866, 16888, 19325, 19336, 19384, 19439

Processing (D4430)
3621, 11084, 12056, 12114, 12222, 12236, 12279, 15927, 16236, 16244, 16311, 16318, 16490, 16746, 16747, 16748, 16749, 16750, 16751, 16752, 16753, 16754, 16755, 16756, 16757, 16758, 16759, 16760, 16761, 16762, 16763, 16764, 16765, 16766, 16767, 16768, 16769, 16770, 16771, 16772, 16773, 16774, 16775, 16776, 16777, 16778, 16779, 16780, 16781, 16782, 16783, 16784, 16785, 16786, 16787, 16788, 16789, 16790, 16791, 16792, 16793, 16794, 16795, 16796, 16797, 16798, 16799, 16800, 16801, 16802, 16803, 16804, 16805, 16806, 16807, 16808, 16809, 16810, 16811, 16812, 16813, 16814, 16815, 16816, 16817, 16818, 16819, 16820, 16821, 16822, 16823, 16824, 16825, 16826, 16827, 16828, 16829, 16830, 16831, 16832, 16833, 16834, 16835, 16836, 16837, 16838, 16839, 16840, 16841, 16842, 16843, 16844, 16845, 16846, 16847, 16848, 16849, 16850, 16851, 16852, 16853, 16854, 16855, 16856, 16857...

Transport and handling (D4440)
16237, 16752, 17106, 17148, 17149, 17150, 17151, 17152, 17153, 17154, 17841

Food composition (D7210)
10741, 11084, 12103, 14152, 16231, 16257, 16765, 16845, 16869, 16886, 16890, 16891, 18831, 19161, 19163, 19190, 19217, 19243, 19286, 19287, 19288, 19289, 19290, 19291, 19292, 19293, 19294, 19295, 19296, 19297, 19298, 19299, 19300, 19301, 19302, 19303, 19304, 19305, 19306, 19307, 19308, 19309, 19310, 19311, 19312, 19313, 19314, 19315, 19316, 19317, 19318, 19319, 19320, 19321, 19322, 19323, 19324, 19325, 19326, 19327, 19328, 19329, 19330, 19331, 19332, 19333, 19334, 19335, 19336, 19337, 19338, 19339, 19340, 19341, 19342, 19343, 19344, 19345, 19346, 19347, 19348, 19349, 19350, 19351, 19352, 19353, 19354, 19355, 19356, 19357, 19358, 19359, 19360, 19361, 19362, 19363, 19364, 19365, 19366, 19367, 19368, 19369, 19370, 19371, 19372, 19373, 19374, 19375, 19376, 19377, 19378, 19379, 19380, 19381, 19382, 19383, 19384, 19385, 19386, 19387, 19388, 19389, 19390, 19391, 19392...

B6862 Eggs and egg products

10412, 11269, 11272, 11274, 11275, 11276, 11278, 11282, 11283, 11284, 11306, 11314, 11315, 11316, 11317, 11318, 11319, 11324, 11325, 11351, 11353, 12579, 12596, 12598, 12607, 12611, 12612, 12613, 12629, 12631, 12634, 12639, 12655, 12710, 12713, 13358, 13359, 13360, 13383, 13386, 13387, 13393, 13394, 13408, 13409, 13411, 13415, 14792, 14797, 14806, 14875, 14883, 15691, 17537, 17751, 17796, 18122, 18479, 18506, 18508, 18532, 18628, 18630, 18788, 18791, 19655, 19656, 20164, 20409, 20467

Storage and conservation (D4420)
19404

Processing (D4430)
12597, 16811, 16948, 16949

Transport and handling (D4440)
17155, 17156

Food composition (D7210)
11281, 12594, 13875, 18822, 19056, 19163, 19164, 19169, 19402, 19403, 19404, 19405

B6869 Other dairy products
18468

Storage and conservation (D4420)
16232

Processing (D4430)
16555, 16950, 16951

Food composition (D7210)
19289, 19406

B6870 Other foods
2663, 3897, 5033, 11545, 17394, 18805, 20621

Storage and conservation (D4420)
16189, 16190, 16261, 16345

Processing (D4430)
5035, 16189, 16302, 16304, 16305, 16309, 16310, 16311, 16345, 16346, 16952, 16953, 16954, 16955, 16956, 16957, 16958, 16959, 16960, 16961, 16962, 16963, 16964, 16965, 17098

Food composition (D7210)
7928, 19407, 19408, 19409, 19410, 19411, 19412, 19413, 19414, 19415, 19416

B6880 Drinking water
1358, 1619, 1620, 1621, 1693, 17749, 19514, 19641, 19642, 19731

Transport and handling (D4440)
17157

Food composition (D7210)
18934, 19417, 19418, 19419

B6890 Alcoholic liquors, coffee, tea, tobacco and other table luxuries
3010, 3011, 3012, 3035, 4409, 4455, 6211, 6213, 7366, 7407, 7408, 7423, 7425, 7428, 7431, 7437, 7438, 7439, 7442, 7443, 7444, 7445, 7446, 7450, 7456, 7738, 8798, 8947, 17218, 17284, 17286, 17518, 17575, 17580, 17582, 17707, 17713, 17714, 17743, 17744, 17748, 17754, 17759, 17764, 17790, 17791, 17792, 17830, 17833, 17834, 17838, 17872, 17908, 17912, 17913, 17925, 17996, 18082, 18283, 18478, 19478, 19517, 19636, 19762, 19982, 20199, 20205, 20237, 20295, 20296, 20662, 20671

Harvesting (D4410)
4450, 15817, 15818, 15819, 15820, 17006

Storage and conservation (D4420)
4392, 15825, 16262, 16263, 19439, 19461

Processing (D4430)
6156, 6223, 6225, 7413, 7435, 9884, 16262, 16631, 16966, 16967, 16968, 16969, 16970, 16971, 16972, 16973, 16974, 16975, 16976, 16977, 16978, 16979, 16980, 16981, 16982, 16983, 16984, 16985, 16986, 16987, 16988, 16989, 16990, 16991, 16992, 16993, 16994, 16995, 16996, 16997, 16998,

20119, 20173, 20329

Storage and conservation (D4420)
15926

Processing (D4430)
12222, 12236, 12279, 12585, 16293, 16295, 16315, 16433, 16439, 16480, 16623, 16737, 16922, 16951, 16981, 17093, 17094, 17095, 17096, 17097, 17098, 17099, 17100, 17101

Transport and handling (D4440)
17100, 17125

B8100 People
1231, 1805, 1957, 2039, 2097, 2135, 2200, 2304, 2307, 2387, 2388, 2416, 2417, 2425, 2435, 2436, 2458, 2466, 2469, 2470, 2477, 2478, 2492, 2493, 2500, 2513, 2515, 2574, 2575, 2579, 2583, 2584, 2592, 4493, 11102, 12387, 12887, 13493, 13542, 13543, 13608, 13624, 13682, 13784, 13866, 13867, 13868, 13888, 13901, 15091, 15106, 15282, 15286, 15301, 15305, 15501, 15502, 15513, 15514, 15523, 15730, 16488, 17170, 17171, 17177, 17178, 17180, 17182, 17187, 17190, 17201, 17203, 17218, 17223, 17237, 17252, 17257, 17258, 17266, 17335, 17367, 17376, 17383, 17395, 17401, 17404, 17410, 17440, 17456, 17466, 17535, 17610, 17639, 17730, 17758, 17766, 17782, 17789, 17800, 17857, 17891, 17943, 17945, 18039, 18078, 18095, 18145, 18150, 18193, 18196, 18200, 18209, 18212, 18216, 18219, 18220, 18225, 18226, 18227, 18228, 18229, 18231, 18232, 18235, 18239...

B8200 Business enterprises in general
1218, 17851, 17895, 17937, 17944, 17947, 17965, 17993, 18000, 18003, 18066, 18081, 18140, 18206, 18307, 18308, 18341, 18405, 19649, 19724

Work management (D5100)
17168, 17169, 17170, 17171

Farm management (D5200)
17330, 17331, 17332, 17333, 17334, 17335, 17336, 17337, 17338, 17339, 17340, 17341, 17342

B8210 Farms in general
23, 289, 1119, 1322, 1431, 1989, 2406, 2443, 2446, 2450, 2457, 2458, 2462, 2472, 2473, 2474, 2508, 2638, 5265, 8305, 10633, 15289, 15323, 15534, 17064, 17813, 17898, 17907, 17909, 17914, 17922, 17964, 17981, 17998, 18006, 18010, 18022, 18024, 18025, 18027, 18028, 18035, 18038, 18048, 18054, 18061, 18062, 18068, 18100, 18113, 18144, 18148, 18156, 18165, 18167, 18184, 18186, 18189, 18190, 18196, 18207, 18209, 18210, 18212, 18229, 18236, 18246, 18250, 18252, 18259, 18260, 18264, 18269, 18272, 18275, 18278, 18280, 18292, 18295, 18296, 18298, 18343, 18344, 18345, 18346, 18347, 18352, 18353, 18355, 18404, 18410, 18429, 18431, 19521, 19644, 19650, 19651, 19652, 19653, 19688, 19703, 19783, 19785, 19837, 19853, 19860, 20212, 20225, 20306, 20326, 20337, 20400, 20706, 20707, 20938, 21047

Work management (D5100)
15305, 17163, 17172, 17173, 17174, 17175, 17176, 17177, 17178, 17179, 17180, 17181, 17182, 17183, 17184, 17185, 17186, 17187, 17188, 17189, 18289

Farm management (D5200)
2449, 15308, 15577, 15715, 15748, 17184, 17292, 17301, 17343, 17344, 17345, 17346, 17347, 17348, 17349, 17350, 17351, 17352, 17353, 17354, 17355, 17356, 17357, 17358, 17359, 17360, 17361, 17362, 17363, 17364, 17365, 17366, 17367, 17368, 17369, 17370, 17371, 17372, 17373, 17374, 17375, 17376, 17377, 17378, 17379, 17380, 17381, 17382, 17383, 17384, 17385, 17386, 17387, 17388, 17389, 17390, 17391, 17392, 17393, 17394, 17395, 17396, 17397, 17398, 17399, 17400, 17401, 17402, 17403, 17404, 17405, 17406, 17407, 17408, 17409, 17410, 17411, 17412, 17413, 17414, 17415, 17416, 17417, 17418, 17419, 17420, 17421, 17422, 17423, 17424, 17425, 17426, 17427, 17428, 17429, 17430,

17431, 17432, 17433, 17434, 17435, 17436, 17437, 17438, 17636, 17999, 18031, 18060, 18097, 18837, 19847

B8211 Arable farms
5324, 18142, 18179, 18370, 18394, 19784, 20439, 20947

Work management (D5100)
17190, 17191, 17192

Farm management (D5200)
2710, 2711, 17439, 17440, 17441, 17442, 17443, 17444, 17445, 17446, 17447, 17448, 17449, 17450, 17451, 17452, 17453, 17454, 17455, 17456, 17457, 17458, 17459, 17460, 17461, 17462, 17463, 17464, 17465, 17466, 17467, 17468, 17469, 17470, 17471, 17472, 17473, 17588

B8212 Animal and grassland farms
328, 497, 1576, 2461, 3534, 3618, 5328, 10399, 10405, 10554, 10729, 10862, 10895, 10908, 11268, 11962, 12989, 14142, 15380, 15448, 15470, 15676, 15678, 15695, 15764, 15831, 17829, 17863, 17881, 17985, 17986, 17987, 17988, 18023, 18036, 18037, 18049, 18063, 18119, 18130, 18151, 18172, 18174, 18188, 18197, 18199, 18201, 18316, 19680, 19684, 19704, 19784, 19940, 19941, 19942, 20331, 20440, 20441, 20442

Work management (D5100)
15696, 17162, 17193, 17194, 17195, 17196, 17197, 17198, 17199, 17200, 17201, 17202, 17203, 17204, 17205, 17206, 17207, 17208, 17209, 17210, 17211, 17212, 17213, 17214, 17215, 17216

Farm management (D5200)
10480, 10717, 10728, 10870, 10878, 11055, 11569, 13144, 13282, 14366, 15416, 16783, 17207, 17208, 17213, 17215, 17295, 17296, 17329, 17439, 17449, 17450, 17474, 17475, 17476, 17477, 17478, 17479, 17480, 17481, 17482, 17483, 17484, 17485, 17486, 17487, 17488, 17489, 17490, 17491, 17492, 17493, 17494, 17495, 17496, 17497, 17498, 17499, 17500, 17501, 17502, 17503, 17504, 17505, 17506, 17507, 17508, 17509, 17510, 17511, 17512, 17513, 17514, 17515, 17516, 17517, 17518, 17519, 17520, 17521, 17522, 17523, 17524, 17525, 17526, 17527, 17528, 17529, 17530, 17531, 17532, 17533, 17534, 17535, 17536, 17537, 17538, 17539, 17540, 17541, 17542, 17543, 17544, 17545, 17546, 17547, 17548, 17549, 17550, 17551, 17552, 17553, 17554, 17555, 17556, 17557, 17558, 17559, 17873, 18408

B8213 Mixed farms
3661, 5328, 18109, 19784, 20338

Work management (D5100)
17216

Farm management (D5200)
17449, 17450, 17560, 17561, 17562, 17563

B8214 Horticultural holdings
164, 464, 865, 870, 1190, 1371, 2407, 2454, 2464, 2484, 2606, 2826, 3705, 4059, 4372, 4385, 4386, 4387, 4407, 4408, 4409, 4411, 5853, 5856, 7389, 8108, 8792, 8794, 8796, 8799, 8801, 8803, 8841, 8993, 9722, 9769, 13520, 15132, 17100, 17834, 17896, 17928, 17974, 17975, 17992, 18082, 18170, 18171, 18175, 18176, 18179, 18182, 18183, 18271, 18394, 19784, 20307, 20321, 20411, 20412, 20439, 20636

Work management (D5100)
3144, 3151, 3882, 4350, 15268, 15376, 15482, 15485, 17161, 17166, 17217, 17218, 17219, 17220, 17221, 17222, 17223, 17224, 17225, 17226, 17227, 17228, 17229, 17230, 17231, 17232, 17233, 17234, 17235, 17236, 17237, 17238, 17239, 17240, 17241, 17242, 17243, 17244, 17245, 17246, 17247, 17596, 17623

Farm management (D5200)
1500, 4063, 4235, 17217, 17230, 17231, 17244, 17245, 17293, 17301, 17459, 17564, 17565, 17566, 17567, 17568, 17569, 17570, 17571, 17572, 17573, 17574, 17575, 17576,

17577, 17578, 17579, 17580, 17581, 17582, 17583, 17584,
17585, 17586, 17587, 17588, 17589, 17590, 17591, 17592,
17593, 17594, 17595, 17596, 17597, 17598, 17599, 17600,
17601, 17602, 17603, 17604, 17605, 17606, 17607, 17608,
17609, 17610, 17611, 17612, 17613, 17614, 17615, 17616,
17617, 17618, 17619, 17620, 17621, 17622, 17623, 17624,
17625, 17626, 17627, 17628, 17629, 17630, 17631, 17632,
17633, 17634, 17635, 17830

B8215 Forest enterprises

872, 1322, 1336, 1869, 1930, 1989, 2376, 2383, 2384, 2423,
2438, 2443, 2444, 2607, 2612, 2635, 2636, 4748, 4749, 4766,
4771, 4787, 4845, 5950, 5982, 10320, 10321, 10419, 10420,
15135, 15742, 17972, 18043, 18044, 18064, 18065, 18071,
18072, 18073, 18074, 18076, 18130, 18143, 18211, 18255,
18256, 18300, 19521, 19657, 19658, 19659, 19789, 19797,
20056, 20209, 20318, 20556, 20559, 20648

Work management (D5100)

16426, 17164, 17165, 17248, 17249, 17250, 17251, 17252,
17253, 17254, 17255, 17256, 17257, 17258, 17259, 17260,
17261, 17262, 17263, 17264, 17265, 17266, 17267, 17268,
17269, 17270, 17271, 17272, 17273, 17274, 17275, 17276

Farm management (D5200)

2439, 2608, 8865, 15789, 17261, 17262, 17268, 17270,
17272, 17273, 17275, 17736, 17637, 17638, 17639, 17640,
17641, 17642, 17643, 17644, 17645, 17646, 17647, 17648,
17649, 17650, 17651, 17652, 17653, 17654, 17655, 17656,
17657, 17658, 17659, 17660, 17661, 17662, 17663, 17664,
17665, 17666, 17667, 17668, 17669, 17670, 17671, 17672,
17673, 17674, 17675, 17676, 17677, 17678, 17679, 17680,
17681, 17682, 17683, 17684, 17685, 17686, 17687, 17688

B8216 Fisheries

1933, 10422, 11391, 11398, 12741, 17878, 17897, 18032,
18149, 18173, 20963

Work management (D5100)

17277, 17278, 17279, 17280

Farm management (D5200)

11412, 17280, 17689, 17690, 17691, 17692, 17693, 17694,
17695, 17696, 17697, 17698

B8219 Other farms

14998

Work management (D5100)

17281

B8220 Service and supply firms

1921, 2028, 2029, 2490, 9524, 15310, 15523, 16465, 17956,
17989, 17997, 18118, 18199, 18284, 18294, 18369, 18432,
18476, 18814, 19519, 19662, 19686

Work management (D5100)

17282, 17283, 18906

Farm management (D5200)

15284, 17347, 17423, 17699, 17700, 17701, 17702, 17703,
17704, 17705, 17706

B8230 Marketing and processing firms

166, 239, 1632, 2032, 10354, 10355, 10385, 11232, 11781,
15407, 15467, 15741, 15991, 16275, 16365, 16588, 16622,
16626, 16781, 16838, 16920, 16951, 17744, 17752, 17776,
17780, 17790, 17791, 17792, 17850, 17872, 17892, 17893,
17908, 17912, 17913, 17994, 18041, 18042, 18084, 18088,
18100, 18119, 18121, 18122, 18139, 18246, 18293, 18386,
19571, 19600, 19632, 19683, 19697, 19698, 19758, 19759,
19835, 20254, 20363, 20529, 20801, 20822, 21089

Work management (D5100)

17179, 17218, 17284, 17285, 17286, 17287, 17288, 17289,
19183

Farm management (D5200)

17149, 17294, 17295, 17394, 17503, 17505, 17506, 17580,
17582, 17706, 17707, 17708, 17709, 17710, 17711, 17712,

17713, 17714, 17715, 17716, 17717, 17718, 17719, 17720,
17721, 17722, 17723, 17724, 17725, 17726, 17727

B8240 Recreational enterprises

18976

Work management (D5100)

17290, 17291

Farm management (D5200)

17728

B8290 Other business enterprises

1123, 1552, 1931, 2144, 2413, 18117, 18447, 18448, 18976,
19061, 19577, 19776, 20220, 20251

Farm management (D5200)

17402, 17403, 17729, 17730

B8300 Marketing systems and sectors thereof

2735, 17102, 17333, 17571, 17572, 17731, 17735, 17737,
17738, 17739, 17743, 17745, 17747, 17749, 17750, 17751,
17752, 17759, 17760, 17767, 17770, 17771, 17781, 17783,
17784, 17785, 17800, 17811, 17812, 17813, 17823, 17827,
17829, 17830, 17831, 17832, 17833, 17834, 17835, 17836,
17837, 17838, 17839, 17840, 17842, 17843, 17846, 17857,
17864, 17865, 17866, 17867, 17868, 17869, 17870, 17871,
17877, 17878, 17881, 17887, 17889, 18172, 18188, 18194,
18195, 18316, 18361

B8400 Communities, areas and regions in general

338, 368, 376, 409, 413, 468, 607, 888, 1024, 1125, 1159,
1260, 1269, 1316, 1324, 1343, 1345, 1590, 1637, 1643, 1674,
1699, 1710, 1725, 1781, 1792, 1793, 1797, 1800, 1803, 1855,
1938, 1939, 1994, 2089, 2112, 2135, 2249, 2372, 2373, 2375,
2377, 2378, 2380, 2385, 2389, 2391, 2396, 2397, 2402, 2403,
2408, 2409, 2411, 2412, 2413, 2425, 2427, 2428, 2442, 2445,
2456, 2470, 2472, 2475, 2476, 2478, 2480, 2481, 2482, 2483,
2485, 2486, 2487, 2491, 2494, 2497, 2498, 2500, 2504, 2505,
2507, 2508, 2509, 2510, 2513, 2514, 2530, 2543, 2545, 2552,
2557, 2558, 2569, 2570, 2571, 2573, 2574, 2578, 2580, 2581,
2582, 2583, 2585, 2586, 2588, 2709, 2874, 5186, 5187, 6100,
10117, 10330, 10343, 15323, 17746, 17749, 17909, 17910,
17943, 17994, 18047, 18048, 18091, 18236...

B8410 The european community in general

374, 1002, 1372, 2004, 2019, 2257, 2458, 3011, 6210, 6231,
7399, 7421, 7841, 8401, 10298, 13120, 14101, 17458, 17502,
17561, 17679, 17734, 17756, 17757, 17795, 17796, 17797,
17798, 17799, 17804, 17859, 17930, 17990, 18073, 18109,
18173, 18334, 18961, 19113, 19147, 19218, 19219, 20280,
20408

B8411 Belgium and luxembourg

1, 1038, 1113, 10622, 12972, 17173, 17739, 17897, 17898,
17899, 18240, 19251, 19785, 19853

B8412 Denmark

180, 472, 754, 2090, 2092, 2323, 2384, 3706, 4787, 4993,
10405, 13184, 13291, 13936, 14188, 15703, 17165, 20324,
20325, 20329, 20333

B8413 United kingdom

2449, 2503, 2551, 6205, 13291, 17406, 17851, 18003, 19817

B8414 France

181, 182, 192, 201, 203, 206, 210, 211, 212, 213, 216, 218,
220, 221, 222, 226, 232, 238, 240, 241, 243, 244, 484, 485,
490, 639, 641, 646, 648, 649, 760, 835, 910, 911, 914, 1035,
1124, 1126, 1127, 1128, 1129, 1130, 1131, 1133, 1136, 1138,
1145, 1405, 1406, 1418, 1424, 1426, 1430, 1431, 1432, 1436,
1438, 1439, 1445, 1447, 1763, 1806, 1868, 2003, 2096, 2100,
2103, 2104, 2105, 2108, 2109, 2110, 2111, 2386, 2387, 2388,
2503, 2542, 2563, 2564, 2613, 2734, 2735, 2751, 2924, 3102,
3174, 3175, 3329, 3539, 3542, 3544, 3545, 3546, 3662, 3709,
3846, 3905, 4067, 4213, 4233, 4804, 4867, 5247, 5259, 5444,
5461, 5517, 5589, 5592, 5596, 5659, 5858, 5859, 5860, 5971,
6362, 6364, 6494, 6894, 7056, 7408, 7465, 8024, 8159...

B - SUBJECT AREAS

B8415 Federal republic of germany
10, 12, 13, 112, 116, 117, 118, 119, 124, 127, 133, 136, 143,
144, 153, 155, 163, 166, 167, 171, 172, 366, 439, 442, 446,
447, 448, 451, 464, 572, 573, 575, 576, 588, 591, 592, 599,
606, 608, 825, 828, 872, 874, 879, 880, 882, 990, 991, 1011,
1122, 1203, 1208, 1211, 1230, 1231, 1238, 1239, 1242, 1254,
1257, 1259, 1262, 1268, 1270, 1274, 1278, 1289, 1299, 1301,
1302, 1304, 1305, 1306, 1322, 1329, 1331, 1333, 1335, 1336,
1342, 1343, 1347, 1348, 1349, 1350, 1351, 1352, 1359, 1363,
1364, 1366, 1367, 1371, 1730, 1741, 1742, 1745, 1801, 1804,
1846, 1847, 1848, 1849, 1850, 1851, 1852, 1853, 1859, 1866,
1867, 1868, 1869, 1870, 1877, 1891, 1892, 1897, 1902, 1907,
1908, 1911, 1913, 1916, 1917, 1920...

B8416 Republic of ireland
93, 94, 251, 253, 256, 258, 259, 260, 261, 262, 263, 264,
265, 266, 512, 2121, 2124, 2129, 2131, 2132, 2134, 2139,
2140, 2141, 2142, 2143, 2144, 2147, 2149, 2150, 2393, 2394,
2446, 2830, 2995, 3663, 4431, 6707, 6795, 8305, 8993,
11030, 11391, 11441, 11442, 11453, 11749, 11925, 12164,
12946, 13237, 13944, 13945, 14367, 15438, 15756, 15757,
15968, 16124, 17279, 17412, 17858, 17860, 18144, 18149,
18343, 18344, 18345, 18347, 18411, 18431, 19099, 19100,
19526, 19702, 19707, 20358, 20360, 20875

B8417 Italy
269, 270, 275, 276, 277, 278, 284, 286, 287, 288, 289, 291,
312, 316, 330, 518, 525, 533, 685, 691, 692, 795, 926, 927,
930, 1070, 1179, 1180, 1183, 1184, 1185, 1467, 1501, 1502,
1503, 1508, 1517, 1518, 1540, 1552, 1815, 1817, 2151, 2156,
2157, 2158, 2160, 2162, 2169, 2177, 2178, 2447, 2448, 2449,
2450, 2845, 2848, 3346, 3350, 3355, 3371, 3372, 3487, 3492,
3605, 3606, 3607, 3608, 3612, 3686, 3767, 4033, 4088, 4096,
4100, 4101, 4195, 4360, 4433, 4441, 4450, 4451, 4459, 4464,
4482, 4828, 4829, 4831, 4832, 4834, 4894, 4948, 4981, 4999,
5000, 5013, 5308, 5454, 6015, 6205, 6249, 6269, 6270, 6391,
6392, 6400, 6401, 6404, 6409, 6812, 7072, 7080, 7084, 7144,
7146, 7147, 7151, 7322, 7323, 7352, 7429, 7432, 7433, 7436,
7440...

B8418 The netherlands
334, 349, 350, 352, 354, 360, 366, 371, 378, 379, 384, 385,
388, 391, 392, 393, 394, 408, 419, 421, 422, 423, 424, 425,
426, 427, 565, 708, 709, 712, 724, 939, 942, 953, 954, 957,
979, 1106, 1557, 1558, 1559, 1562, 1563, 1583, 1596, 1600,
1604, 1606, 1619, 1620, 1621, 1622, 1626, 1630, 1631, 1639,
1641, 1646, 1648, 1651, 1653, 1655, 1657, 1658, 1665, 1668,
1670, 1673, 1676, 1677, 1684, 1686, 1687, 1688, 1691, 1692,
1693, 1694, 1695, 1696, 1697, 1698, 1700, 1701, 1703, 1704,
1705, 1707, 1709, 1711, 1712, 1714, 1715, 1774, 1782, 1819,
2183, 2188, 2191, 2193, 2194, 2195, 2196, 2198, 2219, 2225,
2235, 2238, 2242, 2246, 2250, 2258, 2262, 2267, 2274, 2275,
2277, 2278, 2279, 2280, 2286, 2288, 2290, 2293, 2298...

B8420 Developing countries in general
401, 461, 717, 971, 974, 978, 1227, 1862, 2294, 2499, 2874,
2893, 3039, 3617, 5349, 5365, 5479, 5526, 8419, 9423, 9516,
10445, 10451, 11070, 11071, 11457, 11780, 12631, 12633,
12885, 13538, 13681, 14587, 15283, 15310, 15315, 15731,
15732, 15733, 16019, 16029, 16060, 16083, 16508, 16517,
16579, 16602, 16714, 17341, 17741, 17742, 17750, 17809,
17893, 17903, 17904, 17921, 17930, 17959, 17962, 18010,
18016, 18040, 18056, 18059, 18090, 18205, 18206, 18208,
18368, 18374, 18375, 18391, 20194, 20398, 20403, 20410,
20807, 20960, 20963, 21049

B8421 Developing countries in africa
110, 114, 125, 179, 412, 417, 440, 617, 721, 1071, 1229,
1283, 1313, 1825, 1863, 1893, 1915, 1922, 1943, 2306, 2310,
2316, 2318, 2374, 2871, 3665, 4362, 4885, 5029, 5132, 6067,
6422, 6429, 9373, 9538, 9539, 10058, 10064, 10091, 10095,
10620, 10641, 10642, 10913, 10928, 11072, 11812, 12281,
12768, 12984, 13581, 13594, 13973, 14101, 15644, 17084,
17172, 17351, 17361, 17384, 17442, 17729, 17762, 17781,
17900, 17919, 17923, 17924, 17944, 17946, 17985, 17986,
18009, 18116, 18245, 18248, 18260, 18288, 18310, 18356,
18371, 18373, 18395, 18399, 18401, 18404, 21033, 21034,
21035, 21036, 21037, 21038, 21040, 21041, 21042

B8422 Developing countries in asia
402, 418, 428, 438, 443, 996, 1221, 1287, 1733, 1734, 1899,
2305, 2406, 2544, 2604, 2890, 3256, 3257, 4162, 4362, 4851,
5471, 5841, 5997, 10449, 11072, 11086, 11460, 11968,
13594, 15316, 17351, 17356, 17770, 17892, 17901, 17928,
17935, 17957, 17960, 17961, 17976, 17977, 18004, 18005,
18210, 18212, 18213, 18214, 18221, 18241, 18244, 18246,
18247, 18249, 18250, 18251, 18253, 18254, 18264, 18265,
18266, 18273, 18274, 18275, 18280, 18360, 18363, 18367,
18369, 18400, 18417, 19694, 20649, 20656

B8423 Developing countries in south america
120, 151, 462, 1119, 1121, 1241, 1328, 1894, 1900, 1903,
1909, 1912, 2034, 2035, 2204, 2873, 4362, 4745, 4846, 4849,
5966, 5967, 5996, 6233, 7106, 7452, 8575, 9517, 10298,
11072, 12244, 14101, 17351, 17473, 17474, 17507, 17747,
17920, 17928, 17958, 17978, 17987, 18017, 18018, 18232,
18257, 18370, 18372, 19541

B8424 Developing countries in oceania
1925, 11072

B8490 Other countries, communities, areas and regions
128, 129, 135, 228, 416, 461, 582, 729, 764, 917, 978, 1049,
1050, 1227, 1305, 1799, 1854, 1881, 1892, 1895, 2003, 2004,
2014, 2015, 2016, 2017, 2018, 2019, 2021, 2022, 2137, 2263,
2264, 2265, 2266, 2267, 2269, 2270, 2271, 2272, 2273, 2283,
2294, 2303, 2323, 2893, 2903, 3011, 3039, 3071, 3617, 4400,
4881, 4914, 4959, 5481, 5841, 6067, 6210, 7841, 8223, 8224,
8401, 8510, 8940, 9672, 10447, 10448, 10640, 10927, 11365,
11368, 11407, 12633, 13540, 13553, 13554, 13555, 13936,
14101, 15283, 15396, 16029, 16083, 16413, 16579, 17165,
17331, 17360, 17502, 17708, 17748, 17757, 17780, 17809,
17998, 18042, 18133, 18173, 18331, 18334, 19147, 19297,
19766, 20324, 20385, 20408, 20495, 20538, 20960, 21049

B8500 Agricultural economy in general
17150, 17765, 17773, 17793, 17905, 17909, 17937, 17953,
17966, 17969, 18014, 18053, 18057, 18060, 18097, 18107,
18115, 18120, 18129, 18133, 18138, 18139, 18153, 18215,
18338, 20182, 20319

B8510 Agricultural economy of the european community
1814, 17426, 17748, 17772, 17774, 17775, 17802, 17851,
17877, 17887, 17894, 17916, 17938, 17963, 17967, 17982,
17984, 18013, 18020, 18028, 18033, 18052, 18058, 18087,
18098, 18131, 18146, 18147, 18160, 18162, 18166, 18185,
18197, 18200, 18201, 18349

B8511 Agricultural economy of belgium and luxembourg
17292, 17732, 17736, 17895, 18239

B8512 Agricultural economy of denmark
17385, 17386, 17387, 17389, 17822, 18055, 18061, 18062,
18063, 20331

B8513 Agricultural economy of the united kingdom
2395, 18083, 18161, 18199

B8514 Agricultural economy of france
17981, 18067, 18069, 18070, 18074, 18075, 18076, 18077,
18078, 18080, 18083, 18084, 18085, 18089, 18090, 18095,
18096, 18098, 18102, 18103, 18104, 18105, 18106, 18108,
18114, 18118, 18124, 18126, 18127, 18130, 18134, 18135,
18136, 18137, 18306, 18340

B8515 Agricultural economy of the federal republic of germany
1923, 17777, 17779, 17782, 17801, 17803, 17902, 17907,
17908, 17914, 17917, 17918, 17922, 17929, 17931, 17932,

17933, 17936, 17939, 17940, 17941, 17942, 17968, 17974, 17975, 17979, 17980, 17983, 17988, 17992, 17995, 17996, 17999, 18001, 18002, 18007, 18008, 18011, 18019, 18021, 18022, 18024, 18025, 18030, 18034, 18049, 18083, 18267, 18283

B8516 Agricultural economy of the republic of ireland
18145, 18147, 18148

B8517 Agricultural economy of italy
2395, 17416, 17869, 18152, 18154, 18155, 18156, 18157, 18158, 18161, 18162, 18163, 18164, 18350

B8518 Agricultural economy of the netherlands
17612, 18167, 18168, 18169, 18175, 18176, 18177, 18178, 18179, 18180, 18182, 18183, 18184, 18185, 18187, 18191, 18192, 18193, 18194, 18195, 18196, 18197, 18198, 18202, 18203, 18207, 18364, 18416

B8590 Agricultural economy of the other countries
17426, 17911, 17919, 17920, 17934, 17935, 17948, 17949, 17950, 17951, 17952, 17954, 17955, 17962, 17976, 17985, 17986, 17987, 18004, 18009, 18049, 18050, 18051, 18056, 18087, 18092, 18110, 18116, 18123, 18128, 18132, 18160, 18161, 18204, 18214, 18232, 18274, 18285, 18363, 18366, 18371

B8900 Other subject areas related to human resources, organizations, institutes
1309, 1928, 2111, 2748, 6298, 6300, 6301, 6867, 7660, 12946, 16436, 18125, 18224, 18235, 18237, 18309, 18333, 18334, 18335, 18336, 18407, 18417, 18421, 18423, 18424, 18425, 18442, 18443, 18867, 19525, 19795, 19796, 19812, 20373, 20537, 20824

B9100 Experimental outfit in general
11171, 19834, 20142, 20377, 20378, 20691

B9110 Measuring implements
292, 351, 362, 858, 946, 1226, 1263, 1425, 1572, 1584, 1587, 1591, 1664, 1690, 1708, 1721, 1723, 1737, 1787, 1823, 3788, 5166, 5168, 8388, 10722, 11113, 11165, 11777, 14128, 15297, 15322, 15414, 15579, 15624, 16044, 16710, 16907, 18914, 19166, 19187, 19254, 19295, 19398, 19797, 19840, 19841, 19842, 20037, 20038, 20043, 20053, 20054, 20074, 20081, 20085, 20116, 20165, 20169, 20174, 20234, 20254, 20345, 20346, 20347, 20349, 20432, 20433, 20653, 21005

B9120 Laboratory animals
2345, 15932, 15933, 18900, 18970, 18974, 18978, 19244, 19312, 19468, 19491, 19499, 19670, 19676, 19750, 19964, 19972, 20091, 20106, 20431, 20487, 20575, 20601, 20609, 20624, 20660, 20670, 20694, 20695, 20697, 20721, 20728, 20936

Animal management general and animal husbandry (D3100)
10668, 11531, 11532, 11533, 11534, 11535, 11536, 11537, 12860, 12866

Animal nutrition (D3200)
11647, 11703, 12308, 12309, 12390, 12391, 12404, 12460, 12648, 12651, 12652, 12836, 12837, 12838, 12839, 12840, 12841, 12842, 12843, 12844, 12845, 12846, 12847, 12848, 12849, 12850, 12851, 12852, 12853, 12854, 12855, 12856, 12857, 12858, 12859, 12860, 12861, 12862, 12863, 12864, 12865, 12866, 12867, 12868, 12869, 14887, 15119, 15120, 18899

Animal breeding (D3300)
12883, 13001, 13279, 13287, 13461, 13462, 13463, 13464, 13465, 13466, 13467, 13468, 13469, 13470

Animal diseases, veterinary medicine (D3400)
12858, 12861, 13561, 13563, 13627, 13878, 13884, 14126, 14443, 14632, 14633, 14653, 14657, 14887, 15009, 15023, 15083, 15084, 15085, 15086, 15087, 15088, 15089, 15090, 15091, 15092, 15093, 15094, 15095, 15096, 15097, 15098,

15099, 15100, 15101, 15102, 15103, 15104, 15105, 15106, 15107, 15108, 15109, 15110, 15111, 15112, 15113, 15114, 15115, 15116, 15117, 15118, 15119, 15120, 15121, 15122, 15123, 15124, 15125, 15126, 15127, 15128, 15129, 15130

B9130 Plants for experimental purposes
20099, 20100, 20107

Plant production general and crop husbandry (D2100)
2700

Plant nutrition and fertilization (D2200)
5189

B9190 Other experimental outfit
386, 545, 1266, 1269, 1278, 1279, 2203, 4880, 5713, 6077, 6078, 6209, 6434, 6561, 6571, 6613, 6829, 7781, 9002, 9070, 9653, 13654, 13981, 15920, 16019, 16488, 16879, 18871, 19314, 19823, 19824, 19975, 20055, 20082, 20174, 20248, 20341, 20351, 20372, 20385, 20431, 20433, 20470, 20472, 20473, 20485, 20496, 20509, 20760, 20802

B9200 General methodology and experimental techniques
143, 217, 318, 335, 339, 343, 353, 359, 362, 364, 368, 373, 377, 379, 383, 401, 407, 413, 553, 554, 562, 566, 568, 701, 702, 707, 710, 711, 736, 803, 811, 813, 815, 817, 823, 857, 858, 861, 891, 970, 1090, 1091, 1114, 1187, 1243, 1263, 1266, 1273, 1290, 1307, 1310, 1319, 1320, 1325, 1330, 1413, 1555, 1562, 1579, 1581, 1584, 1588, 1595, 1597, 1598, 1624, 1633, 1634, 1643, 1644, 1646, 1649, 1652, 1659, 1664, 1666, 1667, 1669, 1675, 1682, 1708, 1720, 1722, 1792, 1793, 1795, 1796, 1798, 1807, 1818, 1819, 1820, 1821, 1822, 1823, 1824, 1864, 1865, 2012, 2218, 2224, 2239, 2249, 2253, 2285, 2402, 2405, 2413, 2429, 2456, 2481, 2482, 2483, 2491, 2494, 2495, 2498, 2504, 2507, 2509, 2514, 2530, 2545, 2565, 2569...

B9300 Documentation materials
182, 206, 210, 211, 212, 221, 222, 232, 243, 246, 330, 349, 374, 376, 377, 385, 387, 389, 390, 391, 392, 1118, 1372, 1493, 1792, 1796, 1798, 1800, 1801, 1803, 1815, 1816, 1818, 1822, 1939, 1992, 1999, 2003, 2004, 2009, 2177, 2178, 2215, 2378, 2385, 2386, 2405, 2428, 2556, 2562, 2566, 2594, 4057, 4768, 5320, 5330, 5354, 6083, 7655, 7841, 8386, 8401, 9234, 9424, 9853, 11444, 13298, 17450, 17661, 17664, 17690, 17882, 18003, 18043, 18101, 18169, 18209, 18405, 18412, 18430, 19781, 19782, 19787, 19790, 19791, 19792, 19793, 19794, 19795, 19796, 19798, 19799, 19800, 19801, 19802, 19803, 19804, 19805, 19806, 19807, 19808, 19810, 19811, 19817, 19819, 19821, 19822, 19864, 20049, 20125, 20202, 20208, 20211, 20218, 20219, 20220, 20221, 20222, 20223, 20233, 20243, 20246, 20257, 20280, 20283...

B9900 Research which cannot be allocated elsewhere
25, 26, 27, 30, 32, 33, 34, 35, 36, 37, 38, 39, 40, 41, 42, 43, 44, 45, 47, 48, 49, 50, 51, 52, 53, 54, 55, 56, 57, 58, 59, 60, 61, 62, 63, 66, 69, 70, 71, 72, 73, 74, 75, 77, 78, 80, 81, 82, 83, 84, 85, 86, 87, 88, 89, 90, 91, 92, 1157, 1460, 1809, 1810, 1811, 1812, 1813, 2616, 2619, 2630, 2631, 5075, 5076, 5078, 7759, 7760, 7761, 7993, 8001, 8018, 8020, 8988, 8989, 8990, 8991, 8992, 10407, 10410, 10413, 10415, 13509, 13514, 13517, 15146, 15148, 15154, 15155, 15159, 15173, 15183, 15184, 15185, 15186, 15187, 15188, 15191, 15192, 15193, 15194, 15195, 15196, 15197, 15198, 15200, 15221, 15225, 15228, 15229, 15230, 15235, 15247, 15248, 15250, 15251, 15257, 15262, 15264...

LIST OF RESEARCH ORGANISATIONS

BELGIUM

BE.01.00.00 Faculté des Sciences agronomiques de l'Etat.
Passage des Déportés 2; 5800 Gembloux.

BE.01.00.01   Chaire de Technologie forestiére.

BE.01.00.02   Chaire d'Horticulture.
Avenue de la Faculté 2; 5800 Gembloux.

BE.01.00.03   Chaire de Zootechnie.

BE.01.00.04   Chaire des Plantes ornementales.
Avenue de la Faculté 2; 5800 Gembloux.

BE.01.00.05   Chaire des Sciences de la Terre.

BE.01.00.06   Chaire de Chimie générale et organique.

BE.01.00.07   Chaire d'Economie rurale.

BE.01.00.08   Chaire de Phytotechnie des Régions tempérées.

BE.01.00.09   Chaire de Génie rural II.

BE.01.00.10   Chaire de Statistique.
Avenue de la Faculté 59; 5800 Gembloux.

BE.01.00.11   Chaire de Zoologie appliquée.

BE.01.00.12   Chaire de Mécanisation.
Avenue de la Faculté d'Agronomie 13; 5800 Gembloux.

BE.01.00.13   Chaire de Génie rural I.
Avenue de la Faculté d'Agronomie 16; 5800 Gembloux.

BE.01.00.14   Chaire de Génétique.

BE.01.00.15   Chaire de Technologie agricole et Alimentaire.

BE.01.00.16   Chaire de la Science du Sol.

BE.01.00.17   Chaire de Zoologie appliquée et de Phytiâtrie.

BE.01.00.18   Chaire de Physique et Chimie physique.

BE.01.00.19   Chaire de Phytotechnie des Régions chaudes.

BE.01.00:20   Chaire de Zoologie générale et de Faunistique.

BE.01.00.21   Laboratoire de Chimie analytique.

BE.01.00.22   Laboratorie de Pathologie végétale.
Avenue Maréchal Juin 8; 5800 Gembloux.

BE.01.00.23   Centre d'étude des Légumineuses.

BE.01.00.24   Centre d'Ecologie forestiére et Centre de Recherches sur l'Elevage et les Productions fourragéres en Haute Belgique.

BE.02.00.00 Faculté des Sciences agronomiques de l'Université catholique de Louvain.
Place Croix du Sud 1 á 3; 1348 Louvain–la–Neuve.

BE.02.01.00 Département de Physique et de Chimie appliquées.

BE.02.01.01   Laboratoire de Chimie organo–minérale.

BE.02.01.02   Unité des Cristaux imparfaits.

BE.02.01.03   Unité de Chimie du Solide et de Catalyse.

BE.02.01.04   Unité de Physico–Chimie de Surface.

BE.02.01.05   Unité de Biochimie physiologique.

BE.02.02.00 Département d'Economie rurale.

BE.02.03.00 Département de Biologie végétale appliquée.

BE.02.03.01   Unité des Eaux et Foréts.

BE.02.03.02   Unité de Phytopathologie.

BE.02.03.03   Unité de Phytotechnie tropicale et subtropicale.

BE.02.03.04   Unité de Génétique.

BE.02.04.00 Département de Génie rural.

BE.02.05.00 Département de Microbiologie et des Industries agricoles.

BE.02.05.01   Unité des Industries agricoles.

BE.02.05.02   Unité de Microbiologie.

BE.02.06.00 Département de Biologie animale appliquée.

BE.02.06.01   Unité de Biochimie de la Nutrition.

BE.02.06.02   Unité d'Ecologie des Prairies.

BE.02.06.03   Unité de Nutrition animale.

BE.02.07.00 Département des Sciences du Sol.

BE.02.07.01   Unité de Pédologie générale.

BE.02.07.02   Centre d'Etude des Sols tropicaux.

BE.02.08.00 Institut de Sciences naturelles appliques. Place Croix du Sud 1; 1348 Louvain–la–Neuve.

BE.03.00.00 Faculteit van de Landbouwwetenschappen van de Rijksuniversiteit te Gent.
Coupure Links 533; 9000 Gent.

BE.03.00.01   Seminarie voor Landhuishoudkunde.

BE.03.00.02   Seminarie voor Tuinbouweconomie.

BE.03.00.03   Seminarie voor Toegepaste Wiskunde en Biometrie.

BE.03.00.04   Seminarie voor Marktkunde van Land– en Tuinbouwprodukten.

BELGIUM

BE.03.00.05   Laboratorium voor Organische Scheikunde.

BE.03.00.06   Laboratorium voor Biochemie.

BE.03.00.07   Laboratorium voor Algemene en Industriële Microbiologie.

BE.03.00.08   Laboratorium voor Analytische– en Agrochemie.

BE.03.00.09   Laboratorium voor Fysische en Radiobiologische Chemie.

BE.03.00.10   Laboratorium voor Fytofarmacie.

BE.03.00.11   Laboratorium voor Fytopathologie en voor Fytovirologie.

BE.03.00.12   Laboratorium voor Industriële Gistingen.

BE.03.00.13   Laboratorium voor Bodemfysica, Bodemconditionering en Tuinbouwbodemkunde.

BE.03.00.14   Laboratorium voor Agrarische Bodemkunde.

BE.03.00.15   Laboratorium voor Tuinbouwplantenteelt (met tuinbouwproefbedrijf).

BE.03.00.16   Laboratorium voor Plantenteelt van de Warme Streken en Landhuishoudkunde van de Ontwikkelingslanden.

BE.03.00.17   Laboratorium voor Landbouwplantenteelt.

BE.03.00.18   Laboratorium voor Herbologie.

BE.03.00.19   Laboratorium voor Plantecologie.

BE.03.00.20   Laboratorium voor Land– en Tuinbouwmachines.

BE.03.00.21   Laboratorium voor Landbouwwerktuigkunde.

BE.03.00.22   Laboratorium voor Hydraulica, Kultuurtechniek en Topografie.

BE.03.00.23   Laboratorium voor Fundamentele en Toegepaste Technologie.

BE.03.00.24   Laboratorium voor Dierkunde.

BE.03.00.25   Laboratorium voor Levensmiddelenchemie.

BE.03.00.26   Leerstoel voor Voeding.en Hygiëne.

BE.03.00.27   Laboratorium voor Biologie en Technologie van het Hout.

BE.03.00.28   Onderzoekcentrum voor Veeteelt.

BE.03.00.29   Onderzoekcentrum voor Boerderijbouwkunde.

BE.03.00.30   Onderzoekcentrum voor Bosbouw, Bosbedrijfsvoering en bospolitiek (met Proefbos).

BE.04.00.00   Faculteit der Landbouwwetenschappen van de Katholieke Universiteit te Leuven.

Kardinaal Mercierlaan 92; 3030 Leuven (Heverlee).

BE.04.01.00 **Afdeling voor Oppervlakte Scheikunde en Colloidale scheikunde.**

BE.04.01.01   Centrum voor Oppervlakte en Colloidale Scheikunde.
de Croylaan 42; 3030 Leuven (Heverlee).

BE.04.02.00 **Afdeling voor Bodem– en planten beheersing.**

BE.04.02.01   Centrum voor Bodemvruchtbaarheid en Bodembiologie.

BE.04.02.02   Studiecentrum voor Tuinbouwgronden.

BE.04.02.03   Laboratorium voor Fytopathologie en Plantenbescherming.

BE.04.02.04   Laboratorium voor Fytotechnie en Plantenveredling.
Pastorijstraat 2; 3040 Bierbeek (Korbeek–Lo).

BE.04.03.00 **Afdeling voor Faunabeheersing en –uitbating.**

BE.04.03.01   F.A. Janssenslaboratorium voor Genetica.

BE.04.03.02   Laboratorium voor Eco–fysiologie der Huisdieren.

BE.04.04.00 **Afdeling Landelijk Genie.**

BE.04.04.01   Laboratorium voor Agrarische Bouwkunde.

BE.04.04.02   Laboratorium voor Landbouwwerktuigkunde.

BE.04.04.03   Laboratorium voor Landtechniek.

BE.04.05.00 **Afdeling voor Levensmiddelentechnologie en Industriële Microbiologie.**

BE.04.05.01   Laboratorium voor Toegepaste Koolhydraatchemie.

BE.04.05.02   Laboratorium voor Industriele Conservering Levensmiddelen.

BE.04.05.03   Laboratorium voor Industriele Microbiologie en Biochemie.

BE.04.05.04   Laboratorium voor Tropische Landbouwindustrieen.

BE.04.06.00 **Afdeling Landbouweconomie.**

BE.04.06.01   Centrum voor Landbouweconomisch onderzoek.

BE.04.06.02   Statistiek en Proeftechniek.

BE.04.07.00 **Afdeling voor Organische en Anorganische Analyse.**

BE.04.07.01   Laboratorium voor Analytische en Minerale Scheikunde.

BE.04.07.02   Laboratorium voor Toegepaste Organische Scheikunde.

BE.04.08.00 **Laboratorium voor Bodemgenese en Bodemgeografie. de Croylaan 42; 3030 Leuven (Heverlee).**

BE.04.09.00 **Laboratorium voor Tropische Plantenteelt.**

BE.05.00.00 **Faculteit van de Diergeneeskunde. Casinoplein 24; 9000 Gent.**

BE.05.01.00 **Leerstoel Heelkunde van de huisdieren.**

BE.05.02.00 **Leerstoel Interne Geneeskunde van de grote Huisdieren.**

BE.05.03.00 **Leerstoel Voortplanting en Verloskunde van de Huisdieren.**

BE.05.04.00 **Leerstoel Anatomie van de Huisdieren.**

BE.05.05.00 **Leerstoel Pathologische Anatomie van de Huisdieren.**

BE.05.06.00 **Leerstoel pluimveepathologie, Bacteriologie, Besmettelijke Ziekten.**

BE.05.07.00 **Leerstoel Parasitologie en Parasitaire Ziekten van de Huisdieren.**

BE.05.08.00 **Leerstoel Veterinaire Fysiologie.**

BE.05.09.00 **Leerstoel Fysiologische Scheikunde.**

BE.05.10.00 **Leerstoel Dierlijke Genetica en Veeteelt. Heidestraat 19; 9220 Merelbeke.**

BE.05.11.00 **Leerstoel voor Diervoedingsleer.   Heidestraat 19; 9220 Merelbeke.**

BE.05.12.00 **Leerstoel Eetwaren van Dierlijke Oorsprong. Wolterslaan 12; 9000 Gent.**

BE.06.00.00 **Faculté de Médecine Vétérinaire de l'Université de Liege. Rue des Vétérinaires 45; 1070 Bruxelles.**

BE.06.00.01   Chaire de'anatomie.

BE.06.00.02   Chaire de bactériologie.

BE.06.00.03   Chaire de génétique.

BE.06.00.04   Chaire d'histologie.

BE.06.00.05   Chaire d'inspection des denrées alimentaires.

BE.06.00.06   Chaire d'obstétrique.

BE.06.00.07   Chaire de pathologie chirurgicale.

BE.06.00.08   Chaire de pathologie générale.

BE.06.00.09   Chaire de pathologie médicale.

BE.06.00.10   Chaire de thérapeutique.

BE.06.00.11   Chaire de physiologie.

BE.06.00.12   Chaire de zootechnie.

BE.07.00.00 **Rijkscentrum voor Landbouwkundig Onderzoek te Gent. Burgemeester van Gansberghelaan 96; 9220 Merelbeke.**

BE.07.01.00 **Rijksstation voor Entomologie en Nematologie.**

BE.07.02.00 **Rijksstation voor Kleinveeteelt. Burgemeester van Gansberghelaan 92; 9220 Lemberge–Merelbeke.**

BE.07.03.00 **Rijksstation voor Landbouwtechniek. Burgemeester van Gansberghelaan 115; 9220 Lemberge–Merelbeke.**

BE.07.04.00 **Rijksstation voor Plantenveredeling. Burgemeester Van Gansberghelaan 109; 9220 Lemberge–Merelbeke.**

BE.07.05.00 **Rijksstation voor Plantenziekten.**

BE.07.06.00 **Rijksstation voor Sierplantenteelt. Caritasstraat 17; 9230 Melle.**

BE.07.07.00 **Rijksstation voor Veevoeding. Scheldeweg 68; 9231 Melle (Gontrode).**

BE.07.08.00 **Rijksstation voor Zeevisserij. Ankersstraat 1; 8400 Oostende.**

BE.07.09.00 **Rijkszuivelstation. Brusselsesteenweg 370; 9230 Melle.**

BE.07.10.00 **Rijksstation voor Populierenteelt. Gaverstraat 35; 9500 Geraardsbergen.**

BE.08.00.00 **Centre de recherches agronomiques de l'Etat à Gembloux. Avenue de la Faculté d'Agronomie 22; 5800 Gembloux.**

BE.08.01.00 **Station de Chimie et de Physique agricoles. Chaussée de Wavre 115; 5800 Gembloux.**

BE.08.02.00 **Station des Cultures Fruitiéres et Maraichéres. Chaussée de Charleroi 234; 5800 Gembloux.**

BE.08.03.00 **Station d'Amélioration des Plantes. Rue du Bordia 4; 5800 Gembloux.**

BE.08.04.00 **Station de Génie rural. Chaussée de Namur 146; 5800 Gembloux.**

BE.08.05.00 **Station Laitiére. Chaussée de Namur 24; 5800 Gembloux.**

BE.08.06.00 **Station de Phytopathologie. Avenue Maréchal Juin 13; 5800 Gembloux.**

BE.08.07.00 **Station de Phytopharmacie. Rue du Bordia 11; 5800 Gembloux.**

BE.08.08.00 Station de Phytotechnie. Chemin de Liroux 11; 5800 Gembloux.

BE.08.09.00 Station de Haute Belgique. Rue du Serpont 48; 6600 Libramont.

BE.08.10.00 Station de Technologie forestière. Avenue Maréchal Juin 1; 5800 Gembloux.

BE.08.11.00 Station de Zoologie appliquée. Chemin de Liroux 8; 5800 Gembloux.

BE.08.12.00 Station de Zootechnie. Chemin de Liroux 11; 5800 Gembloux.

BE.09.00.00 **Jardin botanique de Belgique.**
Brusselsesteenweg; 1860 Meise.

BE.10.00.00 **Institut de Recherches chimiques.**
Museumlaan 5; 1980 Tervuren.

BE.11.00.00 **Institut national de Recherches vétérinaires.**
Groeselenberg 99; 1180 Bruxelles.

BE.12.00.00 **Institut économique agricole.**
Bd. de Berlaimont 18–20 (bte. 5); 1000 Bruxelles.

BE.13.00.00 **Station des Eaux et Forêts.**
Avenue Dubois 2; 1990 Groenendaal.

BE.14.00.00 **Institut pour l'Encouragement de la recherche scientifique dans l'Industrie et l'Agriculture. (Pour ce qui conserne les projets exécutés, ni dans les Facultés d'Agronomie ou Vétériaire, ni dans les Centres de Recherches de l'Etat) (IRSIA).**
Rue de Crayer 6; 1050 Bruxelles.

BE.17.00.00 **Fonds de la recherche fondamentale collective.**
Rue d'Egmont 5; 1050 Bruxelles.

BE.18.00.00 **Centrum voor Toegepaste Biologie v.z.w.**
de Croylaan 6; 3030 Leuven (Heverlee).

FEDERAL REPUBLIC OF GERMANY

DE.101.00.0 Technische Hochschule Aachen

DE.101.05.0 Lehrstuhl für Botanik und Botanisches Institut
5100 Aachen; Hainbuchenstr. 20

**DE.101.10.0 Lehrstuhl für Landschaftsoekologie und
Landschaftsgestaltung. 5100 Aachen; Schinkelstr. 1.**

**DE.101.15.0 Lehrstuhl für Geograhpie 1 und 2 und
Geographisches Institut 5100 Aachen; Templergraben 55.**

**DE.101.20.0 Lehrstuhl für Hygiene. 5100 Aachen;
Hainbuchenstr. 20**

DE.101.20.1 Abteilung Hygiene und Arbeitsmedizin

**DE.101.25.0 Lehrstuhl für Wirtschafts- und Sozialgeshichte.
5100 Aachen; Templergraben 55.**

**DE.101.35.0 Lehrstuhl für Entwerfen von Hoch- und
Industriebauten und Instutut für Schulbau 5100 Aachen;
Schinkelstr. 1.**

DE.101.35.1 Abteiling Planung und Entwurf
Landwirtschaftlicher Betriebsstätten.

DE.104.00.0 Freie Universität Berlin.

**DE.104.05.0 Institut für Veterinae-Anatomie, -Histologie und
-Embryologie. 1000 Berlin 33; Koserstr. 20**

DE.104.05.1 Fachrichtung Veterinaer-Anatomie.

DE.104.05.2 Fachrichtung Veterinaer-Histologie und
-Embryologie.

**DE.104.10.0 Institut fuer Veterinaer-Physiologie, -Biochemie,
-Pharmakologie und -Toxikologie. 1000 Berlin 33; Koserstr.
20**

DE.104.10.1 Fachrichtungen Physilogoie und Biochemie

DE.104.10.2 Fachrichtungen Pharmakologie und
Toxikologie

**DE.104.15.0 Institut für Morphologische und Funktionelle
Veterinär-Pathologie 1000 Berlin 33; Drosselweg 1-3**

DE.104.15.1 Fachrichtung Veterinär-Pathologie

DE.104.15.2 Fachrichtung Experimentelle Therapie und
Funtionelle Pathologie.
1000 Berlin 33; Bitterstr. 14-16

**DE.104.20.0 Institut für Parasitologie und
Tropenveterisärmedizin. 1000 Berlin 37; Koenigsweg 65.**

DE.104.20.1 Abteilung Entomologie.

DE.104.20.2 Abteilung Helminthologie.

DE.104.20.3 Abteilung Protozoologie.

DE.104.25.0 Institut für mikrobiologie Geflügelkrankheiten

Elektronen-Mikroskopie und Tierhygiene. 1000 Berlin 33;
Königin-Luise-Str.49.

DE.104.25.1 Fachrichtung Tierhygiene.

DE.104.25.2 Fachrichtung Geflügelkrankheiten
1000 Berlin 33; Koserstr. 21

DE.104.25.3 Fachrichtung Elektronenmikroskopie.

DE.104.25.4 Fachrichtung Mikrobiologie.

**DE.104.30.0 Institut fuer Lebensmittelhygiene, Fleischhygiene
und -technologie. 1000 Berlin 33; Brümmerstr. 10.**

DE.104.30.1 Fachrichtung Lebensmittelhygiene.
1000 Berlin 33; Bitterstr. 8-12.

DE.104.30.2 Fachrichtung Fleischhygiene.

**DE.104.35.0 Institut für Versuchstierkunde
Versuchstierkrankheiten Biometrie Dokumentation und
Versicherungsveterinärmedizin. 1000 Berlin 33; Thielallee 36.**

DE.104.35.1 Fachrichtung Versuchstierkunde und
Versuchstierkrankheiten
1000 Berlin 45; Tietzenweg 85-87

DE.104.35.2 Fachrichtung Biometrie.
1000 Berlin 45; Tietzenweg 85-87.

**DE.104.40.0 Klinik für Klauentierkrankheiten und
Fortpflanzungskunde, Tierärztliche Aussenklinik. 1000 Berlin
37; Königsweg 65.**

DE.104.40.1 Abteilung Andrologie und Haustierbesamung.

**DE.104.45.0 Klinik für Pferdekrankheiten und Allgemeine
Chirurgie. 1000 Berlin 37; Königsweg 50.**

DE.104.45.1 Abteilung Radiologie.

DE.104.45.2 Abteilung Pferdekrankheiten.

**DE.104.50.0 Klinik und Poliklinik für Kleine Haustiere 1000
Berlin 33; Bitterstr. 8-12.**

**DE.104.55.0 Institut für Tierzucht und Tierernährung 1000
Berlin 33; Brümmerstr. 34.**

**DE.104.60.0 Institut für Angewandte Genetik 1000 Berlin 33;
Albrecht-Thär-Weg 6**

DE.104.60.1 Abteilung für Cytologie und Cytogenetik

**DE.104.65.0 Institut fuer Physische Geographie. 1000 Berlin
33; Altensteinstr. 19.**

DE.104.70.0 Osteuropa-Institut. 1000 Berlin 33; Garystr. 55.

DE.104.70.1 Abteilung Landeskunde

**DE.104.75.0 Institut für Ökonomische und Soziologische
Analyse Poltischer Systeme 1000 Berlin 33; Ihnestr. 21.**

DE.105.00.0 Technische Universität Berlin

DE.105.05.0 Institut fuer Oekologie, Fachgebiet Bodenkunde. 1000 Berlin 33; Englerallee 19.

DE.105.06.0 Institut fuer Oekologie, Fachgebiet Oekosystemforschung und Vegetationskunde. 1000 Berlin 33; Albrecht–Thär–Weg 4.

DE.105.07.0 Institut für Ökologie, Fachgebiet Botanik

DE.105.10.0 Institut für Ökologie, Fachrichtung Botanik. 1000 Berlin 41; Rothenburgstr. 12.

DE.105.11.0 Institut fuer Oekologie, Fachgebiet Zierpflanzenbau. 1000 Berlin 33; Königin Luise Str. 22.

DE.105.12.0 Institut fuer Angewandte Zoologie. 1000 Berlin 41; Grunewaldstr. 34.

DE.105.20.0 Institut für Nutzpflanzenforschung

DE.105.20.1 Fachgebiet Acker– und Pflanzenbau. 1000 Berlin 33; Albrecht–Thär–Weg 5

DE.105.20.2 Fachgebiet Pflanzenernährung. 1000 Berlin 33; Lentzeallee 55/57.

DE.105.20.3 Fachgebiet Gemüsebau. 1000 Berlin 33; Königin–Luise–Str. 22

DE.105.20.4 Fachgebiet Obstbau. 1000 Berlin 33; Albrecht–Thär–Weg 3

DE.105.25.0 Institut für Landschaftsbau. 1000 Berlin 33; Lentzeallee 76

DE.105.25.5 Fachgebiet Kulturtechnik.

DE.105.35.0 Institut für Sozialökonomie der Agrarentwicklung 1000 Berlin 33; Podbielskiallee 64

DE.105.35.1 Abteilung Welternährung.

DE.105.35.2 Abteilung Landwirtschaftliches Förderungswesen.

DE.105.40.0 Institut für Agrarbetriebs– und Standortsökonomie. 1000 Berlin 33; Im Dol 27–29.

DE.105.41.0 Lehrgebiet Landwirtschaftliche Betriebslehre 1000 Berlin 33; Im Dol 27–29.

DE.105.50.0 Institut für Sozialökonomie der Landschaftsentwicklung. 1000 Berlin 33; Königin Luise Str. 22.

DE.105.51.0 Lehrgebiet Gärtnerische Betriebslehre. see 105500.

DE.105.52.0 Lehrgebiet Agrarpolitik und Agrarstatistik. 1000 Berlin 33; Albrecht Thär Weg 2.

DE.105.55.0 Institut für Tierproduktion. 1000 Berlin 33; Lentzeallee 75

DE.105.60.0 Institut für Maschinenkonstruktion 1000 Berlin

33; Zoppoter Str. 35.

DE.105.60.1 Abteilung Landtechnik und Baumaschinen

DE.105.65.0 Institut fuer Landschafts– und Freiraumplanung. 1000 Berlin 10; Franklinstr. 29.

DE.105.65.1 Lehrgebiet für Landschafts– und Gartenplanung

DE.105.70.0 Institut für Geographie. 1000 Berlin 12; Strasse des 17. Juni 135

DE.105.75.0 Lehrstuhl für Anlagen und Verfahrenstechnik für die Lebensmitteltechnologie und Biotechnologie 1000 Berlin 65; Seestr. 13

DE.105.80.0 Institut für Lebensmitteltechnologie. 1000 Berlin 33; Könign–Luise–Str. 22.

DE.105.81.0 Fachgebiet Frucht– und Gemüsetechnologie. 1000 Berlin 33; Königin Luise Str. 22.

DE.105.81.1 Abteilung Lebensmittelchemie.

DE.105.81.2 Abteilung Mikrobiologie.

DE.105.81.3 Abteilung Lebensmitteltechnologie.

DE.105.81.4 Abteilung Chemisch–technische Analyse.

DE.105.81.5 Abteilung Lebensmittelhygiene.

DE.105.82.0 Fachgebiet Getreidetechnologie. 1000 Berlin 65; Seestr. 11.

DE.105.83.0 Fachgebiet Zuckertechnologie. 1000 Berlin 65; Amrumer Str. 32

DE.105.90.0 Institut fuer Lebensmittelchemie. 1000 Berlin 12; Strasse des 17. Juni 175

DE.105.95.0 Institut für Zuckerindustrie, Versuchsanstalt für Zuckertechnologie. 1000 Berlin 65; Amrumer Str. 32

DE.105.95.1 Abteilung Maschinen und Apparate.

DE.105.95.2 Abteilung Analytik und Anwendungstechnik.

DE.105.95.3 Bibliothek.

DE.105.98.0 Institut für Gärungsgewerbe und Biotechnologie (Versuchs– und Lehranstalt für Brauerei in Berlin) 1000 Berlin 65; Seestr. 13.

DE.105.98.1 Forschungsinstitut für Technologie der Trinkbranntwein und Likörfabrikation 1000 Berlin 65; Seestr. 13

DE.105.98.3 Abteilung für Chemisch–Technische Analyse. 1000 Berlin 65; Seestr. 13

DE.105.98.4 Forschungsinstitut für Technologie der Brauerei und Mälzerei. 1000 Berlin 65; Seestr. 13

DE.105.98.5 Institut für Brauereitechnologie.
see 105980.

DE.105.98.7 Wirtschaftliche Abteilung
1000 Berlin 65; Seestr. 13

DE.105.98.8 Lehrstuhl für Biotechnologie
1000 Berlin 65; Seestr. 12/15.

DE.105.98.9 Lehrstuhl für Mikrobiologie.

DE.105.99.0 Forschungsinstitut für Mikrobiologie. 1000
Berlin 65; Seestr. 13.

DE.108.00.0 Universität Bochum

DE.108.05.0 Lehrstuhl für Allgemeine Botanik. 4630 Bochum;
Buscheystr. 132

DE.108.05.1 Abteilung Biologie.

DE.108.10.0 Lehrstuhl für Pflanzenphysiologie. 4630
Bochum; Buscheystr. 132

DE.108.10.1 Abteilung Biologie.

DE.108.15.0 Geographisches Institut. 4630
Bochum–Querenburg; Postfach 2148.

DE.108.15.1 Forschungsabteilung für Raumordnung.

DE.108.15.2 Abteilung Wirtschafts– und Sozialgeographie.

DE.108.20.0 Institut für Entwicklungsforschung und
entwicklungspolitik 4630 Bochum–Querenburg; Overbergstr.
15

DE.108.20.1 Abteilung Wirtschafts– und Sozialgeographie.

DE.108.20.2 Abteilung Wirtschaftswissenschaften

DE.111.00.0 Universität Bonn

DE.111.05.0 Institut für Bodenkunde. 5300 Bonn; Nussallee
13.

DE.111.05.1 Abteilung für Tropische Böden und
Bodenradiometrie.

DE.111.05.2 Abteilung Mikrobiologie.

DE.111.05.3 Abteilung Experimentelle Bodenkunde.

DE.111.10.0 Agrikulturchemisches Institut 5300 Bonn;
Meckenheimer Allee 176

DE.111.10.1 Abteilung Radioagronomie und Spezielle
Analytik.

DE.111.12.0 Botanisches Institut. 5300 Bonn; Meckenheimer
Allee 176.

DE.111.12.1 Abteilung Molekularbiologie.

DE.111.15.0 Institut für Landwirtschaftliche Botanik. 5300
Bonn; Meckenheimer Allee 176

DE.111.15.1 Abteilung Biometrie

DE.111.15.2 Abteilung Landwirtschaftliche Botanik.
5300 Bonn, Meckenheimer Allee 176.

DE.111.20.0 Institut für Landwirtschaftliche Zoologie und
Bienenkunde. 5300 Bonn; Melbweg 42.

DE.111.20.1 Abteilung Bienenkunde

DE.111.20.2 Abteilung Oekologie.

DE.111.25.0 Institut für Pflanzenbau 5300 Bonn;
Katzenburgweg 5

DE.111.25.1 Lehrstuhl für Allgemeinen Pflanzenbau

DE.111.25.2 Lehrstuhl für Speziellen Pflanzenbau und
Pflanzen–Züchtung

DE.111.25.3 Abteilung Pflanzenbau in Tropen und
Subtropen.

DE.111.25.4 Abteilung Spezieller Pflanzenbau.

DE.111.25.5 Abteilung Pflanzenzüchtung.

DE.111.30.0 Institut für Obstbau und Gemüsebau

DE.111.30.1 Abteilung Frischhaltung und Konservierung
von Obst und Gemuese.
see 111300.

DE.111.35.0 Institut für Pflanzenkrankheiten 5300 Bonn;
Nussallee 9

DE.111.35.1 Abteilung Virologie

DE.111.35.2 Abteilung Entomologie und Pflanzenschutz

DE.111.35.3 Abteilung Mykologie

DE.111.40.0 Institut fuer Anatomie, Physiologie und Hygiene
der Haustiere. 5300 Bonn; Katzenburgweg 7–9

DE.111.40.1 Abteilung Biochemie

DE.111.40.2 Geflügelforschungsstation

DE.111.40.3 Abteilung Tierhygiene.

DE.111.40.4 Abteilung Veterinär– und Lebensmittelhygiene.

DE.111.40.5 Abteilung Anatomie und Physiologie

DE.111.45.0 Institut für Tierernährung. 5300 Bonn;
Endenicher Allee 15.

DE.111.45.1 Abteilung Futtermittelkunde

DE.111.50.0 Institut für Tierzucht und Tierfütterung 5300
Bonn; Endenicher Allee 15.

DE.111.50.1 Abteilung für Kleintierzucht und –Haltung

DE.111.50.2 Abteilung Haustiergenetik.

DE.111.50.3 Abteilung Tierhaltungstechnik

DE.111.55.0 Institut für Ernährungswissenschaft 5300 Bonn; Endenicher Allee 13.

DE.111.57.0 Lehrstuhl fuer Angewandte Ernaehrungsphysiologie. 5300 Bonn; Roemerstr. 164.

DE.111.60.0 Lehrstuhl für Lebensmittelwissenschaft 5300 Bonn; Endenicher Allee 11–13

DE.111.65.0 Institut für Agrarpolitik, Marktforschung und Wirtschaftssoziologie. 5300 Bonn; Nussallee 21.

DE.111.65.1 Lehrstuhl für Volkswirtschaftslehre Agrarpolitik und Landwirtschaftliches Informationswesen

DE.111.65.2 Abteilung Agrargeschichte und Agrarstrukturpolitik.

DE.111.65.3 Lehrstuhl für Wirtschaftssoziologie.

DE.111.65.4 Lehr– und Forschungsbereich Welternaehrungswirtschaft.

DE.111.65.5 Lehrstuhl für Marktforschung.

DE.111.65.6 Agrarwissenschaftliche Forschungsstelle für die Oststaaten.

DE.111.68.0 Forschungsstelle für Agrarpolitik und Agrarsoziologie

DE.111.70.0 Institut für Landwirtschaftliche Betriebslehre. 5300 Bonn; Meckenheimer Allee 174 II.

DE.111.70.1 Lehrstuhl für Angewandte Landwirtschaftliche Betriebslehre

DE.111.70.2 Lehrstuhl für Allgemeine Landwirtschaftliche Betriebslehre

DE.111.70.3 Abteilung für Wirtschaftsberatung.

DE.111.70.4 Abteilung Betriebslehre der Ernährungswirtschaft.

DE.111.75.0 Institut für gesellschafts– und Wirtschaftswissenschaften. 5300 Bonn; Adenauerallee 24–42

DE.111.75.1 Wirtschaftspolitische Abteilung

DE.111.80.0 Institut für Wirtschaftsgeographie 5300 Bonn; Franziskanerstr. 2.

DE.111.81.0 Geographisches Institut 5300 Bonn; Franziskanerstr. 2.

DE.111.85.0 Institut fuer Landtechnik 5300 Bonn; Nussallee 5

DE.111.85.1 Abteilung Haushaltechnik.

DE.111.85.2 Abteilung Landwirtschaftliche

Arbeitsverfahren.

DE.111.90.0 Institut fuer Staedtebau, Bodenordnung und Kulturtechnik. 5300 Bonn; Nussallee 1

DE.111.90.1 Lehrstuhl fuer Landwirtschaftlichen Wasserbau und Kulturtechnik.

DE.111.95.0 Forschungsstelle fuer Forstwirtschaft in Entwicklungslaendern und Forstliche. 5300 Bonn; Beethovenstr. 30.

DE.111.97.0 Klinik und Poliklinik für Hals–, Nasen– und Ohrenkranke. 5300 Bonn; Venusberg.

DE.111.99.0 Meteorologisches Institut 5300 Bonn; auf dem Hügel 20

DE.114.00.0 Technische Universität Braunschweig

DE.114.05.0 Botanisches Institut. 3300 Braunschweig; Humboldtstr. 1.

DE.114.10.0 Institut für Lebensmittelchemie. 3300 Braunschweig; Fasanenstr. 3

DE.114.15.0 Geographisches Institut 3300 Braunschweig; Pockelsstr. 2

DE.114.20.0 Institut für Landmaschinen 3300 Braunschweig; Langer Kamp 19a.

DE.114.20.1 Abteilung Pneumatische Förderung.

DE.114.20.2 Abteilung für Landtechnik.

DE.114.20.3 Abteilung Maschinenbau.

DE.114.20.4 Abteilung Landmaschinen.

DE.114.20.5 Abteilung Schlepper und Erdbaumaschinen.

DE.114.20.6 Abteilung Hydraulische Antriebe.

DE.114.25.0 Institut fuer Mechanische Verfahrenstechnik. 3300 Braunschweig; Volkmaroder Str. 4–5

DE.114.30.0 Leichtweissinstitut für Wasserbau 3300 Braunschweig; Pockelsstr. 4

DE.114.30.1 Lehrgebiet Bodenkunde und Kulturtechnik. 3300 Braunschweig; Beethovenstr. 51a.

DE.114.30.2 Abteilung Landwirtschaftlicher Wasserbau 3300 Braunschweig; Pockelsstr. 4

DE.114.35.0 Lehrstuhl für Entwicklungsplanung und Siedlungswesen. 3300 Braunschweig; Gauss–Str. 14.

DE.114.35.1 Lehrstuhl für Landwirtschaftliche Baukunde.

DE.114.40.0 Lehrstuhl fuer Baukonstruktionslehre und Holzbau. 3300 Braunschweig; Schleinitzstrasse Steinbaracke

DE.114.45.0 Wilhelm–Klauditz–Institut für Holzforschung an der Tu 3300 Braunschweig–Kralenriede; Bienroder Weg 54–E

DE.114.50.0 Lehrstuhl und Institut für Werkzeugmaschinen und Fertigungstechnik. 3300 Braunschweig; Langer Kamp 19

DE.114.50.1 Versuchsfeld für Werkzeugmaschinen und Fabrikbetrieb.

DE.114.50.2 Versuchsfeld für Holzbearbeitung.

DE.114.55.0 Lehrstuhl für Städtebau, Städtischen Strassen– und Tiefbau und Institut für Stadtbauwesen.

DE.117.00.0 Technische Hochschule Darmstadt.

DE.117.03.0 Fachgebiet Botanik im Fachbereich Biologie.

DE.117.05.0 Eduard–Zintl–Institut. 6100 Darmstadt; Hochschulstr. 4.

DE.117.10.0 Fachgebiet Mikrobiologie. 6100 Darmstadt; Schnittspahnstr. 7

DE.117.10.1 Abteilung für Ökologie und Physiologie der Bakterien.

DE.117.15.0 Institut für Makromolekulare Chemie 6100 Darmstadt; Alexanderstr. 24

DE.117.15.1 Lehrstuhl I.

DE.117.15.2 Abteilung Cellulose– und Papierchemie.

DE.117.20.0 Fachgebiet Geographie. 6100 Darmstadt; Schnittspahnstr. 9

DE.117.25.0 Fachgebiet Landtechnik. 6100 Darmstadt; Hochschulstr. 1.

DE.117.30.0 Institut für Wasserbau und Wasserwirtschaft. 6100 Darmstadt; Rundeturmstrasse 1

DE.117.30.1 Fachgebiet Konstruktiver Wasserbau und Wasserwirtschaft

DE.117.40.0 Fachbereich Biologie.

DE.120.00.0 Universität Erlangen–Nuernberg.

DE.120.05.0 Botanisches Institut. 8520 Erlangen; Schlossgarten 4

DE.120.10.0 Institut für Mikrobiologie und Biochemie. 8520 Erlangen; Friedrichstrasse 33.

DE.120.10.1 Lehrstuhl für Mikrobiologie.

DE.120.15.0 Physiologisch–Chemisches Institut. 8520 Erlangen; Wasserturmstr. 5.

DE.120.20.0 Institut fuer Geographie. 8520 Erlangen; Kochstr. 4.

DE.120.25.0 Wirtschafts– und Sozialgeographisches Institut 8520 Erlangen; Findelgasse 7–9

DE.120.25.1 Abteilung Wirtschaftsgeographie

DE.120.30.0 Lehrstuhl für Psychologie. 8500 Nürnberg; Unschlittplatz 1

DE.123.00.0 Universität Frankfurt.

DE.123.05.0 Institut für Lebensmittelchemie. 6000 Frankfurt/Main; Georg–Voigt–Str. 16.

DE.123.07.0 Zoologisches Institut. 6000 Frankfurt/Main; Siesmayerstr. 70

DE.123.10.0 Institut für Bienenkunde 6370 Oberursel Taunus; im Rosengärtchen

DE.123.15.0 Geographisches Institut. 6000 Frankfurt–Main; Senckenberganlage 36

DE.123.20.0 Seminar für Wirtschaftsgeographie. 6000Frankfurt/ Main; Bockenheimer Landstr 140.

DE.123.20.5 Abteilung Überseeische Strukturforschung.

DE.123.25.0 Seminar Für Agrarwesen. 6000 Frankfurt/Main; Senckenberganlage 31

DE.123.28.0 Lehrstuhl für Agrarpolitik. 6000 Frankfurt–Main; Senckenberganlage 36

DE.123.28.1 Fachbereich Wirtschaftswissenschaften.

DE.123.30.0 Institut für Ländliche Strukturforschung 6000 Frankfurt/Main; Zeppelinallee 31

DE.123.35.0 Institut für Sozialforschung 6000 Frankfurt/Main; Senckenbergallee 26.

DE.126.00.0 Universität Freiburg

DE.126.05.0 Institut für Bodenkunde und Waldernährungslehre. 7800 Freiburg; Bertoldstr. 17.

DE.126.10.0 Forstbotanisches Institut 7800 Freiburg; Bertoldstr. 17.

DE.126.15.0 Forstzoologisches Institut. 7800 Freiburg; Bertoldstr. 17.

DE.126.15.1 Abteilung Biologische Schädlingsbekämpfung

DE.126.20.0 Meteorologisches Institut 7800 Freiburg; Wintererstr. 84

DE.126.25.0 Institut für Biologische Holzforschung. 7800 Freiburg; Bertoldstr. 17.

DE.126.30.0 Waldbau–Institut. 7800 Freiburg; Bertoldstr. 17.

DE.126.35.0 Arbeitsgruppe Umweltforschung und Landespflege. 7800 Freiburg, Belfortstrasse 18–20.

DE.126.40.0 Forschungsstelle für Experimentelle Landschaftsökologie. 7800 Freiburg; Belfortstr. 18–20.

DE.126.45.0 Institut für Forsteinrichtung und Forstliche Betriebswirtschaft. 7800 Freiburg; Holzmarktplatz 4.

DE.126.45.1 Abteilung Luftbildmessung und –interpretation.

7800 Freiburg; Erbprinzenstr. 17a.

**DE.126.50.0 Institut fuer Forstpolitik und Raumordnung.
7800 Freiburg; Bertoldstr. 17.**

DE.126.50.5 Arbeitsbereich Forstgeschichte.

DE.126.51.0 Arbeitsbereich Holzmarktforschung.
7800 Freiburg Kaiser–Joseph–Str. 239

**DE.126.55.0 Forstgeschichtliches Institut 7800 Freiburg;
Bertoldstr. 17.**

**DE.126.60.0 Institut für Forstliche Ertragskunde. 7800
Freiburg; Bertoldstr. 17.**

DE.126.60.1 Abteilung für Forstliche Biometrie.
7800 Freiburg; Holzmarktplatz 6.

**DE.126.65.0 Institut für Forstbenutzung und Forstliche
Arbeitswissenschaft 7800 Freiburg; Holzmarktplatz 4.**

**DE.126.67.0 Geógraphisches Institut I. 7800 Freiburg;
Werderring 4.**

DE.126.67.5 Lehrstul für Geographie und Hydrologie.

**DE.126.70.0 Geographisches Institut II. 7800 Freiburg;
Werderring 4.**

**DE.126.80.0 Limnologisches Institut
(Walter–Schlienz–Institut) 7550 Konstanz–Egg; Mainaustr.
112**

DE.126.80.1 Abteilung Limnobotanik.

DE.126.80.2 Abteilung Fliesswasserforschung.

DE.126.80.3 Abteilung Zooplankton und
Nahrungskettenprobleme.

DE.126.84.0 Medizinische Universitätsklinik.

DE.126.85.0 Medizinische Poliklinik

DE.129.00.0 Universität Giessen.

**DE.129.02.0 Institut für Bodenkunde und Bodenerhaltung.
6300 Giessen; Ludwigstr. 23.**

**DE.129.04.0 Institut für Pflanzenernährung. 6300 Giessen;
Braugasse 7.**

DE.129.04.1 Abteilung Gewebekultur

DE.129.04.2 Abteilung Gefässversuche

DE.129.04.3 Abteilung Wachstumsregulation

DE.129.04.4 Isotopen Abteilung.

DE.129.04.5 Abteilung Analytik.

DE.129.04.6 Abteilung Vegetionsversuche

DE.129.04.7 Abteilung Pflanzenphysiologie

DE.129.05.0 Fachbereich Biologie.

**DE.129.06.0 Botanisches Institut. 6300 Giessen;
Senckenbergstr. 17–25.**

DE.129.06.1 Lehrstuhl I Mykologie und Zellphysiologie

DE.129.06.2 Lehrstuhl II Pflanzenökologie.

DE.129.06.3 Abteilung Ökologie der Mikroorganismen.

DE.129.06.4 Abteilung Umweltschutz.

**DE.129.08.0 Institut für Landwirtscaftliche Mikrobiologie
6300 Giessen; Landgraf–Philipp–Platz 4–6**

**DE.129.10.0 I.Zoologisches Institut. 6300 Giessen;
Stephanstr. 24**

**DE.129.12.0 Institut für Pflanzenbau und Pflanzenzüchtung.
6300 Giessen; Ludwigstr. 23.**

DE.129.12.1 Abteilung Biometrie
6300 Giessen; Bismarckstr. 20.

DE.129.12.2 Abteilung Pflanzenzuechtung.

DE.129.12.3 Abteilung Ökophysiologie der
Nahrungspflanzen.

**DE.129.14.0 Institut für Grünlandwirtschaft und Futterbau.
6300 Giessen; Ludwigstr. 23.**

DE.129.14.1 Abt Ökophysiologie der Grünlandpflanzen.

DE.129.14.2 Abteilung Qualität der Futterpflanzen.

DE.129.14.3 Abteilung Futterpflanzenzüchtung.

**DE.129.16.0 Institut für Obstbau 6300 Giessen; Ludwigstr.
37.**

DE.129.16.1 Abteilung Obstzüchtung.

DE.129.16.2 Abteilung Physiologie der Obstgewächse.

DE.129.16.3 Abteilung Phytopathologie.

**DE.129.18.0 Institut für Phytopathologie 6300 Giessen;
Ludwigstr. 23.**

DE.129.18.1 Zoologische Abteilung

DE.129.18.2 Abteilung für Virologie.

DE.129.18.3 Abteilung für Mykologie.

DE.129.18.4 Abteilung Entomologie

DE.129.18.5 Abteilung Nematologie.

DE.129.19.5 Abteilung Nematologie

DE.129.20.0 Fachgebiet Vorratsschutz im Fachbereich Umweltsicherung 6300 Giessen; Alter Steinbacher Weg 36

DE.129.22.0 Fachgebiet Rasenforschung im Fachbereich Umweltsicherung. 6300 Giessen; Schlossgasse 7

DE.129.25.0 Institut für Tierzucht und Haustiergenetik. 6300 Giessen; Bismarckstr. 16.

DE.129.25.1 Abteilung Biochemische Genetik und Immunbiologie.

DE.129.25.2 Abteilung Milchwissenschaft

DE.129.25.3 Abteilung Bioklimatologie

DE.129.25.4 Abteilung Zytogenetik.

DE.129.25.5 Abteilung Populationsgenetik.

DE.129.27.0 Institut für Tierernährung. 6300 Giessen; Landgraf–Philipp–Platz 4–6

DE.129.30.0 Institut für Landwirtschaftliche Betriebslehre. 6300 Giessen; Landgraf–Philipp–Platz 4–6

DE.129.30.1 Abteilung Landwirtschaftliche Betriebslehre I

DE.129.30.2 Abteilung Regionalwissenschaften.

DE.129.30.3 Abteilung Wirtschaftsberatung

DE.129.30.4 Abteilung Rechnungswesen und Datenverarbeitung

DE.129.30.5 Abteilung Regionalwissenschaften.

DE.129.31.0 Institut für Landwirtschaftliche Betriebslehre II. 6300 Giessen; Landgraf Philipp–Platz 4.

DE.129.32.0 Institut fuer Agrarpolitik und Marktforschung. 6300 Giessen; Senckenbergstr. 3.

DE.129.34.0 Institut für Agrarsoziologie 6300 Giessen; Eichgärtenallee 3

DE.129.34.1 Abteilung Wirtschafts– und Regionalsoziologie.

DE.129.36.0 Institut für Ländliches Genossenschaftswesen 6300 Giessen; Landgraf–Philipp–Platz 4

DE.129.38.0 Zentrum für Kontinentale Agrar– und Wirtschaftsforschung 6300 Giessen; Rathenaustr. 17.

DE.129.38.1 Abteilung Agrar– und Ernaehrungsoekonomik.

DE.129.38.2 Abteilung Bodenkunde und Bodenerhaltung.

DE.129.38.3 Abteilung Tierzucht und Tierhaltung.

DE.129.38.4 Abteilung Pflanzenbau und Pflanzenzüchtung.

DE.129.38.5 Abteilung Veterinärmedizin

DE.129.40.0 Institut für Landtechnik. 6300 Giessen; Braugasse 7.

DE.129.40.1 Abteilung Aussenwirtschaft – Pflanzliche Produktion

DE.129.40.2 Abteilung Landwirtschaftliche Bauforschung

DE.129.45.0 Institut für Landeskultur. 6300 Giessen; Landgraf–Philipp–Platz 4–6

DE.129.51.0 Professur für Volkswirtschaftslehre und Entwicklungsländerforschung 6300 Giessen; Licher Str. 74.

DE.129.52.0 Geographisches Institut. 6300 Giessen; Landgraf– Philipp– Platz 2

DE.129.52.1 Abteilung Geographie der Tropen.

DE.129.53.0 Abteilung Ernährung in den Tropen 6300 Giessen; Wilhelmstr. 20

DE.129.54.0 Institut für Tropische Veterinärmedizin. 6300 Giessen; Wilhelmstr. 15

DE.129.55.0 Tropeninstitut. 6300 Giessen; Schottstr. 2–4

DE.129.55.1 Abteilung Bodenkunde und Bodenerhaltung der Tropen und Subtropen.

DE.129.55.2 Abteilung Pflanzenbau und Pflanzenzüchtung.

DE.129.55.3 Abteilung Tierzucht und Tierernährung 6300 Giessen; Hardtallee 51.

DE.129.55.4 Abteilung Phytopathologie und Angewandte Entomologie. 6300 Giessen; Schottstr. 2–4

DE.129.60.0 Veterinär–Anatomisches Institut 6300 Giessen; Frankfurter Str. 94

DE.129.60.1 Lehrstuhl I, Veterinär–Anatomie I.

DE.129.60.2 Abteilung Vergleichende Anatomie der Haus– und Wildtiere.

DE.129.60.3 Abteilung für Allgemeine und Klinische Methodik.

DE.129.60.5 Lehrstuhl II, Veterinär–Anatomie II.

DE.129.62.0 Veterinär–Pathologisches Institut 6300 Giessen; Frankfurter Str. 94

DE.129.64.0 Veterinär–Physiologisches Institut. 6300 Giessen; Frankfurter Str. 94

DE.129.66.0 Institut für Pharmakologie und Toxikologie. 6300 Giessen; Frankfurter Str. 107.

DE.129.68.0 Institut fuer Zuchthygiene und Veterinaermedizinische Genetik. 6300 Giessen; Hofmannstr. 10

DE.129.70.0 Institut für Tierärztliche Nahrungsmittelkunde 6300 Giessen; Frankfurter str. 94.

DE.129.70.1 Abteilung fuer Hygiene der Milch, Fische und Eier.

DE.129.72.0 Institut für Virologie 6300 Giessen; Frankfurter Str. 85–87.

DE.129.74.0 Institut für Bakteriologie und Immunologie. 6300 Giessen; Frankfurter Str. 107.

DE.129.76.0 Institut für Biochemie und Endokrinologie. 6300 Giessen; Frankfurter Strasse 100

DE.129.76.1 Abteilung Endokrinologie.

DE.129.76.2 Abteilung Angewandte Biochemie und Klinische LaboratoriumsdiagnosAbteilung Angewandte Biochemie und Klinische Laboratoriumsdiagnostik. tik.

DE.129.78.0 Chirurgische Veterinärklinik und Chirurgische Veterinärpoliklinik 6300 Giessen; Frankfurter Str. 94

DE.129.80.0 Institut und Klinik fuer Gefluegelkrankheiten. 6300 Giessen; Frankfurter Str. 87

DE.129.82.0 Institut für Hygiene und Infektionskrankheiten der Tiere 6300 Giessen; Frankfurter Str. 85–87.

DE.129.82.1 Abteilung Zoonosen.

DE.129.84.0 Institut für Parasitologie und Parasitäre Krankheiten der Tiere. 6300 Giessen; Frankfurter Str. 94

DE.129.86.0 Ambulatorische und Geburtshilfliche Veterinärklinik 6300 Giessen; Frankfurter Str. 94

DE.129.86.1 Seminar für Verhaltensforschung. 6300 Giessen; Frankfurter Str. 94

DE.129.88.0 Medizinische und Gerichtliche Veterinärklinik. 6300 Giessen; Frankfurter Str.102.

DE.129.88.1 Lehrstuhl I.

DE.129.90.0 Institut für Ernährungswissenschaft I. 6300 Giessen; Wilhelmstr. 20

DE.129.92.0 Institut für Ernährungswissenschaft II 6300 Giessen; Wiesenstr. 3–5

DE.132.00.0 Universität Göttingen

DE.132.03.0 Institut für Bodenkunde. 3400 Göttingen; Von Siebold–Str. 4.

DE.132.03.1 Abteilung Angewandte Bodenkunde.

DE.132.03.2 Abteilung Bodenphysik

DE.132.03.3 Abteilung Mineralogie

DE.132.03.4 Abteilung Ökochemie.

DE.132.06.0 Institut für Agrikulturchemie. 3400 Göttingen; Nikolausberger Weg 7

DE.132.06.1 Abteilung für Biochemie und Physiologie

DE.132.06.2 Abteilung Qualitaet Landwirtschaftlicher Produkte.

DE.132.06.3 Abteilung Versuchswesen und Chemische Analytik.

DE.132.06.4 Abteilung Wasserchemie.

DE.132.06.5 Abteilung Düngung und Düngemittel.

DE.132.07.0 Pflanzenphysiologisches Institut. 3400 Göttingen; Untere Karspüle 2

DE.132.09.0 Systematisch–Geobotanisches Institut 3400 Göttingen; Untere Karspüle 2

DE.132.09.1 Lehrstuhl für Pflanzensystematik

DE.132.09.2 Abteilung für Palynologie.

DE.132.09.3 Lehrstuhl für Geobotanik

DE.132.12.0 Institut für Mikrobiologie 3400 Göttingen; Grisebachstr. 8

DE.132.12.1 Lehrstuhl I

DE.132.12.2 Abteilung Bakterienphysiologie.

DE.132.12.5 Institut für Mikrobiologie der Gesellschaft für Strahlen– und Umweltforschung mbH. 3400 Göttingen; Grisebachstr.8.

DE.132.15.0 Physiologisch–Chemisches Institut. 3400 Göttingen; Humboldtallee 7

DE.132.18.0 Institut für Pflanzenbau und Pflanzenzüchtung. 3400 Göttingen; Von Siebold–Str. 8.

DE.132.18.1 Lehrstuhl fuer Acker– und Pflanzenbau.

DE.132.18.2 Lehrstuhl für Pflanzenzüchtung.

DE.132.18.3 Abteilung für Futterpflanzenzüchtung.

DE.132.18.4 Abteilung Pflanzenzüchtung.

DE.132.21.0 Institut für Pflanzenpathologie und Pflanzenschutz. 3400 Göttingen; Grisebachstr. 8

DE.132.21.1 Entomologische Abteilung.

DE.132.21.2 Mykologische Abteilung.

DE.132.21.3 Abteilung Pathophysiologie.

DE.132.24.0 Institut für Tropischen und Subtropischen Pflanzenbau. 3400 Göttingen; Grisebachstr. 6

DE.132.25.0 Institut für Ausländische Landwirtschaft. 3400 Goettingen; Buesgenweg 2.

DE.132.25.1 Abteilung Ausländische Landwirtschaft

DE.132.27.0 Institut für Tierzucht und Haustiergenetik. 3400 Göttingen; Albrecht–Thär–Weg 1.

DE.132.27.1   Abteilung Fortpflanzungsbiologie.

DE.132.27.2   Abteilung Tierhaltung

DE.132.27.3   Abteilung Tierzüchtung

DE.132.27.4   Abteilung Haustiergenetik

DE.132.27.5   Abteilung Tierzüchtung und –haltung an tropischen und subtropischen Standorten.

DE.132.27.6   Abteilung Fortpflanzung und Biotechnik.

DE.132.30.0 Institut für Tierphysiologie und Tierernährung 3400 Göttingen; Oskar–Kellnerweg 6

DE.132.30.1   Abteilung Fuetterungslehre.

DE.132.30.2   Abteilung Leistungsphysiologie.

DE.132.30.3   Abteilung Ernährungsphysiologie.

DE.132.30.4   Abteilung Wiederkauer–Ernährung.

DE.132.30.5   Abteilung Futterverwertung und Futterbewertung.

DE.132.33.0 Tierärztliches Institut. 3400 Göttingen; Groner Landstr. 2.

DE.132.33.1   Abteilung Tierhygiene an Tropischen und Subtropischen Standorten

DE.132.33.2   Abteilung Rinderkrankheiten

DE.132.33.3   Geflügelabteilung

DE.132.33.4   Veterinäruntersuchungslaboratorium.

DE.132.33.5   Abteilung Fortpflanzungsphysiologie

DE.132.33.6   Abteilung Molekular und Immunogenetik.

DE.132.33.7   Abteilung Parasitologie

DE.132.36.0 Institut für Agrarökonomie. 3400 Goettingen; Nikolausberger Weg 9c.

DE.132.36.1   Abteilung Betriebslehre

DE.132.36.2   Abteilung Marktwirtschaft

DE.132.36.3   Abteilung Agrarpolitik.

DE.132.36.4   Abteilung Landwirtschaftliche Marktlehre

DE.132.36.5   Abteilung Sektorale Wirtschaftspolitik

DE.132.36.6   Abteilung Ökonometrie.

DE.132.36.7   Abteilung Strukturforschung.

DE.132.39.0 Institut für Sozialpolitik und Sozialrecht 3400

Göttingen; Nikolausberger Weg 5 C.

DE.132.39.1   Abteilung Sozialpolitik

DE.132.42.0 Volkswirtschaftliches Seminar. 3400 Göttingen; Nikolausberger Weg 5 C.

DE.132.42.1   Volkwirtschaftliches Seminar.

DE.132.45.0 Institut für Landwirtschaftsrecht. 3400 Göttingen; Nikolausberger Weg 9 A.

DE.132.48.0 Ibero–Amerika Institut für Wirtschaftsforschung. 3400 Göttingen; Gosslerstr. 1 b

DE.132.51.0 Geographisches Institut. 3400 Göttingen; Herzberger Landstr. 2.

DE.132.51.1   Lehrstuhl II Anthropogeographie

DE.132.51.2   Abteilung Sozial– und Wirtschaftsgeographie.

DE.132.54.0 Landmaschinen–Institut. 3400 Göttingen; Gutenbergstr. 33.

DE.132.57.0 Abteilung fuer Biophotogrammetrie. 3400 Göttingen; Oskar–Kellner–Weg 6

DE.132.58.0 Interfakultatives Lehrgebiet Chemie. 3400 Göttingen; von–Siebold–Str. 2

DE.132.60.0 Institut für Bodenkunde und Waldernährung. 3400 Göttingen; Büsgenweg 2.

DE.132.60.1   Abteilung Bodenchemie und Waldernährung

DE.132.60.2   Abteilung Tropische Bodenkunde

DE.132.60.3   Abteilung Bodenphysik und Bodenhydrologie.

DE.132.63.0 Forstbotanisches Institut 3400 Göttingen; Büsgenweg 2.

DE.132.66.0 Institut für Forstzoologie. 3400 Göttingen; Büsgenweg 3.

DE.132.69.0 Lehrstuhl für Forstgenetik und Forstpflanzenzüchtung 3400 Göttingen; Büsgenweg 2.

DE.132.69.5   Abteilung für forstliche Biometrie der forstlichen Fakultät. 3400 Goettingen; Buesgenweg 5.

DE.132.72.0 Institut für Forsteinrichtung und Forstliche ertragskunde 3400 Göttingen; Büsgenweg 5.

DE.132.75.0 Institut für Waldbau 3400 Göttingen; Büsgenweg 1.

DE.132.75.1   Lehrstuhl für Waldbau in den Gemässigten Zonen.

DE.132.75.2   Lehrstuhl für Naturwaldforschung und Waldbau der Tropen und Subtropen

DE.132.78.0 Institut für Forstbenutzung. 3400

Göttingen–Weende; Büsgenweg 4.

DE.132.81.0 Institut für Wildforschung und Jagdkunde. 3400 Göttingen; Büsgenweg 3.

DE.132.84.0 Institut für Forstliche Betriebswirtschaftslehre. 3400 Göttingen; Büsgenweg 5.

DE.132.87.0 Institut für Forstpolitik, Holzmarktlehre, Forstgeschichte und Naturschutz. 3400 Göttingen; Büsgenweg 5.

DE.132.90.0 Institut für Waldarbeit und Forstmaschinenkunde 3400 Göttingen; Büsgenweg 6.

DE.132.95.0 Medizinische Klinik und Poliklinik. 3400 Goettingen; Humboldtallee 1.

DE.132.96.0 Zoologisches Institut und Museum. 3400 Goettingen; Berliner Str. 28.

DE.135.00.0 **Universität Hamburg.**

DE.135.05.0 Institut für Angewandte Botanik. 2000 Hamburg 36; Marseiller Str. 7.

DE.135.05.1 Abteilung Kultur– und Versuchstechnik.

DE.135.05.2 Abteilung Pflanzenschutz (Pflanzenschutzamt Hamburg).

DE.135.05.3 Abteilung Nutzpflanzenbiologie.

DE.135.05.4 Abteilung Saatgutprüfung

DE.135.05.5 Abteilung Warenkunde.

DE.135.05.6 Abteilung Landwirtschaftliche Chemie

DE.135.10.0 Zoologisches Institut und Zoologisches Museum. 2000 Hamburg 13; Martin–Luther–King–Platz 3.

DE.135.15.0 Institut für Organische Chemie und Biochemie 2000 Hamburg 13; Papendamm 6

DE.135.15.1 Abteilung Organische Chemie.

DE.135.20.0 Institut für Physiologische Chemie am Universitätskrankenhaus Eppendorf. 2000 Hamburg 20; Martinstr. 52.

DE.135.20.1 Abteilung Corrinoide.

DE.135.20.2 Abteilung Ernährungsphysiologie.

DE.135.25.0 Institut für Hydrobiologie und Fischereiwissenschaft. 2000 Hamburg 50; Olbersweg 24.

DE.135.25.1 Abteilung Fischereiwissenschaft

DE.135.30.0 Institut für Geographie und Wirtschaftsgeographie 2000 Hamburg 13; Rothenbaumchaussee 21–23.

DE.138.00.0 **Universität Hannover**

DE.138.03.0 Institut für Bodenkunde. 3000 Hannover–Herrenhausen;Herrenhäuser Str. 2.

DE.138.03.1 Abteilung Bodenkunde.

DE.138.06.0 Institut für Pflanzenernährung. 3000 Hannover–Herrenhausen; Herrenhäuser Str. 2

DE.138.09.0 Institut für Botanik 3000 Hannover–Herrenhausen; Herrenhäuser Str. 2

DE.138.09.1 Abteilung Biologie und Meteorologie.

DE.138.12.0 Institut für Vegetationskunde. 3000 Hannover 1; Nienburger Str. 17

DE.138.15.0 Institut für Meteorologie und Klimatologie. 3000 Hannover–Herrenhausen; Herrenhäuser Str. 2

DE.138.18.0 Institut für Biophysik 3000 Hannover–Herrenhausen; Herrenhäuser Str. 2

DE.138.18.1 Institut für Strahlenbotanik der Gesellschaft für Strahlen– und Umweltforschung.

DE.138.21.0 Institut für Gemüsebau. 3000 Hannover–Herrenhausen; Herrenhäuser Str. 2

DE.138.24.0 Institut für Obstbau und Baumschule 3203 Sarstedt/Hannover; Haus Steinberg

DE.138.24.1 Abteilung Baumschule.

DE.138.24.2 Abteilung Physiologie

DE.138.27.0 Institut für Zierpflanzenbau 3000 Hannover–Herrenhausen; Herrenhäuser Str. 2

DE.138.30.0 Institut für Angewandte Genetik 3000 Hannover–Herrenhausen; Herrenhäuser Str. 2

DE.138.33.0 Institut für Pflanzenkrankheiten und Pflanzenschutz 3000 Hannover–Herrenhausen; Herrenhäuser Str. 2

DE.138.33.1 Abteilung Virologie

DE.138.36.0 Institut für Gartenbauökonomie. 3000 Hannover–Herrenhausen, Herrenhäuser Str. 2.

DE.138.36.1 Lehrstuhl A Produktion.

DE.138.36.2 Lehrstuhl B Absatz.

DE.138.36.5 Lehrgebiet Berufsdidaktik des Gartenbaues.

DE.138.39.0 Institut für Technik in Gartenbau und Landwirtschaft. 3000 Hannover–Herrenhausen; Herrenhäuser Str. 2

DE.138.45.0 Lehrstuhl und Institut für Wasserwirtschaft, Hydrologie und landwirtschaftlichen Wasserbau. 3000 Hannover; Callinstr. 15.

DE.138.46.0 Institut für Siedlungswasserwirtschaft. 3000 Hannover; Welfengarten 1.

DE.138.50.0 Lehrstuhl für Bautechnik und Holzbau. 3000 Hannover; Callinstr. 15.

DE.138.55.0 Institut für Photogrammetrie und Ingenieurvermessungen. 3000 Hannover; Nienburger Str. 1.

DE.138.60.0 Institut für Landschaftspflege und Naturschutz. 3000 Hannover–Herrenhausen; Herrenhäuser Str. 2.

DE.138.60.1 Abteilung Landespflege.

DE.138.63.0 Lehrstuhl für Ländliches Bau– und Siedlungswesen. 3000 Hannover; Wilhelm– Busch– Str. 8.

DE.138.66.0 Institut für Grünplanung und Gartenarchitektur 3000 Hannover–Herrenhausen; Herrenhäuser Str. 2

DE.138.66.1 Abteilung Landespflege.

DE.138.70.0 Geographisches Institut 3000 Hannover; Schneiderberg 50.

DE.138.70.1 Afrika–Abteilung.

DE.138.70.2 Angew.Phys.Geographie.

DE.138.75.0 Lehrstuhl für Lebensmittelchemie 3000 Hannover; Wunstorfer Str. 14

DE.139.00.0 Tierärztliche Hochschule Hannover

DE.139.02.0 Botanisches Institut. 3000 Hannover; Bischofsholer Damm 15.

DE.139.02.1 Abteilung Biochemie der Pflanzen.

DE.139.05.0 Anatomisches Institut 3000 Hannover; Bischofsholer Damm 15.

DE.139.06.0 Chemisches Institut. 3000 Hannover; Bischofsholer Damm 15.

DE.139.10.0 Institut für Pathologie. 3000 Hannover; Bischofsholer Damm 15.

DE.139.10.1 Abteilung Immunpathologie

DE.139.10.2 Abteilung Pathologie der Versuchstiere.

DE.139.10.3 Abteilung Elektronenmikroskopie.

DE.139.10.4 Abteilung Diagnostik.

DE.139.15.0 Institut für Pharmakologie, Toxikologie und Pharmazie. 3000 Hannover; Bischofsholer Damm 15.

DE.139.15.1 Abteilung Toxikologie

DE.139.20.0 Physiologisches Institut. 3000 Hannover; Bischofsholer Damm 15.

DE.139.20.1 Abteilung Strahlenbiologie.

DE.139.20.2 Abteilung Ernährungsphysiologie.

DE.139.20.3 Abteilung Muskelphysiologie

DE.139.20.4 Abteilung Stoffwechselphysiologie

DE.139.20.5 Abteilung Endokrinologie.

DE.139.25.0 Institut für Physiologische Chemie. 3000 Hannover; Bischofsholer Damm 15.

DE.139.26.0 Institut für Tierhygiene 3000 Hannover; Bischofsholer Damm 15.

DE.139.26.1 Abteilung für Tierhaltung.

DE.139.30.0 Institut für Tierernährung. 3000 Hannover; Bischofsholer Damm 15.

DE.139.30.1 Arbeitsgruppe Futtermittelkunde und Phytochemie.

DE.139.35.0 Institut für Tierzucht und Vererbungsforschung. 3000 Hannover–Kirchrode; Bünteweg 17

DE.139.35.1 Abteilung Haustierökologie

DE.139.35.2 Abteilung Haustiergenetik.

DE.139.40.0 Institut für Haustierbesamung und Andrologie. 3000 Hannover; Bischofsholer Damm 15.

DE.139.40.1 Abteilung Spermatologie

DE.139.45.0 Institut für Mikrobiologie und Tierseuchen. 3000 Hannover; Bischofsholer Damm 15.

DE.139.45.1 Abteilung Mykologie

DE.139.50.0 Institut für Parasitologie 3000 Hannover–Kirchrode; Bünteweg 17

DE.139.50.1 Abteilung Veterinärmedizinische Entomologie

DE.139.50.2 Abteilung Veterinärmedizinische Protozoologie

DE.139.50.3 Abteilung Helminthologie.

DE.139.50.4 Laboratorium für Parasitologie in den Tropen.

DE.139.55.0 Institut für Virologie 3000 Hannover; Bischofsholer Damm 15.

DE.139.55.1 Abteilung Tumorvirologie.

DE.139.55.2 Abteilung Allgemeine Virologie

DE.139.56.0 Institut für Lebensmittelkunde, Fleischhygiene und –technologie 3000 Hannover; Bischofsholer Damm 15.

DE.139.60.0 Institut für Hygiene und Technologie des Fleisches 3000 Hannover; Bischofsholer Damm 15.

DE.139.65.0 Institut für Hygiene und Technologie der Milch 3000 Hannover; Bischofsholer Damm 15.

DE.139.65.1 Abteilung Bakteriologie.

DE.139.70.0 Institut für Geflügelkrankheiten. 3000 Hannover; Bischofsholer Damm 15.

DE.139.70.1 Abteilung Ziervögel.

DE.139.70.2 Vogelklinik.

DE.139.70.3 Abteilung Bakteriologie

DE.139.70.4 Abteilung Virologie

DE.139.70.5 Arbeitsgruppe Zier- und Wildvögel.

DE.139.75.0 Klinik fuer Kleine Klauentiere und Forensische Medizin und Ambulatorische Klinik. 3000 Hannover; Bischofsholer Damm 15.

DE.139.80.0 Klinik für Geburtshilfe und Gynäkologie des Rindes. 3000 Hannover; Bischofsholer Damm 15; Im Richard–Götze–Haus.

DE.139.80.1 Abteilung Herdensterilität

DE.139.80.2 Abt. für Angewandte Endokrinologie.

DE.139.80.3 Abt.Experimentelle Fortpflanzungsbiologie.

DE.139.80.4 Klinisch–Bakteriologisches Labor.

DE.139.80.5 Abteilung Euterkrankheiten.

DE.139.85.0 Klinik für Pferde 3000 Hannover; Bischofsholer Damm 15.

DE.139.90.0 Klinik für Rinderkrankheiten 3000 Hannover; Bischofsholer Damm 15; I Richard–Götze–Haus

DE.139.95.0 Klinik für Kleine Haustiere 3000 Hannover; Bischofsholer Damm 15.

DE.139.96.0 Institut für Statistik und Biometrie. 3000 Hannover; Bischofsholer Damm 15.

DE.142.00.0 Universität Heidelberg

DE.142.10.0 Hygiene–Institut. 6900 Heidelberg; Thibautstr. 2

DE.142.10.1 Abteilung Lebensmittelhygiene

DE.142.10.2 Abteilung Allgemeine Hygiene und Umwelthygiene.

DE.142.15.0 Suedasien–Institut. 6900 Heidelberg; Im Neuenheimer Feld 13.

DE.142.20.0 Institut für Internationale Vergleichende Agrarpolitik und Agrarsoziologie am Südasien–Institut. 6900 Heidelberg; Im Neuenheimer Feld 13.

DE.142.30.0 Geographisches Institut 6900 Heidelberg; Grabengasse

DE.142.30.1 Abteilung Südamerika.

DE.142.40.0 Fachgruppe Innere Medizin I.

DE.142.45.0 Fakultät für Klinische Medizin I.

DE.144.00.0 Universität Hohenheim.

DE.144.03.0 Dokumentationsstelle. 7000 Stuttgart 70; Paracelsusstr. 2.

DE.144.04.0 Abteilung Organische Chemie. 7000 Stuttgart 70, Emil–Wolff–Str. 14.

DE.144.10.0 Abteilung für Mikrobiologie und Molekularbiologie. 7000 Stuttgart 70, Otto–Sander–Str. 5.

DE.144.13.0 Institut fuer Zoologie – Fachgebiet Parasitologie. 7000 Stuttgart 70; Emil–Wolff–Str. 27.

DE.144.14.0 Institut fuer Zoophysiologie. 7000 Stuttgart 70; Emil–Wolff–Str. 27.

DE.144.14.1 Arbeitsgruppe Endokrinologie. 7000 Stuttgart 70; Emil Wolff Str. 27.

DE.144.14.2 Arbeitsgruppe Vegetative Physiologie.

DE.144.15.0 Institut fuer Zoologie. 7000 Stuttgart 70, Emil–Wolff–Str. 27.

DE.144.15.1 Fachgebiet Biologie der Fische.

DE.144.16.0 Institut fuer Organische Chemie. 7000 Stuttgart 70; Postfach 106.

DE.144.16.5 Lehrstuhl fuer Organische Chemie. 7000 Stuttgart 70; Postfach 106.

DE.144.19.0 Institut für Biologische Chemie. 7000 Stuttgart 70; Garbenstr. 30.

DE.144.23.0 Lehrstuhl für Allgemeine Lebensmitteltechnologie. 7000 Stuttgart 70; Garbenstr. 25

DE.144.24.0 Institut für Allgemeine Lebensmitteltechnologie und technische Biochemie. 7000 Stuttgart 70; Garbenstr. 25.

DE.144.24.1 Laboratorium fuer Technische Biochemie.

DE.144.24.2 Laboratorium für Gärungstechnologie.

DE.144.26.0 Institut fuer Spezielle Lebensmitteltechnologie – Fachgebiet Gemuese– und Fruechtetechnologie. 7000 Stuttgart 70; Garbenstr. 25.

DE.144.27.0 Institut fuer Spezielle Lebensmitteltechnologie – Fachgebiet Milchwissenschaft. 7000 Stuttgart 70, Garbenstr. 25.

DE.144.28.0 Abteilung Getreidetechnologie. 7000 Stuttgart 70; Garbenstr. 25.

DE.144.32.0 Institut für Bodenkunde und Standortlehre. 7000 Stuttgart 70, Postfach 106.

DE.144.32.1 Fachgebiet Bodenchemie.

DE.144.32.2 Fachgebiet Allgemeine Bodenkunde mit

Gesteinskunde.

DE.144.32.3   Fachgebiet Biometeorologie.

DE.144.32.4   Fachgebiet Bodenphysik.

DE.144.34.0 Öko–Physiologie und Vegetationskunde. 7000 Stuttgart 70; Emil–Wolff–Str. 27.

DE.144.35.0 Abteilung Landschaftsökologie. 7000 Stuttgart 70; Emil–Wolff–Str. 27.

DE.144.39.0 Abteilung für Pflanzenernährung I. 7000 Stuttgart 70, Emil Wolff Str. 25.

DE.144.40.0 Institut für Pflanzenernährung. 7000 Stuttgart 70; Emil Wolff Str. 25.

DE.144.40.5   Abteilung für Pflanzenernährung III. 7000 Stuttgart 70, Fruwirthstr. 20.

DE.144.41.0 Institut für Pflanzenzüchtung und Populationsgenetik. 7000 Stuttgart 70; Postfach 106.

DE.144.41.1   Fachgebiet Samenkunde und Keimungsphysiologie.

DE.144.42.0 Lehrstuhl für Angewandte Genetik und Pflanzenzüchtung. 7000 Stuttgart 70; Emil Wolff Str. 25.

DE.144.43.0 Institut fuer Pflanzenzuechtung und Populationsgenetik – Fachgebiet Spezielle Pflanzenzuechtung. 7000 Stuttgart 70; Emil Wolff Str. 25.

DE.144.44.0 Institut fuer Obst– Gemuese– und Weinbau. 7000 Stuttgart 70; Emil Wolff Str. 25.

DE.144.44.5   Lehrstuhl fuer Obstbau.

DE.144.45.0 Lehrstuhl fuer Weinbau. 7000 Stuttgart 70; Emil–Wolff–Str. 25.

DE.144.48.0 Institut fuer Pflanzenbau. 7000 Stuttgart 70; Postfach 106.

DE.144.48.5   Lehrstuhl für Acker– und Pflanzenbau.

DE.144.49.0 Institut für Pflanzenbau. 7000 Stuttgart 70; Fruwirthstr. 23.

DE.144.50.0 Lehrstuhl für Grünlandlehre. 7000 Stuttgart 70; Kirchnerstr. 30.

DE.144.51.0 Abteilung Pflanzenbau in den Tropen und Subtropen. 7000 Stuttgart 70; Kirchnerstr. 30

DE.144.52.0 Lehrstuhl für Phytopathologie und Pflanzenschutz. 7000 Stuttgart 70, Otto Sander Str. 5.

DE.144.52.1   Isotopenlabor.

DE.144.53.0 Abteilung für Mikrobiologie und Phytopathologie. 7000 Stuttgart 70; Postfach 106.

DE.144.54.0 Institut fuer Phytomedizin. 7000 Stuttgart 70, Kirchnerstr. 30.

DE.144.54.1   Isotopenlabor.

DE.144.54.2   Fachgebiet Virologie und Bakteriologie.

DE.144.54.3   Fachgebiet Herbologie.

DE.144.55.0 Fachgebiet Angewandte Entomologie. 7000 Stuttgart 70; Kirchnerstr. 30

DE.144.58.0 Lehrstuhl für Anatomie und Physiologie der Haustiere. 7000 Stuttgart 70; Fruwirthstr. 35.

DE.144.58.1   Tierklinik.

DE.144.59.0 Abteilung Futtermittelkunde. 7000 Stuttgart 70; Ochsenhof 2.

DE.144.60.0 Institut für Tiermedizin und Tierhygiene mit Tierklinik. 7000 Stuttgart 70; Ochsenhof 2.

DE.144.60.5   Lehrstuhl für Anatomie und Physiologie der Haustiere. 7000 Stuttgart 70; Postfach 106.

DE.144.61.0 Fachgebiet Tierhygiene. 7000 Stuttgart 70; Ochsenhof 2

DE.144.62.0 Institut fuer Tierernaehrung. 7000 Stuttgart 70; Postfach 106.

DE.144.62.1   Abteilung Tierernährung.

DE.144.62.2   Abteilung Ernährungsphysiologie der Tiere

DE.144.62.5   Fachrichtung Futtermittelkunde.

DE.144.66.0 Abteilung Biotechnik der Grosstiere. 7000 Stuttgart 70; Emil–Wolff–Str. 34.

DE.144.67.0 Institut für Tierhaltung und Tierzüchtung. 7000 Stuttgart 70. Emil Wolff Str. 34.

DE.144.67.5   Lehrstuhl Tierhaltung.

DE.144.68.0 Fachgebiet Tierzüchtung. 7000 Stuttgart 70; Emil Wolff Str. 34.

DE.144.68.5   Lehrstuhl fuer Kleintierzucht.

DE.144.69.0 Fachgebiet Allgemeine Tierhaltung. 7000 Stuttgart 70; Emil Wolff Str. 34.

DE.144.70.0 Institut für Agrartechnik. 7000 Stuttgart 70, Garbenstr. 9.

DE.144.70.1   Lehrstuhl für Grundlagen der Landtechnik.

DE.144.70.5   Lehrstuhl für Landtechnik.

DE.144.71.0 Fachgebiet fuer Verfahrenstechnik in der Tierproduktion und Landwirtschaftliches Bauwesen. 7000 Stuttgart 70; Garbenstr. 9.

DE.144.72.0 Fachgebiet Verfahrenstechnik fuer Intensivkulturen. 7000 Stuttgart 70; Garbenstr. 9.

FEDERAL REPUBLIC OF GERMANY

DE.144.74.0 Institut für Landwirtschaftliche Betriebslehre.
7000 Stuttgart 70; Osthof.

DE.144.74.5 Lehrstuhl für Angewandte Landwirtschaftliche Betriebslehre.

DE.144.75.0 Fachgebiet Planung der Landwirtschaftlichen Produktion. 7000 Stuttgart 70; Osthof.

DE.144.75.5 Institut für Landwirtschaftliche Betriebslehre; Lehrstuhl für Ökonomik der Landwirtschaftlichen Produktion in den Tropen und Subtropen.

DE.144.78.0 Lehrstuhl für Ökonomik der Landwirtschaftlichen Produktion. 7000 Stuttgart 70; Osthof.

DE.144.79.0 Institut fuer Haushalts- u. Konsumoekonomik. 7000 Stuttgart 70; Osthof.

DE.144.79.5 Lehrstuhl fuer Wirtschaftslehre des Haushalts.

DE.144.85.0 Institut fuer Agrarpolitik und Landwirtschaftliche Marktlehre. 7000 Stuttgart 70; Postfach 106.

DE.144.85.5 Fachgebiet Landwirtschaftliche Marktlehre.

DE.144.86.0 Fachgebiet Agrarpolitik. 7000 Stuttgart 70; Postfach 106.

DE.144.86.1 Forschungsstelle für Genossenschaftswesen.

DE.144.86.5 Fachgebiet International Vergleichende Agrarpolitik.
7000 Stuttgart 70; Postfach 106.

DE.144.88.0 Institut für Sozialwissenschaften.
7000 Stuttgart 70; Postfach 106.

DE.144.88.5 Fachgebiet Wirtschafts-, Sozial- mit Agrargeschichte.

DE.144.88.6 Forschungsstelle für Vergleichende Untersuchungen der Landwirtschaft in den Beiden Deutschen Staaten.

DE.144.89.0 Abteilung Wirtschafts-. Sozial- und Agrargeschichte. 7000 Stuttgart 70, Garbenstr. 17.

DE.144.90.0 Institut für Sozialwissenschaften; Fachgebiet Recht. 7000 Stuttgart 70; Postfach 106.

DE.144.91.0 Abteilung für Öffentliches Recht, Agrarrecht und Umweltrecht. 7000 Stuttgart 70; Garbenstr. 17

DE.144.92.0 Lehrstuhl für Zoophysiologie. 7000 Stuttgart 70; Schloss Westflügel.

DE.144.95.0 Abteilung Ländliche Sozialforschung. 7000 Stuttgart 70; Osthof.

DE.144.96.0 Institut für Agrarsoziologie, Landwirtschaftliche Beratung und Angewandte Psychologie. 7000 Stuttgart 70; Garbenstr. 17.

DE.144.96.5 Fachgebiet Ländliche Sozialforschung.

DE.144.97.0 Institut für Landeskultur und Pflanzenökologie. 7000 Stuttgart 70; Postfach 106.

DE.144.98.0 Fachgebiet Pflanzenökologie. 7000 Stuttgart 70, Osthof.

DE.144.99.0 Fachgebiet Landeskultur. 7000 Stuttgart 70, Postfach 106.

DE.144.99.1 Abteilung Landeskultur.

DE.145.00.0 Universität Karlsruhe.

DE.145.05.0 Meteorologisches Institut 7500 Karlsruhe; Kaiserstr. 12.

DE.145.10.0 Institut für Lebensmittelchemie. 7500 Karlsruhe; Kaiserstr. 12.

DE.145.15.0 Institut für Lebensmittelverfahrenstechnik 7500 Karlsruhe; Kaiserstr. 12.

DE.145.20.0 Institut für Kolbenmaschinen 7500 Karlsruhe; Kaiserstr. 12.

DE.145.20.1 Lehrauftrag für Landmaschinen.

DE.145.25.0 Institut fuer Orts-, Regional- und Landesplanung. 7500 Karlsruhe; Postfach 6380.

DE.145.25.1 Abteilung Landwirtschaftliches Bauwesen.

DE.145.30.0 Institut für Wasserbau und Wasserwirtschaft mit 'Theodor-Rehbock-Flussbaulaboratorium'. 7500 Karlsruhe; Kaiserstr. 12.

DE.145.30.1 Lehrgebiet Landwirtschaftlicher Wasserbau.

DE.145.30.2 Abteilung Hydrologie – Wasserwirtschaft.

DE.145.35.0 Geodätisches Institut.
7500 Karlsruhe; Englerstr. 7

DE.145.40.0 Lehrstuhl für Ingenieurholzbau und Baukonstruktionen der Versuchsanstalt für Stahl, Holz und Steine. 7500 Karlsruhe; Kaiserstr. 12.

DE.145.45.0 Institut fuer Wasserbau III. 7500 Karlsruhe; Kaiserstr. 12.

DE.148.00.0 Universität Kiel

DE.148.05.0 Institut für Pflanzenernährung und Bodenkunde. 2300 Kiel; Neue Universität Haus 41

DE.148.05.1 Lehrstuhl Bodenkunde.

DE.148.05.2 Lehrstuhl Pflanzenernährung.

DE.148.10.0 Institut für Pflanzenbau und Pflanzenzüchtung. 2300 Kiel; Olshausenstr. 40–60.

DE.148.10.1 Abteilung Gruenlandwirtschaft, Futterbau und

Landschaftsoekologie.
see 148100.

DE.148.15.0 Institut für Landwirtschaftliche
Verfahrenstechnik. 2300 Kiel; Olshausenstr. 40–60.

DE.148.20.0 Institut für Phytopathologie 2300 Kiel; Neue
Universität Haus C 1

DE.148.30.0 Institut für Tierernährungslehre. 2300 Kiel;
Neue Universität Haus C 1

DE.148.35.0 Institut für Tierzucht und Tierhaltung. 2300
Kiel; Olshausenstr. 40–60.

DE.148.35.1 Abteilung Tierzucht

DE.148.35.2 Abteilung Tierhaltung

DE.148.40.0 Institut für Landwirtschaftliche Betriebs– und
Arbeitslehre. 2300 Kiel; Holtzkoppelweg 14

DE.148.40.1 Lehrstuhl für Angewandte Landwirtschaftliche
Betriebslehre.

DE.148.40.5 Lehrstuhl für Wirtschaftslehre des Landbaues.

DE.148.45.0 Institut für Agrarpolitik und Marktlehre. 2300
Kiel; Holzkoppelweg 14.

DE.148.45.1 Lehrstuhl Agrarpolitik.

DE.148.45.2 Lehrstuhl Marktlehre.

DE.148.45.3 Abteilung Absatz und
Welternährungswirtschaft

DE.148.45.4 Abteilung Wirtschaftslehre des Haushalts.

DE.148.50.0 Soziologisches Seminar. 2300 Kiel; Neue
Universität Haus 38

DE.148.55.0 Institut für Wasserwirtschaft und
Meliorationswesen 2300 Kiel; Neue Universität Haus C 1

DE.148.60.0 Institut für Weltwirtschaft. 2300 Kiel;
Düsternbrookerweg.

DE.148.60.1 Abteilung Struktur und Weltwirtschaft.

DE.148.65.0 Institut für Meereskunde.
2300 Kiel; Düsterbrooker Weg– 120–122.

DE.148.70.0 Abteilung für Marine Mikrobiologie. 2300 Kiel;
Düsterbrooker Weg 120–122

DE.149.45.3 Abteilung Absatz– und
Welternährungswirtschaft.

DE.151.00.0 Universität Köln.

DE.151.05.0 Institut für Biochemie 5000 Köln; an der
Bottmeuhle 2.

DE.151.10.0 Zoologisches Institut der Universität Köln. 5000
Köln 41; Weyertal 119.

DE.151.10.1 Lehrstuhl für Physiologische Ökologie.

DE.151.15.0 Geographisches Institut 5000 Köln 41;
Albertus–Magnus–Platz

DE.151.20.0 Wirtschafts– und Sozialgeographisches Institut
5000 Köln 41; Albertus–Magnus–Platz

DE.151.25.0 Seminar für Genossenschaftswesen 5000 Köln 41;
Universität.

DE.151.30.0 Institut für Handelsforschung.

DE.154.00.0 Universität Mainz.

DE.154.05.0 Institut für Spezielle Botanik und Botanischer
Garten 6500 Mainz; Saarstr. 21.

DE.154.07.0 Hygiene–Institut.
6500 Mainz; Langenbeckstr. 1

DE.154.10.0 Institut für Mikrobiologie und Weinforschung.
6500 Mainz; Ernst–Ludwig–Str. 10

DE.154.20.0 Physiologisch–Chemisches Institut. 6500 Mainz;
Joh.–Joachim–Becker–Weg 13.

DE.154.20.1 II. Lehrstuhl.

DE.157.00.0 Universität Marburg.

DE.157.05.0 Institut für Pharmazeutische Chemie und
Lebensmittelchemie 3550 Marburg; Marbacher Weg 6.

DE.157.07.0 Institut für Genossenschaftswesen in
Entwicklungsländrn. 3550 Marburg A Plan 2

DE.157.10.0 Institut fuer Angewandte Zoologie. 3550
Marburg; Lahnberge.

DE.160.00.0 Universität München

DE.160.03.0 Institut fuer Bodenkunde und Standortslehre der
Forstlichen Forschungsanstalt Muenchen. 8000 München 40;
Amalienstrasse 52.

DE.160.06.0 Forstbotanisches Institut der Forstlichen
Forschungsanstalt Muenchen. 8000 München 40;
Amalienstrasse 52.

DE.160.06.1 Lehrstuhl für Anatomie, Physiologie und
Pathologie der Pflanzen.

DE.160.06.2 Abteilung Ökologie.

DE.160.09.0 Lehrstuhl für Angewandte Zoologie. 8000
München 40, Amalienstr. 52.

DE.160.09.1 Abteilung Forstzoologie

DE.160.09.2 Abteilung Forstschutz

DE.160.12.0 Institut fuer Meteorologie der Forstlichen
Forschungsanstalt Muenchen. 8000 München 40;
Amalienstrasse 52.

FEDERAL REPUBLIC OF GERMANY

DE.160.12.1 Lehrstuhl für Bioklimatologie und Angewandte Meteorologie.

DE.160.15.0 Waldbauinstitut der Forstlichen Forschungsanstalt Muenchen. 8000 München 40; Amalienstrasse 52.

DE.160.15.1 Lehrstuhl für Waldbau und Forsteinrichtung.

DE.160.18.0 Institut fuer Forstpflanzenzuechtung, Samenkunde und Immissionsforschung der Forstlichen Forschungsanstalt Muenchen. 8000 München 40; Amalienstrasse 52.

DE.160.18.1 Lehrstuhl für Saatgut, Genetik und Züchtung der Waldbäume.

DE.160.21.0 Institut fuer Waldwachstumskunde der Forstlichen Forschungsanstalt Muenchen. 8000 München 40; Amalienstrasse 52.

DE.160.21.1 Lehrstuhl für Waldwachstumskunde. 8000 München 40; Amalienstrasse 52.

DE.160.27.0 Institut für Wildforschung und Jagdkunde. 8103 Oberammergau; Ettaler Str. 3

DE.160.30.0 Institut fuer Forstpolitik und Forstliche Betriebswirtschaftslehre der Forstlichen Forschungsanstalt Muenchen. 8000 München 40; Amalienstrasse 52.

DE.160.30.1 Abteilung Forstpolitik und Forstgeschichte.

DE.160.30.2 Abteilung Betriebswirtschaft und Holzmarkt.

DE.160.31.0 Institut fuer Forstliche Arbeitswissenschaft und Verfahrenstechnik der Forstlichen Forschungsanstalt Muenchen. 8000 München 80; Hohenlindener Str. 5.

DE.160.31.1 Lehrstuhl für Forstliche Arbeitswissenschaft und Verfahrenstechnik.

DE.160.33.0 Institut fuer Holzforschung. 8000 München 13; Winzererstr. 45.

DE.160.33.1 Abteilung Technologie.

DE.160.33.2 Abteilung Holzchemie und Ultrastrukturforschung.

DE.160.33.3 Abteilung Forstnahe Holzforschung.

DE.160.33.4 Abteilung Holzphysik und Verfahrenstechnik.

DE.160.33.5 Abteilung Anatomie und Pathologie des Holzes.

DE.160.33.6 Abteilung Mechanik und Prüfung von Holz.

DE.160.40.0 Institut für Tieranatomie. 8000 München 22; Veterinärstr. 13.

DE.160.43.0 Lehrstuhl für Makroskopische Anatomie der Tiere. 8000 München 22; Veterinärstr. 13.

DE.160.46.0 Lehrstuhl für Histologie und Embryologie der Tiere. 8000 München 22; Veterinärstr. 13.

DE.160.49.0 Institut für Tierpathologie. 8000 München 22; Veterinärstr. 13

DE.160.49.1 Lehrstuhl für Allgemeine Pathologie und Pathologische Anatomie.

DE.160.49.5 Lehrstuhl für Allgemeine Pathologie und Neuropathologie.

DE.160.55.0 Institut für Pharmakologie Toxikologie und Pharmazie 8000 münchen 22; Veterinärstr. 13

DE.160.55.1 Abteilung Toxikologie und Radiologie.

DE.160.55.2 Abteilung Pharmakologie und Toxikologie.

DE.160.58.0 Institut für Pharmazie und Lebensmittelchemie 8000 München 2; Sophienstr. 10.

DE.160.58.1 Abteilung Lebensmittelchemie.

DE.160.60.0 Institut für Physiologie, Physiologische Chemie und Ernährungsphysiologie. 8000 München 22; Veterniärstr. 13.

DE.160.61.0 Lehrstuhl für Tierphysiologie und Physiologische Chemie. see 160600.

DE.160.61.1 Isotopenabteilung.

DE.160.62.0 Lehrstuhl für Ernährungsphysiologie. 8000 München 22; Veterniärstr. 13.

DE.160.62.1 Abteilung Immunologie

DE.160.64.0 Lehrstuhl für Haustiergenetik. 8042 Oberschleissheim; St. Hubertusstr. 2

DE.160.70.0 Institut für Tierzucht und Tierhygiene. 8000 München 22; Veterinärstr. 13.

DE.160.70.1 Lehrstuhl für Tierhygiene.

DE.160.70.5 Lehrstuhl für Tierzucht.

DE.160.71.0 Lehrstuhl für Haustiergenetik. 8042 Oberschleissheim; St.Hubertusstr. 2.

DE.160.73.0 Institut für Hygiene und Technologie der Lebensmittel Tierischen Ursprungs. 8000 München 22; Veterinärstr. 13.

DE.160.73.1 Lehrstuhl für Hygiene und Technologie der Milch.

DE.160.73.2 Lerhstuhl für Hygiene und Technologie der Lebensmittel Tierischen Ursprungs.

DE.160.73.3 Abteilung Umweltkontamination. see 160730.

DE.160.73.4 Abteilung Serologie.

DE.160.73.5   Abteilung Virologie.

DE.160.73.6   Abteilung Radiologie.

DE.160.76.0 Institut für Medizinische Mikrobiologie, Infektions– und Seuchenmedizin. 8000 München 22; Veterinärstr. 13.

DE.160.76.1   Abteilung Bakteriologie und Mykologie.

DE.160.76.2   Abteilung Epidemologie.

DE.160.76.3   Abteilung Infektions– und Seuchenmedizin

DE.160.76.4   Abteilung Virologie

DE.160.76.5   Abteilung Immunbiologie.

DE.160.77.0 Institut für Zoologie und Hydrobiologie. 8000 München 22; Kaulbachstr. 37.

DE.160.77.1   Abteilung Fischereibiologie und Fischkrankheiten

DE.160.77.2   Abteilung Teichwirtschaft

DE.160.79.0 Lehrstuhl für Paläoanatomie, Domestikationsforschung und Geschichte der Tiermedizin. 8000 München 22; Veterinärstr. 13.

DE.160.82.0 Institut für Vergleichende Tropenmedizin und Parasitologie. 8000 Muenchen 23; Kaulbachstr. 37.

DE.160.82.1   Abteilung Virologie.

DE.160.82.2   Abteilung Protozoologie

DE.160.82.3   Abteilung Parasitologie.

DE.160.82.4   Abteilung Fischerei.

DE.160.82.5   Abteilung für Infektions– und Tropenmedizin

DE.160.85.0 Klinik fuer Innere Krankheiten der Tiere. 8042 Oberschleissheim; Mittenheimer Str. 178.

DE.160.85.5 Lehrstuhl fuer Krankheiten des Hausgeflügels, der Zier– und Zoovoegel.

DE.160.88.0 Medizinische Tierklinik 8000 München 22; Veterinärstr. 13.

DE.160.88.2 Lehrstuhl II.

DE.160.91.0 Gynäkologische und Ambulatorische Tierklinik 8000 München 22; Königinstr. 12.

DE.160.91.1 Lehrstuhl für Physiologie und Pathologie der Fortpflanzung, Insbesondere Andrologie und Künstliche Besamung

DE.160.91.2 Gynaekologische und Ambulatorische Tierklinik, Abteilung fuer Krankheiten der Milchdruese und der Jungtiere.

DE.160.91.3 abt. Aufzuchtkrankheiten

DE.160.94.0 Chirurgische Tierklinik 8000 München 22; Veterinärstr. 13.

DE.160.96.0   Medizinische Poliklinik

DE.160.97.0 Zoologisches Institut. 8000 München 2; Luisenstr. 14.

DE.161.00.0 Technische Unveristät München.

DE.161.01.0 Institut fuer Bodenkunde, Pflanzenernaehrung und Phytopathologie. 8050 Freising–Weihenstephan.

DE.161.02.0 Lehrstuhl fuer Bodenkunde. 8050 Freising–Weihenstephan.

DE.161.04.0 Lehrstuhl für Pflanzenernährung. 8050 Freising–Weihenstephan.

DE.161.05.0 Lehrstuhl fuer Phytopathologie. 8050 Freising–Weihenstephan.

DE.161.06.0 Lehrstuhl für Chemie und Landwirtschaftliche Technologie. 8050 Freising–Weihenstephan.

DE.161.06.1   Abteilung Chemische Pflanzenphysiologie.

DE.161.08.0 Institut für Lebensmittelchemie. 8046 Garching; Lichtenbergstr. 4.

DE.161.10.0 Lehrstuhl für Chemisch–Technische Analyse und Chemische Lebensmitteltechnologie. 8050 Freising–Weihenstephan.

DE.161.11.0 Lehrstuhl für Allgemeine Lebensmitteltechnologie. 8050 Freising–Weihenstephan.

DE.161.12.0 Lehrstuhl fuer Physik. 8050 Freising–Weihenstephan.

DE.161.14.0 Lehrstuhl für Gemüsebau. 8050 Freising–Weihenstephan.

DE.161.16.0 Lehrstuhl für Obstbau. 8050 Freising–Weihenstephan.

DE.161.16.1   Abt. Weinbau.

DE.161.18.0 Lehrstuhl für Zierpflanzenbau. 8050 Freising–Weihenstephan.

DE.161.20.0 Lehrstuhl für Phytopathologie. 8050 Freising–Weihenstephan.

DE.161.22.0 Bayerische Hauptversuchsanstalt für Landwirtschaft. 8050 Freising–Weihenstephan.

DE.161.24.0 Institut fuer Landwirtschaftlichen Gaertnerischen Pflanzenbau. 8050 Freising–Weihenstephan.

DE.161.24.1   Abteilung Ackerbau und Versuchswesen.

DE.161.25.0   Lehrstuhl fuer Pflanzenbau und

Pflanzenzuechtung.
8050 Freising–Weihenstephan.

DE.161.25.5 Lehrstuhl fuer Gruenlandlehre.
8050 Freising–Weihenstephan.

DE.161.25.6 Abteilung Zytogenetik

DE.161.26.0 Lehrstuhl fuer Gemuesebau. 8050
Freising–Weihenstephan.

DE.161.26.1 Abteilung Ackerbau und Versuchswesen.

DE.161.26.5 Lehrstuhl für Obstbau.
8050 Freising–Weihenstephan.

DE.161.26.6 Abteilung Weinbau.

DE.161.27.0 Lehrstuhl fuer Ziepflanzenbau.
8050 Freising–Weihenstephan.

DE.161.28.0 Institut für Tierernährung 8050
Freising–Weihenstephan.

DE.161.30.0 Lehrstuhl für Tierzucht. 8050
Freising–Weihenstephan.

DE.161.30.5 Abteilung Tierhaltung

DE.161.32.0 Lehrstuhl für Physiologie der Fortpflanzung und
Laktation. 8050 Freising–Weihenstephan.

DE.161.33.0 Institut für Milchwissenschaft und
Lebensmittelverfahrenstechnik.
8050 Freising–Weihenstephan.

DE.161.34.0 Lehrstuhl für Milchwissenschaft. 8050
Freising–Weihenstephan.

DE.161.35.0 Lehrstuhl für Lebensmittelverfahrenstechnik
und Molkereitechnologie.
8050 Freising–Weihenstephan.

DE.161.36.0 Lehrstuhl für Tierhygiene und Nutztierkunde.
8050 Freising–Weihenstephan; Hohenbachernstr. 15.

DE.161.36.1 Abteilung Parasitologie und Fischbiologie.

DE.161.40.0 Lehrstuhl für Angewandte Landwirtschaftliche
Betriebslehre. 8050 Freising–Weihenstephan.

DE.161.42.0 Lehrstuhl für Agrarpolitik und
Landwirtschaftliches Marktwesen. 8050
Freising–Weihenstephan.

DE.161.42.1 Abteilung Landwirtschaftliche Marktlehre.

DE.161.42.2 Abteilung Agrarpolitik.

DE.161.42.3 Abt. für Ländliche Soziologie.

DE.161.44.0 Lehrstuhl für Wirtschaftslehre des Landbaues.
8050 Freising–Weihenstephan.

DE.161.46.0 Lehrstuhl für Wirtschaftslehre des Gartenbaues.
8050 Freising–Weihenstephan.

DE.161.48.0 Lehrstuhl für Wirtschaftslehre der Brauerei.
8050 Freising–Weihenstephan.

DE.161.50.0 Institut für Landmaschinen 8000 München 2;
Arcisstr. 21.

DE.161.52.0 Institut für Landtechnik 8050
Freising–Weihenstephan.

DE.161.52.1 Abteilung Technische Grundlagen.

DE.161.52.2 Abteilung Technik in Gartenbau und
Landschaftspflege.

DE.161.52.3 Abteilung Arbeitslehre.

DE.161.52.4 Abteilung Technik in der Tierischen
Produktion.

DE.161.52.5 Abteilung Technik in der Pflanzlichen
Produktion.

DE.161.54.0 Lehrstuhl für Mechanik. 8050
Freising–Weihenstehpan.

DE.161.56.0 Lehrstuhl für Entwerfen und Ländliches
Bauwesen. 8000 München 13; Isabellastr. 13.

DE.161.58.0 Lehrstuhl für Baukonstrukton und Holzbau.
8000 München 2; Arcisstr. 21.

DE.161.58.1 Abteilung Bauwesen.

DE.161.58.2 Baunigenieurabteilung II.

DE.161.65.0 Lehrstuhl für Landschaftsarchitektur. 8050
Freising–Weihenstephan.

DE.161.67.0 Lehrstuhl für Landschaftsökologie. 8050
Freising–Weihenstephan.

DE.161.69.0 Lehrstuhl für Raumforschung, Raumordnung
und Landesplanung am Institut für Städtebau und
Raumplanung. 8000 München 2; Gabelsberger Str. 30

DE.161.69.1 Fachbereich Architektur.

DE.161.71.0 Geographisches Institut 8000 München 2;
Postfach 202420

DE.161.77.0 Lehrstuhl für Technische Mikrobiologie und
Technologie der Brauerei II. 8050 Freising–Weihenstephan.

DE.161.79.0 Lehrstuhl für Technologie der Brauerei I. 8050
Freising–Weihenstephan.

DE.161.81.0 Lehrstuhl für Maschinenwesen und
Energiewirtschaft der Brauerei. 8050
Freising–Weihenstephan.

DE.161.82.0 Lehrstuhl für Brauereianlagen.
8050 Freising–Weihenstephan.

DE.161.86.0 Lehrstuhl und Pruefamt für
Wasserguetewirtschaft und Gesundheitsingenieurwesen.

8000 München 2; Arcisstr. 21.

DE.161.86.1　Abteilung Bauingenieurwesen

DE.164.00.0　**Universität Münster**

DE.164.05.0　**Institut für Angewandte Botanik　4400 Münster; Hindenburgplatz 55.**

DE.164.05.1　Abt.Ökologie und Umweltschutz

DE.164.10.0　**Institut für Mikrobiologie　4400 Münster; Piusallee 7**

DE.164.15.0　**Institut für Lebensmittelchemie.　4400 Münster; Piusallee 7**

DE.164.20.0　**Hygiene–Institut.　4400 Münster; Westring 10**

DE.164.20.1　Abteilung Lebensmittelhygiene

DE.164.30.0　**Institut für Genossenschaftswesen.　4400 Münster; am Stadtgraben 9.**

DE.164.30.1　Abteilung Allgemeine Kooperationsforschung

DE.164.35.0　**Institut für Siedlungs und Wohnungswesen.　4400 Münster; am Stadtgraben 9.**

DE.164.35.1　abteilung Landentwicklung und Agrarstruktur

DE.164.40.0　**Institut fuer Geographie.　4400 Münster ; Robert–Koch–Str. 26.**

DE.164.40.1　Lehrstuhl Landschaftökologie.

DE.164.45.0　**Lehrstuhl für Biochemie der Pflanzen.　4400 Münster; Hindenburg Platz 55.**

DE.167.00.0　**Universität des Saarlandes**

DE.167.05.0　**Fachrichtung für Organische Chemie.　6600 Saarbrücken; Universität, Bau 23.**

DE.167.05.1　Arbeitskreis Neunhoeffer.

DE.167.10.0　**Fachrichtung Ernährungs– und Haushaltswissenschaften im Fachbereich Analytische und Biologische Chemie.　6600 Saarbrücken; Universität Bau 12.**

DE.167.10.1　Abteilung Ernährungswissenschaft.

DE.167.10.2　Abteilung Analytische Phytochemie.

DE.167.15.0　**Fachrichtung Hygiene und Mikrobiologie.　6550 Hamburg/Saar; Haus 5**

DE.167.15.1　Abteilung Ökologische und Angewandte Mikrobiologie

DE.167.20.0　**Institut für Biogeographie.　6600 Saarbrücken 15; Universität, Bau 11.**

DE.170.00.0　**Universität Stuttgart.**

DE.170.05.0　**Institut für Lebensmittelchemie　7000 Stuttgart**

80; Pfaffenwaldring 55.

DE.170.10.0　**Institut für Siedlungswasserbau und Wassergütewirtschaft.　7000 Stuttgart 80; Bandtäle.**

DE.170.10.1　Abteilung Biologie.

DE.170.15.0　**Institut für Ländliche Siedlungsplanung.　7000 Stuttgart 1; Kepplerstr. 11**

DE.170.20.0　**Institut für Photogrammetrie　7000 Stuttgart 1; Kepplerstr. 11**

DE.170.25.0　**Geographisches Institut　7000 Stuttgart; Silcher Str. 9.**

DE.170.25.1　Abteilung Wüstenforschung.

DE.170.25.5　Lehrstuhl fuer Wirtschaftsgeographie.

DE.170.30.0　**Amtliche Forschungs und Materialprüfungsanstalt für das Bauwesen, Otto Graf Institut an der Universität.　7000 Stuttgart 80; Pfaffenwaldring**

DE.170.30.1　Abteilung für Holz und Holzverbindungen

DE.170.35.0　**Institut für Tragkonstruktionen und Konstruktives Entwerfen.　7000 Stuttgart 1; Kepplerstr. 11**

DE.173.00.0　**Universität Tübingen.**

DE.173.05.0　**Institut für Biologie I.　7400 Tübingen; Auf der Morgenstelle 1.**

DE.173.05.1　Lehrstuhl für Spezielle Botanik

DE.173.10.0　**Wirtschaftswissenchaftliches Seminar.　7400 Tübingen; Mahlstr. 36.**

DE.173.10.1　Abteilung Volkswirtschaft.

DE.173.10.5　Lehrstuhl für Volkswirtschaftslehre, insbesondere Wirtschafts– und Finanztheorie.

DE.173.15.0　Pharmakologisches Institut. 7400 Tuebingen; Wilhelmstr. 56.

DE.173.15.5　Abteilung fuer Molekularpharmakologie.

DE.173.20.0　**Tropenmedizinisches Institut.　7400 Tuebingen; Wilhelmstr. 11.**

DE.173.25.0　Institut für Organische Chemie. 7400 Tuebingen; Auf der Morgenstelle 18

DE.173.40.0　**Lehrstuhl für Populationsgenetik**

DE.176.00.0　**Universität Würzburg.**

DE.176.05.0　**Zoologisches Institut.　8700 Würtzburg; Röntgenring 10.**

DE.176.05.1　Lehrstuhl III.

DE.176.10.0　**Institut für Hygiene und Mikrobiologie　8700 Würzburg; Josef– Schneider– Str. 2; Bau 17.**

DE.176.10.1   Bakteriologische Forschungsabteilung

DE.176.15.0 Institut für Pharmazie und Lebensmittelchemie.
8700 Würzburg; Klinikstr. 3

DE.176.15.1   Abteilung Lebensmittelchemie.

DE.176.25.0 Institut für Zoologie I.

DE.176.30.0 Institut für Pharmakologie und Toxikologie.

DE.201.00.0 Bundesforschungsanstalt fuer
Landwirtschaft Braunschweig–Voelkenrode.
3300 Braunschweig; Bundesallee 50.

DE.201.01.0 Institut für Biochemie des Bodens

DE.201.02.0 Institut für Bodenbiologie

DE.201.03.0 Institut für Grünlandwirtschaft, Futterbau und
Futterkonservierung.

DE.201.04.0 Institut für Pflanzenbau und Saatgutforschung

DE.201.05.0 Institut für Tierernährung.

DE.201.06.0 Institut für Landtechnische Grundlagenforschung

DE.201.07.0 Institut für Landmaschinenforschung.

DE.201.08.0 Institut für Landwirtschaftliche Bauforschung

DE.201.09.0 Institut für Betriebstechnik

DE.201.10.0 Institut für Betriebswirtschaft.

DE.201.11.0 Institut für Landwirtschaftliche marktforschung.

DE.201.12.0 Institut für Strukturforschung

DE.201.20.0 Institut für Tierzucht und Tierverhalten. 3051
Mariensee über Wunstorf/Hann.

DE.201.30.0 Institut für Kleintierzucht. 3100 Celle,
Dörnbergstr. 25/27.

DE.202.00.0 Bundesforschungsanstalt für Forst– und
Holzwirtschaft, Reinbek.
2050 Hamburg 80; Leuschnerstr. 91.

DE.202.01.0 Institut für Weltforstwirtschaft 2057 Reinbek;
Bei Hamburg Schloss.

DE.202.02.0 Institut für Forstgenetik und
Forstpflanzenzüchtung. 2070 Schmalenbeck Bei Ahrensburg;
Sieker Landstr. 2

DE.202.03.0 Institut für Arbeitswissenschaft 2057 Reinbek Bei
Hamburg; Vorwerksbusch.

DE.202.04.0 Institut für Holzbiologie und Holzschutz. 2050
Hamburg 80; Leuschnerstr. 91.

DE.202.05.0 Institut für Holzchemie und Chemische
Technologie des holzes 2050 Hamburg 80; Leuschnerstr. 91.

DE.202.06.0 Institut für Holzphysik und Mechanische
Technologie des Holzes. 2050 Hamburg 80; Leuschnerstr. 91.

DE.204.00.0 Bundesforschungsanstalt für
Rebenzuechtung Geilweilerhof.
6741 Siebeldingen/ Pfalz

DE.205.00.0 Bundesforschungsanstalt fuer
Naturschutz und Landschaftsoekologie.
5300 Bonn 2; Konstantinstr. 110.

DE.205.01.0   Institut für Landschaftspflege und
Landschaftsökologie.
5300 Bonn–Bad; Godesberg Heerstrasse 110

DE.205.02.0   Institut für Naturschutz und Tieroekologie.
5300 Bonn–Bad; Godesberg Heerstr. 110.

DE.205.03.0   Institut fuer Vegetationskunde.
5300 Bonn 2; Konstantinstr. 110.

DE.206.00.0 Bundesforschungsanstalt für
Gartenbauliche Pflanzenzüchtung Ahrensburg.
2070 Ahrensburg; Bornkampsweg.

DE.207.00.0 Bundesanstalt für Milchforschung Kiel
2300 Kiel; Hermann– Weigmann– Str. 1–27.

DE.207.01.0 Institut für Milcherzeugung.

DE.207.02.0 Institut für Hygiene

DE.207.03.0 Institut für Mikrobiologie

DE.207.04.0 Institut für Chemie.

DE.207.05.0 Institut für Physik.

DE.207.06.0 Institut für Verfahrenstechnik

DE.207.07.0 Institut für Betriebswirtschaft und
Marktforschung.

DE.207.08.0 Bibliothek und Archiv.

DE.208.00.0 Bundesforschungsanstalt für Fischerei
Hamburg
2000 Hamburg 50; Palmaille 9

DE.208.01.0 Institut für Seefischerei.

DE.208.02.0 Institut für Küsten– und Binnenfischerei

DE.208.03.0 Institut für Fangtechnik

DE.208.04.0 Institut für Biochemie und Technologie.

DE.208.05.0 Isotopenlaboratorium 2000 Hamburg 55;
Wüstland 2

DE.208.06.0 Informations– und Dokumentationsstelle. 2000
Hamburg 50; Palmaille 9

DE.209.00.0 Bundesforschungsanstalt für Getreide–
und Kartoffelverarbeitung.

4930 Detmold; Am Schützenberg 12

DE.209.01.0 Institut für Biochemie und Analytik.

DE.209.02.0 Institut für Müllereitechnologie.

DE.209.03.0 Institut für Bäckereitechnologie.

DE.209.04.0 Institut für Stärke- und Kartoffeltechnologie.

DE.209.05.0 Zentrale Einrichtung Information und Dokumentation.

DE.209.06.0 Zentrale Einrichtung Radiochemie.

DE.210.00.0 Bundesanstalt für Fleischforschung Kulmbach Oskar-von-Miller-Str. 20.

DE.210.01.0 Institut für Fleischerzeugung.

DE.210.02.0 Institut für Technologie

DE.210.03.0 Institut für Bakteriologie und Histologie

DE.210.04.0 Institut für Chemie und Physik.

DE.210.05.0 Sekretariat für Codex Committee on Meat.

DE.210.06.0 Isotopenlabor.

DE.211.00.0 Bundesforschungsanstalt für Ernährung. 7500 Karlsruhe; Engesserstr. 20.

DE.211.01.0 Institut für Biologie.

DE.211.02.0 Institut für Lebensmittelchemie.

DE.211.03.0 Zentrallabor für Isotopentechnik.

DE.211.04.0 Institut für Verfahrenstechnik.

DE.211.05.0 Institut für Biochemie.

DE.211.10.0 Institut für Hauswirtschaft. 7000 Stuttgart-Hohenheim; Garbenstrasse 13.

DE.213.00.0 Bundesanstalt für Fettforschung Münster. 4400 Münster/Westfalen; Piusallee 76

DE.213.01.0 Institut für Allgemeine und Analytische Chemie

DE.213.02.0 Institut für Biochemie und Technologie.

DE.215.00.0 Biologische Bundesanstalt für Land- und Forstwirtschaft, Berlin und Braunschweig 3300 Braunschweig; Messeweg 11/12 and 1000 Berlin 33; Koenigin-Luise-Str. 19.

DE.215.01.0 Institut für Pflanzenschutzmittelforschung 1000 Berlin 33; König- Louise-Str. 19

DE.215.03.0 Institut für Vorratsschutz 1000 Berlin 33; Königin-Louise- str. 19.

DE.215.04.0 Institut fuer Mikrobiologie. 1000 Berlin 33; Königin-Louise-Str. 19.

DE.215.06.0 Institut für Nichtparasitaere Pflanzenkrankheiten 1000 Berlin 33; Königin-Louise- Str. 19

DE.215.07.0 Institut fuer Pflanzenschutz im Zierpflanzenbau. 1000 Berlin 33; Königin-Louise-Str. 19.

DE.215.08.0 Fachgruppe fuer Chemische Mittelpruefung. 3300 Braunschweig; Messeweg 11/12.

DE.215.09.0 Fachgruppe fuer Botanische Mittelpruefung. 3300 Braunschweig; Messeweg 11/12.

DE.215.10.0 Fachgruppe fuer Zoologische Mittelpruefung. 3300 Braunschweig; Messeweg 11/12.

DE.215.11.0 Fachgruppe fuer Anwendungstechnik. 3300 Braunschweig; Messeweg 11/12.

DE.215.12.0 Institut fuer Pflanzenschutz in Ackerbau und Gruenland. 3300 Braunschweig; Messeweg 11/12.

DE.215.13.0 Institut für Unkrautforschung. 3300 Braunschweig; Messeweg 11/12.

DE.215.14.0 Institut für Biochemie 3300 Braunschweig; Messeweg 11/12.

DE.215.15.0 Institut fuer Viruskrankheiten der Pflanzen. 3300 Braunschweig; Messeweg 11/12.

DE.215.17.0 Institut für Biologische Schädlingsbekaempfung 6100 darmstadt; Heinrichstr. 43

DE.215.18.0 Institut fuer Pflanzenschutz in Ackerbau und Gruenland/Aussenstelle Kiel-Kitzeberg. 2300 Kiel-Kitzeberg; Schlosskoppelweg 8

DE.215.19.0 Institut fuer Nematologie. 4400 Münster; Toppheideweg 88

DE.215.19.1 Aussenstelle 5153 Elsdorf/Rhld; Dürener Str. 1

DE.215.20.0 Institut fuer Pflanzenschutz im Gemuesebau. 5035 Fischenich (Krs.Koeln); Marktweg 60

DE.215.21.0 Institut fuer Pflanzenschutz im Obstbau. 6901 Dossenheim über Heidelberg; Schwabenheimer Strasse; Postfach 73

DE.215.22.0 Institut fuer Pflanzenschutz im Weinbau. 5550 Bernkastel-Küs; Brüningstr. 84

DE.215.23.0 Institut fuer Pflanzenschutz im Forst. 3510 Hann.-Münden; Kasseler str. 22.

DE.215.24.0 Dienstelle fuer Wirtschaftliche Fragen und Rechtsangelegenheiten im Pflanzenschutz. 1000 Berlin 33; Koenigin-Luise-Str. 19.

DE.215.30.0 Dokumentationsschwerpunkt Pflanzenkrankheiten und Pflanzenschutz 1000 Berlin 33;

Königin–Louise– str. 19.

DE.216.00.0 Bundesforschungsanstalt für Viruskrankheiten der Tiere Tübingen 7400 Tübingen; Paul–Ehrlich– str. 28

DE.216.01.0 Institut fuer Mikrobiologie.

DE.216.02.0 Institut fuer Immunologie.

DE.216.03.0 Abteilung fuer Impfstoffe.

DE.217.00.0 Bundesanstalt für Materialprüfung Berlin–Dahlem 1000 Berlin 45; Unter den Eichen 87

DE.217.01.0 Abteilung Biologische Materialprüfung

DE.217.02.0 Abteilung Bautenschutz.

DE.217.03.0 Abteilung Sondergebiete der Materialprüfung.

DE.218.00.0 Bundesanstalt für Gewaesserkunde 5400 Koblenz; Kaiserin–Augusta–Anlagen 15; Postfach 309

DE.301.00.0 Deutscher Wetterdienst.

DE.301.01.0 Deutscher Wetterdienst, Abteilung Agrarmeteorologie des Zentralamtes. 6050 Offenbach/m.; Frankfurter str. 135; Postfach 185

DE.301.02.0 Agrarmeteorologische Beratungsstelle Bonn. 5300 Bonn– Bad Godesberg; Mittelstr. 121

DE.301.03.0 Agrarmeteorologische Forschungsstelle Braunschweig Völkenrode. 3300 Braunschweig–Fal; Bundesallee 50

DE.301.04.0 Wetteramt Freiburg Dezernat Agrarmeteorologie. 7800 Freiburg; Stefan–Meier–Str. 4–6.

DE.301.05.0 Agrarmeteorologische Forschungsstelle Geisenheim 6222 Geisenheim/Rhg.; Kreuzweg 17; Postfach 1233

DE.301.06.0 Agrarmeteorologische Forschungsstelle Ahrensburg. 2000 Hamburg 4; Bernhard–Nocht–Str. 76.

DE.301.07.0 Wetteramt Schleswig Dezernat Agrarmeteorologie. 2380 Schleswig; Regenpfeiferweg 9.

DE.301.08.0 Agrarmeteorologische Beratungsstelle Stuttgart–Hohenheim. 7000 Stuttgart 70; Nordstr. 12

DE.301.09.0 Wetteramt Trier Dezernat Agrarmeteorologie. 5500 Trier/Mosel Petrisberg; Sickingerstr. 41.

DE.301.10.0 Agrarmeteorologische Forschungsstelle Weihenstephan. 8050 Weihenstephan B. Freising.

DE.301.11.0 Wetterwarte und Agrarmeteorologische Beratungsstelle Würzburg–Stein. 8700 Würzburg 2; Oberer Steinberg, Postfach 1093.

DE.301.12.0 Agrarmeteorologische Forschungsstelle Giessen.

6300 Giessen; Bergstr. 21.

DE.305.00.0 Bundesgesundheitsamt Berlin 1000 Berlin 33; Postfach.

DE.305.01.0 Max–von–Pettenkofer–Institut 1000 Berlin 33 (Dahlem); Unter Den Eichen 82–84; (Postanschrift: 1000 Berlin 33; Postfach)

DE.305.01.1 Abteilung Lebensmittelchemie.

DE.305.01.2 Abteilung Ernaehrungsmedizin.

DE.305.02.0 Institut fuer Wasser–, Boden– und Lufthygiene. 1000 Berlin 33 (Dahlem); Corrensplatz 1; (Postanschrift: 1000 Berlin 33; Postfach)

DE.305.03.0 Institut für Veterinärmedizin – Robert von Ostertag–Institut–. 1000 Berlin 33; Postfach.

DE.305.03.1 Abt. Zoonosen– und Tierseuchenforschung.

DE.305.03.2 Abt. Arzneimittel, Tierernaehrung und Rueckstandsforschung.

DE.305.03.3 Fachgebiet Milchhygiene.

DE.400.00.0 Max–Planck–institute, Max–Planck–Gesellschaft (generalverwaltung und Präsidialbüro. Residezstr. 1A, 8000 München 2 (Postanschrift: 8000 München 1; Postfach 647)

DE.401.00.0 Max–Planck–Institut für Ernährungsphysiologie 4600 Dortmund; Rheinlanddamm 201

DE.401.01.0 Abteilung Epidemiologie

DE.402.00.0 Max–Planck–Institut Für Züchtungsforschung (Erwin–Baur–Institut); 5000 Köln 30; Vogelsang

DE.404.00.0 Max–Planck–Institut Für Limnologie. 2320 Plön/Holstein; August–Thienemann–Str. 2; Postf.165

DE.404.01.0 Abteilung Tropenökologie

DE.500.00.0 Institutionen der Länder

DE.501.00.0 Baden–Würtemberg

DE.501.01.0 Staatliche Landwirtschaftliche Untersuchungs– und Forschungsanstalt Augustenberg. 7500 Karlsruhe–Durlach; Nesslerstr. 23.

DE.501.01.1 Abt.Futtermitteluntersung und Mikrobiologie.

DE.501.01.2 Abt.Versuchswesen und Umwelt

DE.501.01.3 Abt.Spurenelement und Radio Aktivitätsuntersuchung

DE.501.01.4 Abteilung Saatgutprüfung und Angewandte Botanik.

DE.501.05.0 Landesanstalt für Pflanzenschutz. 7000 Stuttgart–W; Reinsburgstr. 107.

DE.501.05.1 Abteilung Integrierter Pflanzenschutz.

DE.501.05.2 Abteilung Phytopathologie

DE.501.05.3 Abteilung Landwirtschaft.

DE.501.05.4 Rückstandslabor

DE.501.05.5 Abteilung Biologische Schädlingsbekämpfung

DE.501.10.0 Staatliches Weinbauinstitut Versuchs– und Forschungsanstalt für Weinbau und Weinbehandlung Freiburg 7800 Freiburg i. b. Merzhauser Str. 119

DE.501.10.1 Abteilung Botanik

DE.501.10.2 Abteilung Rebenzüchtung und Standortforschung

DE.501.10.3 Abteilung Betriebs– Und Arbeitswirtschaft

DE.501.10.4 Abteilung Weinchemie.

DE.501.10.5 Abteilung Mikrobiologie.

DE.501.10.6 Abteilung Bodenkunde.

DE.501.10.7 Abteilung Zoologie.

DE.501.11.5 Landesanstalt für Tabakbau und Tabakforschung Forchheim.
7501 Forchheim/Bahnhof; Kutschenweg 10.

DE.501.12.0 Landesanstalt fuer Tabakbau und Tabakforschung.
7501 Forchheim/Bahnhof; Kutschenweg 10.

DE.501.12.1 Abteilung Pflanzenbau.

DE.501.12.2 Abteilung Tabakchemie.

DE.501.20.0 Staatliche Versuchsanstalt für Grünlandwirtschaft Und Futterbau Aulendorf 7960 Aulendorf/KreisRavensburg; Lehmgrubenweg 5.

DE.501.20.1 Abteilung Futterkonservierung.

DE.501.20.2 Abteilung Versuchswesen.

DE.501.20.3 Abteilung Pflanzensoziologie.

DE.501.25.0 Landesstelle Für Naturschutz und Landschaftspflege Baden–Württenberg. 7140 Ludwigsburg Favoritenschloss.

DE.501.25.1 Bezirksstelle Für Naturschutz und Landschaftspflege Nordwürttemberg.
7000 Stuttgart; Dillmannstr. 3

DE.501.30.0 Staatliches Tierärztliches Untersuchungsamt Aulendorf/Württ. 7960 Aulendorf/Württbg; Löwenbreitstr. 20

DE.501.30.1 Abteilung Bakteriologie

DE.501.32.0 Tierhygienisches Institut Freiburg 7800 Lehen/Post Freiburg; Hugstetter Landstrasse.

DE.501.32.1 Abteilung Angewandte Ethologie

DE.501.32.2 Chemische Abteilung

DE.501.32.3 Bakteriologische Abteilung.

DE.501.32.4 Virologische Abteilung.

DE.501.32.5 Serologische Abteilung.

DE.501.34.0 Staatliches Tierärztliches Untersuchungsamt Stuttgart. 7000 Stuttgart/N; Azenbergstr. 16

DE.501.34.1 Abteilung Virologie.

DE.501.34.2 Abteilung Bakteriologie

DE.501.40.0 Chemische Landesuntersuchungsanstalt Stuttgart 7000 Stuttgart–N; Kienestr. 18.

DE.501.50.0 Forstliche Versuchs– und Forschungsanstalt Baden–Wuerttemberg. 7800 Freiburg/Brsg.; Sternwaldstr. 16

DE.501.50.1 Abteilung Ertragskunde.
7800 Freiburg/Brsg.; Schweighofstr. 16.

DE.501.50.2 Abteilung Botanik und Standortkunde.
7000 Stuttgart–Weilimdorf; Fasanengarten.

DE.501.50.3 Sektion Bodenchemie und Pflanzenernährung.
7800 Freiburg/Brsg.; Sternwaldstr. 16.

DE.501.50.4 Abteilung Landespflege.
7800 Freiburg–brsg.; Sternwaldstr. 16

DE.501.50.5 Abteilung Arbeitswirtschaft und Forstbenutzung.
7800 Freiburg/Brsg.; Schweighofstr. 6.

DE.501.50.6 Abteilung Betriebswirtschaft.
7800 Freiburg/Brsg.; Sternwaldstr. 16

DE.501.50.7 Abteilung Biometrie und Informatik.
7800 Freiburg/Brsg., Sternwaldstr. 16.

DE.501.50.8 Abteilung Waldschutz.
7806 Wittental Post Ebnet bei Freiburg/Brsg.

DE.501.50.9 Abteilung Waldwachstum.

DE.501.60.0 Fachhochschule Esslingen. 7300 Esslingen A. N.; Kanalstr. 33

DE.501.70.0 Landessaatzuchtanstalt an der Universität Hohenheim. 7000 Stuttgart 70; Postfach 106.

DE.501.80.0 Institut fuer Seenforschung und Fischereiwesen. 7994 Langenargen/Bodensee; Untere Seestr. 81.

DE.501.80.5 Abteilung Fischereiwesen.

DE.502.00.0 **Bayern**

DE.502.05.0 **Bayerische Landesanstalt fuer Bodenkultur und Pflanzenbau, Freising–Muenchen. 8000 München 19; Menzinger Str. 54**

DE.502.05.1 Abteilung Pflanzenschutz.

DE.502.05.3 Abt. Versuchs– und Untersuchungswesen, Information.

DE.502.05.5 Abteilung Boden– und Landschaftspflege

DE.502.05.8 Abteilung Pflanzenbau und Pflanzenzüchtung.

DE.502.05.9 Abschnitt Hopfen–Wolnzach.

DE.502.06.3 Aussenstelle Pflanzenschutz Würzburg.

DE.502.10.0 **Fachhochschule Weihenstephan. 8050 Freising–Weihenstephan**

DE.502.10.1 Institut für Bodenkunde und Pflanzenernährung.

DE.502.10.2 Institut für Botanik und Pflanzenschutz

DE.502.10.3 Institut für Gärtnerische Betriebslehre.

DE.502.10.4 Institut für Gemüsebau.

DE.502.10.5 Institut fuer Obstbau und Obstbaumschulwesen.

DE.502.10.6 Institut für Zierpflanzenbau unter Glas.

DE.502.10.7 Institut für Stauden, Gehölze und Angewandte pflanzensoziologie

DE.502.10.8 Institut für Obst– und Gemüseverarbeitung.

DE.502.10.9 Institut für Freiraumplanung.

DE.502.11.0 **Institut für Technik Im Gartenbau.**

DE.502.15.0 **Bayerische Landesanstalt für Weinbau und Gartenbau. 8700 Würzburg; Residenzplatz 3.**

DE.502.15.1 Abt. Rebenzüchtung.

DE.502.15.2 Abteilung Weinchemie.

DE.502.15.3 Abt. Weinbau Veitshöchheim.

DE.502.15.4 Sachgebiet Weinbaulicher Pflanzenschutz.

DE.502.15.5 Abteilung Mikrobiologie.

DE.502.15.6 Abteilung Gartenbau.

DE.502.15.7 Lehr– und Versuchskellerei.

DE.502.20.0 **Staatliche Versuchs– und Lehrwirtschaft für Gartenbau. 8600 Bamberg; Äussere Galgenfuhr 21.**

DE.502.25.0 **Bayerisches Landesamt für Wasserwirtschaft.**

8000 Muenchen 22; Lazarettstr. 67.

DE.502.25.1 Sachgebiet Ingenieurbiologie.

DE.502.30.0 **Bayerische Landesanstalt für Tierzucht. 8011 Grub; Post Poing/Oberbayern**

DE.502.30.1 Abt. Haustiergenetik.

DE.502.30.2 Abt. Tierernährung und Futterkonservierung.

DE.502.30.3 Abt. Rinderproduktion.

DE.502.30.4 Abt. Schweineproduktion.

DE.502.30.5 Abt. Schafzucht.

DE.502.30.6 Abteilung Stallbau und Tierhaltung.

DE.502.35.0 **Bayerische Landesanstalt für Bienenzucht. 8520 Erlangen; Burgbergstr. 70**

DE.502.40.0 **Landesanstalt für Tierseuchenbekämpfung. 8042 Oberschleissheim; Veterinaerstr. 2.**

DE.502.50.0 **Deutsche Forschungsanstalt für Lebensmittelchemie 8046 Garching; Lichtenbergstr.**

DE.502.50.1 Abteilung Lebensmittelanalytik.

DE.502.50.2 Abteilung Chemie Biochemie und Mikrobiologie.

DE.502.55.0 **Süddeutsche Versuchs– und Forschungsanstalt für Milchwirtschaft 8050 Freising–Weihenstephan.**

DE.502.55.1 Bakteriologisches Institut.

DE.502.55.2 Institut für Betriebswirtschaft.

DE.502.55.3 Chemisches und Physikalisches Institut.

DE.502.55.4 Institut für Physiologie.

DE.502.60.0 **Bayerische Landesanstalt für Landtechnik. 8050 Freising– Weihenstephan**

DE.502.60.1 Abteilung Technik der Pflanzlichen Produktion

DE.502.60.2 Abteilung Technik und Bauwesen in der Tierischen Produktion.

DE.502.60.3 Abteilung Landwirtschaftliches Bauwesen.

DE.502.60.4 Abteilung Futterbau und Konservierung.

DE.502.60.5 Abteilung Technisch–Physikalische Grundlagen.

DE.502.65.0 **Bayerische Landesanstalt fuer Wasserforschung. 8000 München 22; Kaulbachstr. 37.**

DE.502.65.1 Abt. Abwasserchemie.

DE.502.65.2 Abt. Hydrobiologie.

DE.502.65.3 Abt. Teichwirtschaft.

DE.502.65.4 Abt. Fischuntersuchung.

DE.502.65.5 Abt. Abwasserbiologie.

DE.502.65.6 Abt. Radiologie.

DE.502.65.7 Abwasserversuchsfeld.

DE.505.10.0 Veterinäruntersuchungsanstalt. 2000 Hamburg 6; Lagerstr. 36.

DE.506.00.0 Hessen

DE.506.05.0 Hessisches Landesamt für Bodenforschung 6200 Wiesbaden; Leberweg 9

DE.506.05.1 Abteilung Landesaufnahme Ref Bodenkunde

DE.506.10.0 Forschungsanstalt für Weinbau, Gartenbau, Getränketechnologie und Landespflege. 6222 Geisenheim; Postfach 1180.

DE.506.10.1 Institut für Weinbau

DE.506.10.2 Institut für Rebenzüchtung und Rebenveredlung.

DE.506.10.3 Institut für Kellerwirtschaft und Verfahrenstechnik

DE.506.10.4 Institut für Weinchemie und Getränkeforschung.

DE.506.10.5 Institut für Mikrobiologie und Biochemie.

DE.506.10.6 Institut für Obstbau

DE.506.10.7 Institut für Gemüsebau.

DE.506.10.8 Institut für Zierpflanzenbau

DE.506.10.9 Institut für Gartenarchitektur und Landschaftspflege.

DE.506.11.0 Institut für Bodenkunde und Pflanzenernährung. 6222 Geisenheim; Postfach 1180.

DE.506.11.1 Institut für Botanik

DE.506.11.2 Institut für Pflanzenkrankheiten

DE.506.11.3 Institut für Technik

DE.506.11.4 Institut für Betriebswirtschaft und Marktforschung.

DE.506.11.5 Institut für Landschaftsbau.

DE.506.15.0 Hessische Lehr und Forschungsanstalt für Grünlandwirtschaft und Futterbau Eichhof 6430 Bad Hersfeld/ Hessen

DE.506.15.1 Institut für Grünland–Botanik.

DE.506.15.2 Institut für Futterkonservierung und Tierernährung

DE.506.15.3 Institut für Futterbau.

DE.506.15.4 Institut für Grünlandwirtschaft

DE.506.15.5 Institut für Pflanzenzüchtung und Saatguterzeugung

DE.506.15.6 Institut für Agrikulturchemie und Bodenkunde.

DE.506.20.0 Staatliche Vogelschutzwarte für Hessen, Rheinland–Pfalz und Saarland Institut für Angewandte Vogelkunde. 6000 Frankfurt/Main–Fechenheim; Steinhauerstr. 44.

DE.506.22.0 Hessische Landesanstalt für Leistungsprüfungen in der Tierzucht, New–Ulrichstein. 6313 Homberg (ohm) 1.

DE.506.22.1 Abt. Kleintierprüfungen.

DE.506.22.2 Abt. Mastleistungsprüfungen.

DE.506.25.0 Hessische Landesanstalt für Leistungsprüfungen in der Tierzucht – Aussenstelle für Bienenzucht. 3570 Kirchhain; Erlenstr. 15

DE.506.30.0 Staatliches Veterinäruntersuchungsamt. 6000 Frankfurt/Main; Deutschordenstr. 48.

DE.506.30.1 Virusabteilung

DE.506.30.2 Schweinegesundheitsdienst.

DE.506.30.3 Diagnostische Abteilung.

DE.506.32.0 Staatliches Veterinäruntersuchungsamt. 6300 Giessen; Marburger str. 54.

DE.506.34.0 Staatliches Veterinäruntersuchungsamt. 3500 Kassel; Druseltalstr. 61.

DE.506.40.0 Hessische Forsteinrichtungsanstalt. 6300 Giessen; Moltkestr. 10.

DE.506.40.1 Abteilung Forsteinrichtung.

DE.506.40.2 Abteilung Betriebswirtschaft.

DE.506.40.3 Abteilung Ertragskunde.

DE.506.40.4 Sachgebiet Landesplanung.

DE.506.45.0 Hessische Forstliche Versuchsanstalt 3510 Hann.–Münden; Prof. Ölkers–str. 6

DE.506.45.1 Institut für Forstpflanzenzüchtung.

DE.506.45.2 Institut für Forsthydrologie

DE.506.45.3 Institut für Forstproduktion

DE.507.00.0 Niedersachen

DE.507.05.0 Landwirtschaftliche Untersuchngs– und

Forschungsanstalt der Landwirtschaftskammer Hannover.
3250 Hameln; Finkenborner Weg 1a.

DE.507.05.1   Abteilung Anorganische Chemie.

DE.507.10.0 Niedersächsisches Landesamt für
Bodenforschung 3000 Hannover– Buchholz; Postfach 54.

DE.507.10.1   Torfinstitut Hannover

DE.507.10.2   Abteilung Bodenkunde.

DE.507.10.3   Ausseninstitut für Moorforschung und
Angewandte Bodenkunde
2800 Bremen; Friedrich– Missler–str.46–48

DE.507.15.0 Niedersächsisches Landesinstitut für Marschen–
und wurtenforschung 2940 Wilhelmshaven; Viktoriastr. 26–28

DE.507.15.1   Naturwissenschaftliche Abteilung.

DE.507.20.0 Pflanzenschutzamt als Landesstelle fuer
Bisambekaempfung/Ahlem der Landwirtschaftskammer
Hannover. 3110 Ahlem/Hann.; Wunstorfer Landstr. 9.

DE.507.25.0 Pflanzenschutzamt der Landwirtschaftskammer
Weser–ems. 2900 Oldenburg i.o. Mars–La–Tour–Str.9–11

DE.507.30.0 Obstbauversuchsanstalt der
Landwirtschaftskammer Hannover. 2155 Jork Bez.Hamburg;
Niederelbe 1

DE.507.30.1   Abteilung Bodenpflege und Pflanzenernährung.

DE.507.30.2   Abteilung Technik.

DE.507.30.3   Abteilung Botanik

DE.507.30.4   Abteilung Pflanzenschutz.

DE.507.30.5   Obstbauversuchsring des Alten Landes E.V.

DE.507.30.6   Abteilung Kernobst.

DE.507.30.7   Abt. Steinobst.

DE.507.30.8   Abteilung Lagerung.

DE.507.35.0 Grünlandlehranstalt und Marschversuchsstation
für Niedersachsen. 2890 Indfeld Post Abbehausen/Kreis
Wesermarsch.

DE.507.35.1   Abteilung Landeskultur und Bodenkunde.

DE.507.40.0 Landwirtschaftliche Untersuchungs– und
Forschungsanstalt der Landwirtschaftskammer Weser–Ems.
2900 Oldenburg I.O.; Mars–La–Tour–Str. 4.

DE.507.60.0 Tiergesundheitsamt Der Landwirtschaftskammer
Weser–Ems 2900 Oldenburg I.O.; Mars–La–Tour–Str. 1.

DE.507.65.0 Niedersächsische Forstliche Versuchsanstalt 3400
Göttingen; Grätzelstr.2

DE.507.65.1   Abteilung A – Ertragskunde.

DE.507.65.2   Abteilung B – Waldschutz.

DE.507.65.3   Abteilung C – Forstpflanzenzuechtung.
Staufenberg 1; OT 3513 Escherode.

DE.507.70.0 Fachhochschule Osnabrück 4500
Osnabrück–Haste; Oldenburger Landstr. 24

DE.507.70.1   Fachbereich Gartenbau

DE.507.70.2   Fachbereich Landbau
4500 Osnabrück; am Krümpel 31.

DE.507.70.3   Fachbereich Landespflege.
4500 Osnabrück–Haste; Oldenburger Landstr. 24

DE.507.75.0 Werkkunstschule Hildesheim. 3200 Hildesheim;
Dammstr. 45.

DE.507.75.1   Forschungsstelle für Oberflächentechniken

DE.507.80.0 Staatliches Naturhistorisches Museum zu
Braunschweig, Forschungsanstalt fuer Zoologie. 3300
Braunschweig; Pockelsstr. 10a.

DE.507.85.0   Institut für Meeresforschung.
2850 Bremerhaven; Am Handelshafen 12.

DE.508.00.0 Nordrhein–Westfalen

DE.508.05.0 Landesanstalt für Immissions– und
Bodennutzungsschutz Des Landes Nrw 4300 Essen;
Wallneyerstr. 6.

DE.508.05.1   Abteilung Bodennutzungsschutz

DE.508.10.0 Geologisches Landesamt Nordrhein–Westfalen
4150 Krefeld; De–Greiff–Str. 195

DE.508.10.1   Abt. IV– Bodenkunde

DE.508.20.0 Lehr– und Versuchsanstalt für Zierpflanzenbau,
Baumschulen und Floristik der Landwirtschaftskammer
Rheinland. 5300 Bonn–Bad Godesberg; Langer Grabenweg 68.

DE.508.20.1   Abt. Zierpflanzenbau.

DE.508.24.0   Landwirtschaftskammer Westfalen–Lippe.
4400 Münster; von Esmarch–Str. 12.

DE.508.25.0 Institut für Pflanzenschutz, Saatgutunterschung
und Bienenkunde der Landwirtschaftskammer
Westfalen–Lippe 4400 Münster; van Esmarch–Str. 12.

DE.508.30.0 Landesanstalt für Oekologie,
Landschaftsentwicklung und Forstplanung NRW. 4000
Duesseldorf 3ø; Prinz–Georg–Strasse 126.

DE.508.30.1   Abteilung Grünland– und Futterbauforschung.

DE.508.30.2   Abteilung Boden und Bodennutzung.

DE.508.30.3   Abteilung Oekologie.

DE.508.35.0 Lehr– und Versuchsanstalt Haus Düsse der

Landwirtschaftskammer Westfalen–Lippe. 4772 Bad Sassendorf; Ostinghausen

DE.508.35.1 Referat Schweinehaltung

DE.508.45.0 Staatliches Veterinäruntersuchungsamt. 5770 Arnsberg/Westf.; Zur Taubeneiche 10–12.

DE.508.45.1 Abt.Serologie

DE.508.45.2 Abt.Virologie

DE.508.45.3 Abteilung Milch.

DE.508.45.4 Abteilung Lebensmittel.

DE.508.47.0 Staatliches Veterinärunterschungsamt 4150 Krefeld; Deutscher Ring 100.

DE.508.47.1 Abteilung Virologie

DE.508.47.2 Abteilung Lebensmittel.

DE.508.55.0 Lehr– und Versuchsanstalt für Garten– und Landschaftsbau und Friedhofsgärtnerei der Landwirtschaftskammer Rheinland. 4300 Essen; Külskammer Weg 40.

DE.508.60.0 Fachhochschule Köln. 5000 Köln; Ubierring 48

DE.508.60.1 Fachbereich Landmaschinentechnik.

DE.508.65.0 Staatliche Materialprüfungsanstalt Nordrhein–Westfalen 4600 Dortmund–Aplerbeck; Marsbruchstr. 186.

DE.508.65.1 Abteilung Chemie.

DE.508.70.0 Fachhochschule Wuppertal. 5600 Wuppertal.

DE.508.70.1 Fachbereich Bau– und Verkehrstechnik

DE.508.75.0 Landesanstalt für Fischerei des Landes Nordrhein–Westfalen.
5942 Kirchhundem 1; Heinsberger Str. 51–53.

DE.509.00.0 Rheinland–Pfalz

DE.509.05.0 Landespflanzenschutzamt Rheinland–Pfalz. 6500 Mainz–Bretzenheim; Essenhaimer Str. 144

DE.509.10.0 Landes–Lehr– und Versuchsanstalt für Weinbau Gartenbau und Landwirtschaft. 4583 Bad Neuenahr/Ahrweiler 2; Walporzheimer Str. 48.

DE.509.10.1 Abteilung Landwirtschaft.

DE.509.15.0 Landes–Lehr– und Forschungsanstalt für Wein und Gartenbau 6730 Neustadt A. D. Weinstrasse; Maximilianstr. 43–45

DE.509.15.1 Botanische Abteilung.

DE.509.15.2 Abteilung für Chemie

DE.509.15.3 Abteilung Phytomedizin.

DE.509.15.4 Abteilung Botanik und Wiederaufbau

DE.509.15.5 Abteilung Weinbau

DE.509.15.6 Gartenbauabteilung Einschl. Imkerei.

DE.509.15.7 Abteilung Virologie

DE.509.20.0 Landes–Lehr– und Versuchsanstalt für Weinbau, Gartenbau und Landwirtschaft Trier. 5500 Trier; Egbertstr. 18–19

DE.509.20.1 Institut für Bodenkunde.

DE.509.20.2 Institut für Weinchemie und Gärungsphysiologie

DE.509.20.3 Zentralstelle für Klonenselektion.

DE.509.30.0 Landes–Veterinäruntersuchungsamt für Rheinland–Pfalz. 5400 Koblenz; Blücherstr. 34.

DE.511.00.0 Schleswig– Holstein.

DE.511.05.0 Landwirtschaftliche Untersuchungs– und Forschungsanstalt der Landwirtschaftskammer Schleswig–Holstein 2300 Kiel; Gutenbergstr. 75–77

DE.511.05.1 Abteilung Mikronährstoff

DE.511.05.2 Abteilung Ackerbau.

DE.511.10.0 Pflanzenschutzamt des landes Schleswig–Holstein. 2300 Kiel; Westring 383.

DE.511.15.0 Lehr– und Versuchsanstalt für Grünlandwirtschaft, Futterbau und Landeskultur 2257 Bredstedt; Theodor– Storm– Str. 2

DE.511.25.0 Lehr– und Versuchsanstalt für Kleintierzucht Kiel Steenbek der Landwirtschaftskammer Schleswig– Holstein 2300 Kiel–Steenbek; Steenbeker Weg 151.

DE.900.00.0 Private Institutionen

DE.901.00.0 Landwirtschaftliche Versuchsanstalt der Thomasphosphatfabriken
4300 Essen– Bredeney; Prinz– Adolf– Str. 2.

DE.902.00.0 Landwirtschaftliche Forschungsanstalt Büntehof.
Hannover– Kirchrode; Buentehof 8.

DE.902.00.1 Abteilung Bodenkunde.

DE.902.00.2 Abteilung Pflanzenphysiologie

DE.902.00.3 Abteilung Pflanzenernaehrung.

DE.902.00.4 Abteilung Bodenmikrobiologie.

DE.903.00.0 Landwirtschaftliche Versuchsstation der Kali– Chemie A. G. Hannover.
3000 Hannover; Hans– Böckler– Allee 20.

DE.903.50.0 Forschungsstelle von Sengbusch GmbH. 2070 Wulfsdorf; Hamburger Str. 250.

DE.904.00.0 **Institut für Umweltschutz und Agrikulturchemie.**
Dr. Helmut Berge; 5628 Heiligenhaus/ Bez. Düsseldorf; Am Vogelsang 14

DE.906.00.0 **Institut für Hopfenforschung der Deutschen Gesellschaft für Hopfenforschung e.V.**
8069 Hüll, Post Wolnzach/Obb.

DE.907.00.0 **Institut für Zuckerrübenforschung (An der Universität Göttingen)**
3400 Göttingen; Holtenser Landstr. 77.

DE.907.01.0 Abteilung Pflanzenphysiologie

DE.907.02.0 Abteilung Feldversuchswesen und Auswertung

DE.907.03.0 Abt. Pflanzenpathologie und Pflanzenschutz.

DE.909.00.0 **Wissenschaftliche Station für Brauerei in München E. V.**
8000 München 19; Romanstr. 41

DE.910.00.0 **Verfahrenskontrolle der H. Bahlsens Keksfabrik KG Forschungslabor.**
3000 Hannover; Podbielskistr. 11

DE.910.00.1 Chemisch–Analytische Abteilung.

DE.911.00.0 **Institut für Lebensmitteltechnologie und Verpackung E. V. An der Technischen Universität München.**
8000 München 50; Schragenhofstr. 35

DE.912.00.0 **Forschungsinstitut des Foerdervereins fuer schnellwachsende Baumarten.**
3510 Hann.–Münden; Prof. Ölkers Str. 6.

DE.913.00.0 **Forschungsinstitut für Holzwerkstoffe und Holzleime E. V.**
7500 Kalsruhe– Durlach; Dieselstr. 6

DE.914.00.0 **Agrarsoziale Gesellschaft E. V.**
3400 Göttingen; Kurze Geismarstr 23–25; Postf. 667.

DE.914.00.1 Arbeitsgruppe Regionalentwicklung

DE.914.00.2 Arbeitsgruppe Ländliche Sozialpolitik

DE.914.00.3 Arbeitsgruppe Landwirtschaft.

DE.914.00.4 Arbeitsgruppe Agrarstrukturpolitik.

DE.914.00.5 Referat Bildung und Ausbildung

DE.914.00.6 Abteilung Ländliche Sozialfragen.

DE.915.00.0 **Ava Arbeitsgemeinschaft zur Verbesserung der Agrarstruktur in Hessen E. V.**
6200 Wiesbaden; Alexanderstr. 2.

DE.915.50.0 Forschungsgesellschaft für Agrarpolitik und

Agrarsoziologie. 5300 Bonn;Meckenheimer allee 125

DE.916.00.0 **Forschungsstelle für Internationale Agrarentwicklung E. V.**
6900 Heidelberg; Kurfürstenanlage 59.

DE.916.50.0 Forschungsstelle für Gesamtdeutsche Wirtschaftliche und Soziale Fragen –Bereich Agrarforschung –. 1000 Berlin 31; Berliner Str. 154.

DE.916.60.0 Kommission fuer Erforschung der Agrar– und Wirtschaftshaeltnisse des Europaeischen Ostens e.V. 6300 Giessen.

DE.917.00.0 **I F O Institut für Wirtschaftsforschung E. V.**
8000 München 86; Poschinger Str 5.

DE.917.00.1 Abteilung Landwirtschaft.

DE.917.00.2 Abteilung Afrika–Studienstelle.

DE.917.00.3 Abteilung Finanzwirtschaft.

DE.917.50.0 Deutsches Institut für Tropische und Subtropische Landwirtschaft GmbH. 3430 Witzenhausen; Steinstr. 19.

DE.918.00.0 **Institut für Fenstertechnik E.V.**
8201 Aisingerwies über Rosenheim.

DE.919.00.0 **Landtechnischer Verein in Bayern.**
8050 Freising–Weihenstephan.

DE.920.00.0 **Institut für Umweltforschung Villingen.**
Villingen.

DE.933.00.0 **Battelle–Institut E.V**
6000 Frankfurt/Main 90; Wiesbadener Str.

DE.934.00.0 **Dornier System Gmbh.**
7990 Friedrichshafen.

DE.935.00.0 **Versuchsstation Dethlingen des Kuratoriums für Technik und Bauwesen in der Landwirtschaft.**
Dethlingen.

DE.936.00.0 **Zoologisches Forschungsinstitut und Museum A. König**
5300 Bonn; Adenauerallee 150

DE.936.00.1 Abteilung Säugetiere.

DE.938.00.0 Kuratorium fuer Waldarbeit und Forsttechnik. 6079 Buchschlag.

DE.938.00.1 Chemisch–technische Abteilung.

DE.939.00.0 Forschungsstelle fuer Jagdkunde und Wildschadenverhuetung. Forsthaus Hardt; 5300 Bonn–Beuel.

DE.940.00.0 **Forschungsinstitut für Futtermitteltechnik der IFF Braunschweig.**
3301 Thune über Braunschweig.

DENMARK

DK.01.00.00 Landbrugets Samråd for forskning og forsøg [The Joint Committee for Agricultural Research and Experiments].
Axelborg; Vesterbrogade 4a, 4.sal, vær. 19; 1620 København V.

DK.01.01.00 Statens Planteavisforsøg (Research Service for Soil and Plant Sciences). Virumgaard; Kongevejen 83; 2800 Lyngby.

DK.01.01.01 Askov Forsøgsstation.
6600 Vejen.

DK.01.01.02 Blangstedgård Forsøgsstation.
5220 Odense Sø

DK.01.01.03 Borris Forsøgsstation
6900 Skjern

DK.01.01.04 Hornum Forsøgsstation
9600 Års.

DK.01.01.05 Statens Marskforsøg
6280 Højer.

DK.01.01.06 Roskilde Forsøgsstation
Ledreborg Alle; 4000 Roskilde

DK.01.01.07 Rønhave Forsøgsstation.
6400 Sønderborg

DK.01.01.08 Silstrup Forsøgsstation.
7700 Thisted.

DK.01.01.09 St.Jyndevad Forsøgsstation
6360 Tinglev.

DK.01.01.11 Tylstrup Forsøgsstation
9380 Vestbjerg.

DK.01.01.12 Tystofte Forsøgsstation
4230 Skælskør

DK.01.01.13 Statens Væksthusforsøg.
Kirstinebjergvej 10; 5792 Årslev.

DK.01.01.14 Ødum Forsøgsstation.
8370 Hadsten.

DK.01.01.15 Årslev Forsøgsstation
Kirstinebjergvej 6; 5792 Årslev.

DK.01.01.16 Statens Plantepatologiske Forsøg
Lottenborgvej 2; 2800 Lyngby

DK.01.01.17 Statens Planteavls–Laboratorium.
Lottenborgvej 24; 2800 Lyngby.

DK.01.01.18 Dataanalytisk Laboratorium.
Lottenborgvej 24; 2800 Lyngby.

DK.01.01.19 Statens Ukrudtforsøg.
Flakkebjerg; 4200 Slagelse

DK.01.02.00 Statens Husdyrbrugsforsøg [Goverment Resarch in animal Husbandry]. Rolighedsvej 25; 1958 København V.

DK.01.02.01 Afdeling for forsøg med kvæg og får.

DK.01.02.02 Afdelingen for forsøg med svin og heste.

DK.01.02.03 Afdeling for forsøg med fjerkræ og kaniner.

DK.01.02.04 Afdelingen for forsøg med pelsdyr.

DK.01.02.06 Afdelingen for dyrefysiologi, biokemi og analytisk kemi.

DK.01.03.00 Statens Forsøgsmejeri. 3400 Hillerød.

DK.01.04.00 Statens Jordbrugstekniske Forsøg. Bygholm; 8700 Horsens.

DK.01.05.00 Jordbrugsøkonomisk Institut. Valby Langgade 19; 2500 Valby.

DK.01.06.00 De Landbrugstekniske Undersøgelser. Ørritslevgaard; 5450 Otterup.

DK.01.07.00 De Landsøkonomiske Foreninger.

DK.01.07.01 Landskontoret for Planteavl.
Kongsgårdsvej 28; 8260 Viby J..

DK.01.07.02 Landskontoret for Kvæg.
Kongsgårdsvej 28; 8260 Viby J..

DK.01.07.03 Landboorganisationernes driftsøkonomiudvalg.
Engdalsvej 65; 8220 Brabrand.

DK.01.08.00 Bioteknisk Institut. Holbergsvej 10; 6000 Kolding.

DK.01.09.00 Det danske Hedeselskab. 8800 Viborg.

DK.01.10.00 Statens Byggeforskningsinstitut. SBI Postboks 119; 2970 Hørsholm.

DK.02.00.01 Statens Forstlige Forsøgsvæsen.
Møllevangen, Springforbivej 4, 2930 Klampenborg.

DK.02.01.00 Statens Veterinære serumlaboratorium. Bülowsvej 27; 1870 København V.

DK.02.02.02 Statens veterinære Serumlaboratorium; Afd. i Ringsted.
Odinsvej 4; 4100 Ringsted.

DK.02.03.00 Institut for Fjerkræsygdomme. Rypevej 1; 8870 Langå.

DK.02.04.00 Statens Veterinære Serumlaboratorium; Afdeling for Jylland. Hangøvej 2; 8200 Århus N.

DK.02.05.00 Statsfrøkontrollen. Skovbrynet 20; 2800 Lyngby.

DK.02.06.00 Statens Skadedyrlaboratorium. Skovbrynet 14; 2800 Lyngby.

DENMARK

DK.02.07.00 Landbrugets Kartoffelfond; Forædlingsstationen.
Grindstedvej 55; 7184 Vandel.

DK.03.01.00 Den Kongelige Veterinær og Landbohøjskole.
Bülowsvej 13; 1870 København V.

DK.03.01.01 Afdelingen for almen Genetik.

DK.03.01.02 Arboretet.
2970 Hørsholm.

DK.03.01.03 Botanisk Institut.
Rolighedsvej 23; 1958 København V.

DK.03.01.04 Afdelingen for Fysiologisk Botanik.
Thorvaldsensvej 40; 1871 København V.

DK.03.01.05 Fysisk Laboratorium.
Thorvaldsensvej 40; 1871 København V.

DK.03.01.06 Kemisk Institut.
Thorvaldsensvej 40; 1871 København V.

DK.03.01.07 Institut for Matematik og Statistik.
Thorvaldsensvej 40; 1871 København V.

DK.03.01.08 Afdelingen for almindelig Mikrobiologi og
Mikrobiel Økologi.
Rolighedsvej 21; 1958 København V.

DK.03.01.09 Zoologisk Institut.

DK.03.01.21 Ambulatorisk klinik og klinisk
Centrallaboratorium.

DK.03.01.23 Afdelingen for Farmakologi og Toksikologi.

DK.03.01.25 Afdelingen for Fodringslære.

DK.03.01.26 Afdelingen for Fysiologi, Endokrinologi og
Blodtypeforskning.

DK.03.01.27 Afdelingen for Husdyrgenetik.

DK.03.01.28 Institut for Husdyrenes Reproduktion.

DK.03.01.29 Institut for intern Medicin.

DK.03.01.30 Institut for Kirurgi.

DK.03.01.31 Afdelingen for Levnedsmiddelkonservering.
Howitzvej 13; 2000 København F.

DK.03.01.32 Afdelingen for normal Anatomi.

DK.03.01.33 Afdeling for Almindelig Patologi og Patologisk
Anatomi.

DK.03.01.34 Afdelingen for praktisk Kødkontrol.
Kongensgade 24; 4100 Ringsted.

DK.03.01.35 Afdelingen for Retsmedicin.

DK.03.01.36 Stationær klinik for mindre Husdyr.

DK.03.01.37 Institut for Veterinær Mikrobiologi og

Hygiejne.

DK.03.01.38 Afdelingen for Veterinær Virologi og
Immunologi.

DK.03.01.41 Dyrefysiologisk afdeling.
Rolighedsvej 25; 1958 København V.

DK.03.01.42 Husdyrbrugsinstituttet.
Rolighedsvej 23; 1958 København V.

DK.03.01.43 Hydroteknisk laboratorium og klimastation.
Bülowsvej 23, 1870 København V.

DK.03.01.44 Jordbrugsteknisk Institut.
Rolighedsvej 23; 1958 København V.

DK.03.01.45 Afdeling for Landbrugets Plantekultur.
Thorvaldsensvej 40; 1871 København V.

DK.03.01.46 Afdelingen for Landbrugsforsøg.
Forsøgsanlæg Risø; 4000 Roskilde.

DK.03.01.47 Plantepatologisk afdeling.
Thorvaldsensvej 40; 1871 København V.

DK.03.01.48 Afdelingen for Planternes ernæring.
Thorvaldsensvej 40; 1871 København V.

DK.03.01.49 Økonomisk Institut.
Thorvaldsensvej 40; 1871 København V.

DK.03.01.61 Institut for Land– og Byplanlægning.

DK.03.01.71 Havebrugsinstituttet.
Rolighedsvej 23; 1958 København V.

DK.03.01.72 Afdelingen for Frugt, Grønsager og
Planteskole.
Rolighedsvej 23; 1958 København V.

DK.03.01.73 Afdelingen for Have og Landskab.
Rolighedsvej 23; 1958 København V.

DK.03.01.81 Skovbrugsinstituttet.
Thorvaldsensvej 57; 1871 København V.

DK.03.01.91 Afdelingen for Kødteknologi.
Howitzvej 11; 2000 København F.

DK.03.01.92 Mejeriafdelingen.

DK.03.01.93 Mælkerilaboratoriet.

DK.03.01.94 Afdelingen for Vegetabilske levnedsmidlers
Teknologi.
Thorvaldsensvej 40; 1871 København V.

DK.03.01.95 Danmarks Veterinær–og Jordbrugsbibliotek.

DK.03.02.00 Danmarks Tekniske Højskole. 2800 Lyngby.

DK.03.02.01 Laboratoriet for Levnedsmiddelindustri.
Bygning 221; 2800 Lyngby.

DK.03.02.02 Laboratoriet for Teknisk Hygiejne.

DENMARK

Bygning 115; 2800 Lyngby.

**DK.03.03.00 Isotopcentralen, ATV. Skelbækgade 2; 1717
København V.**

**DK.03.04.00 Odense Universitet. Niels Bohrs Alle; 5000
Odense.**

DK.03.04.01 Biokemisk Institut.

**DK.03.05.00 Teknologisk Institut. Gregersensvej; 2630
Tåstrup.**

DK.03.06.00 Aarhus Universitet.

DK.03.06.01 Institut for Genetik og økologi.
Ny Munkegade; 8000 Århus C.

DK.03.06.02 Botanisk Institut.
Nordlandsvej 68; 8240 Risskov.

DK.03.06.03 Institut for Zoologi og zoofysiologi.
Ole Worms Allé; 8000 Århus C.

**DK.03.07.00 Roskilde Universitetscenter. Postboks 260; 4000
Roskilde.**

DK.03.08.00 Københavns Universitet.

DK.03.08.01 Geografisk Institut.
Haraldsgade 68; 2100 København Ø.

FRANCE

FRANCE

FR.01.00.00 Institut National de la Recherche Agronomique Services Centraux.
149, Rue de Grenelle; 75007 Paris.

FR.01.01.00 Centre National de Recherches Agronomiques
Etoile de choisy, Route de Saint – Cyr; 78000 Versailles
(Yvelines).

FR.01.01.01 Station de Bioclimatologie.

FR.01.01.02 Laboratoire de Métrologie.

FR.01.01.03 Station de Science du Sol.

FR.01.01.04 Station de Génétique et d'Amélioration des
Plantes.

FR.01.01.05 Laboratoire d'bl Amélioration des Plantes. bl
Faculté des Sciences d'Orsay

FR.01.01.06 Station de Zoologie.

FR.01.01.07 Station de Pathologie Végétale

FR.01.01.08 Station de Physiologie Végétale.

FR.01.01.09 Laboratoire de Malherbologie.

FR.01.01.10 Station de Phytopharmacie

FR.01.01.11 Laboratoire de Biométrie.

FR.01.01.12 Section Centrale du Service d' Experimentation
et d'Information.

FR.01.01.13 Service d' Etude des Sols et de la Carte
Pédologique de France

FR.01.01.14 Station de Recherches sur l' Abeille et les
Insectes Sociaux
Domaine de La Guyonnerie; 91440 Bures–sur–yvette
(Essonne).

FR.01.01.15 Domaine expérimental du Vieux–Pin
Le Pin–au–Haras; 61310 EXMES.

FR.01.01.16 Laboratoires de Recherches des Chaires de
L'Ecole Nationale Supérieure d'Horticulture (ENSH)
(Botanique et Physiologie Végétale, Chimie, Cultures florales,
Cultures marai chéreset génétique).

FR.01.02.00 Centre National de Recherches Zootechniques
Domaine de Vilvert; 78350 Jouy–en–Josas (Yvelines).

FR.01.02.01 Station de Physiologie Animale.

FR.01.02.02 Station de Génétique Quantitative et Appliquée

FR.01.02.03 Laboratoire de Physiologie des Poissons.

FR.01.02.04 Laboratoire de Nutrition des Poissons.

FR.01.02.05 Station de Recherches de Nutrition

FR.01.02.06 Station de Recherches sur l' Elevage des Porcs.

FR.01.02.07 Station Centrale de Recherches laitières et de
Technologie des Produits Animaux.

FR.01.02.08 Laboratoire de Physiologie Acoustique.

FR.01.02.09 Laboratoire des Petits Vertébrés

FR.01.02.13 Laboratoire de Physiologie de la Lactation

FR.01.02.14 Laboratoire de Biometrie.

FR.01.02.15 Laboratoire de Radiobiologie Appliquée.

FR.01.02.16 Laboratoire d'Écologie Microbienne.

FR.01.02.17 Laboratoire de Physiologie de la Nutrition

**FR.01.03.00 Centre National de Recherches Forestières de
Nancy et Ecole Nationale du Genie Rural, des Eaux et des
Forêts. Champenoux, 54280 Seichamps (Meurthe–et–Moselle)
14, Rue Girardet, 54000 Nancy (Meurthe–et–Moselle).**

FR.01.03.01 Station de Sylviculture et de Production
Champenoux – 54280 Seichamps

FR.01.03.02 Station d' Amélioration des Arbres Forestièrs
Champenoux – 54280 Seichamps

FR.01.03.03 Station de Recherches sur les Sols Forestiers et
la Fertilisation.
Champenoux – 54280 Seichamps

FR.01.03.04 Station de Recherches sur la Qualité des Bois.
Champenoux – 54280 Seichamps

FR.01.03.05 Laboratoire de Pathologie Forestière.
Champenoux – 54280 Seichamps

FR.01.03.06 Laboratoire de Recherches de la Chaire des
Produits Forestierers.
Engref 14 Rue Girardet; 54042 Nancy.

FR.01.03.07 Laboratoire de Phytoecologie Forestiere.

FR.01.03.08 Laboratoire d'Economie Forestière
Engref 14 Rue Girardet; 54000 Nancy.

FR.01.03.09 Domaine Expérimental Forestier.
de L'Hermitage, Pierroton – Cestas 33610 Gazinet.

FR.01.03.10 Centre de Recherches Forestiéres.
Ardon – 45160 Olivet

FR.01.03.12 Station de Biométrie

FR.01.04.00 Centre de Recherches Agronomiques d'Angers.
Beauconze – 49000 Angers.

FR.01.04.01 Station de Recherches d'Arboriculture
Fruitière.

FR.01.04.02 Station de Pathologie Végétale et de
Phytobacté–riologi.

FRANCE

FR.01.04.03 Station d'Agronomie.

FR.01.04.04 Station de Recherches Oenologiques.
Route de St. Clément – 49000 Qugers

FR.01.04.05 Domaine Expérimental Viticole de
Montreuil–Bellay
49260 Montreuil–Bellay

**FR.01.05.00 Centre de Recherches d'Antibes 37, Boulevard
du Cap; 06602 Antibes (Alpes–Maritimes)**

FR.01.05.01 Station de Pathologie Végétale.
Villa Thuret – B.B. 78, 06602 Antibes (Alpes–Maritimes).

FR.01.05.02 Station d'Agronomie et de Physiologie
Végétale.
45 Boulevard Du Cap; 06602 Antibes (Alpes–Maritimes).

FR.01.05.03 Station de Zoologie et de Lutte Biologique.

FR.01.05.04 Station de Recherches sur les Nématodes
123 Boulevard Du Cap; 06602 Antibes (Alpes–Maritimes).

FR.01.05.05 Station d'Amélioration des Plantes Florales.
Domaine de la Gaudine; 83600 Fréjus (Var)

FR.01.05.06 Laboratoire d'Amélioration des Plantes et
Domaine Experimental de la Baronne.
Commune de la Caude; (Route Nationale No. 2209); b.p. No.
55; 06702 St–Laurent–du Var; (Alpes–Maritimes).

**FR.01.06.00 Centre de Recherches d'Avignon Domaine
Saint–Paul; 84140 Montfavet (Vaucluse).**

FR.01.06.01 Station de Science du Sol.

FR.01.06.02 Station de Zoologie

FR.01.06.03 Station d'Amélioration des Plantes
Maraichères.
Domaine Saint–Maurice

FR.01.06.04 Station de Pathologie Végétale

FR.01.06.05 Station de Technologie des Produits Végétaux.

FR.01.06.06 Station de Bioclimatologie.

FR.01.06.07 Laboratoire d'Etude de la pollution
atmostohérique.

FR.01.06.08 Station Expérimentale d'Apiculture.

FR.01.06.09 Station de Zoologie Forestière d'Avignon.
Avenue Antoine Vivaldi; 84000 Avignon.

FR.01.06.10 Section Réqionale d'Avignon ('u Service
d'Expérimentation et d'Information

FR.01.06.11 Station Séricicole
28 Quai Boissier de Sauvage;. 30101 Alès (Gard).

FR.01.06.12 Station de Recherches de Pathologie Comparée
30380 Saint–Christol–les–Alès (Gard).

FR.01.06.13 Domaine Expérimental des Vignères.
Domaine d'Olonne; 84250 Le Thor (Vaucluse).

FR.01.06.15 Domaine de Gotheron

FR.01.06.16 Domaine d'Alenya.

**FR.01.07.00 Centre de Recherches de Bordeaux Domaine de
"la Grande Ferrade"; Point–de–la–Maye 33140 villenave
d'Ornon (Gironde).**

FR.01.07.01 Station de Recherches d'Arboriculture
Fruitière.

FR.01.07.02 Station de Recherches de Viticulture

FR.01.07.03 station de Recherches sur les Champignons

FR.01.07.04 Station d'Agronomie

FR.01.07.05 Station de Biochimie et Physiologie Végétales

FR.01.07.06 Station de Pathologie Végétale

FR.01.07.07 Station de Zoologie

FR.01.07.08 Laboratoire de Pathologie Apicole.
40630 – Sabres (Landes).

FR.01.07.09 Laboratoire d'Analyses.
64 – Montardon (Pyrénées–Atlantiques)

FR.01.07.10 Domaine Expérimental de Chateau Couhins et
Chateau Pont–de–Lagon.
33140 – Villenave d'Ornon.

FR.01.07.11 Station d'Hydrobiologie Continentale
Bp no. 79; 64200 Biarritz (Pyrénées–Atlantiques)

FR.01.07.12 Laboratoire de Recherches de la Chaire de
Zootechnie de l'Ecole Nationale des Ingénieurs des Travaux
Agricoles.
(Enita) de Bordeaux –1 Cours du Général de Gaulle –33170
Gradignan

FR.01.07.13 Laboratoire de Recherches de la Chaire
d'Immunologie de la Faculté des Sciences.

FR.01.07.14 Institut d'Œnologie
Université de Bordeaux ii – 351, cours de la libération –33405
Talence.

**FR.01.08.00 Centre de Recherches de Clermont–Ferrand. 1
–Recherches Agronomiques (Milieu et Productions Végétales);
Domaine de Mon Désir"; 63100 Clermont–Ferrand (Puy–de
Dôme); 2 – Recherches Zootechniques et Vétérinaires; Domaine
de Theix; Saint–Genes–Champaelle 63110 Beaumont.**

FR.01.08.01 Station d'Amélioration des Plantes.
Domaine de Crouelle; 63100 Clermont–Ferrand (Puy
de–Dôme.

FR.01.08.02 Station d'Agronomie

FR.01.08.03 Station de Pathologie Végétale

FRANCE

FR.01.08.04 Domaines Expérimentaux de Marcenat – Domaine de la Boue–.
15330 Marcenat (Cantal).

FR.01.08.05 Station de Recherches sur l'Elevage des Ruminants et du Cheval
Domaine Expérimental du Roc; 63210 Rochefort. Montagne – Orival, Domaine Expérimental des Razats; 63820 Lagueuille

FR.01.08.06 Laboratoire de Bioclimatologie.

FR.01.08.07 Laboratoire des maladies nutritionnelles

FR.01.08.08 Station de Recherches sur la Viande.

FR.01.08.09 Laboratoire de Microbiologie.

FR.01.08.10 Laboratoire d'Economie de l' Elevage

FR.01.08.11 Station de Recherches Vétérinaires

FR.01.08.12 Laboratoire de la Production Laitière

FR.01.08.13 Laboratoire de la Production de Viande

FR.01.08.14 Laboratoire des Aliments.

FR.01.08.15 Laboratoire d'Etudes du Métabolisme Azoté.

FR.01.08.16 Laboratoire d'Etudes du Métabolisme Energétique

FR.01.08.17 Laboratoire des Maladies Métaboliques

FR.01.09.00 **Centre de Recherches de Colmar 28 Rue de Herrlisheim; 68021 Colmar (Haut–Rhin) Bp No. 507.**

FR.01.09.01 Station d'Agronomie

FR.01.09.02 Station de Recherches Viticoles et Œnologiques.
8 Rue Kleber; 68021 Colmar (Haut–Rhin).

FR.01.09.03 Station de Pathologie Végétale

FR.01.09.04 Station de Zoologie

FR.01.09.05 Domaine Expérimental de Colmar.

FR.01.09.06 Section Régionale de Colmar du Service d'Expérimentation et d'Information.

FR.01.10.00 **Centre de Recherches de Dijon. 17, Rue Sully; 21034 Dijon (Côte d'Or) Bv No. 1540**

FR.01.10.01 Station de Science du Sol.

FR.01.10.02 Laboratoire de Microbiologie des Sols.

FR.01.10.03 Station d'Amélioration des Plantes.
Domaine d'Epoisses Bretenéres – 21110 Genlis

FR.01.10.04 Station de Physio Pathologie Végétale
Domaine d'Epoisses – Bretenières 21110 Genlis

FR.01.10.05 Laboratoire de Malherbologie.

Domaine d'Epoisses – Bretenières 21110 Genlis

FR.01.10.06 Station de Recherches sur la Flore Pathogéné dans le Sol

FR.01.10.07 Station de Recherches sur la Faune du Sol

FR.01.10.08 Station de Recherches sur la Qualité des Aliments de l'Homme.

FR.01.10.09 Station de Technologie des Produits Végétaux.

FR.01.10.10 Station de Génie Microbiologique.

FR.01.10.11 Station OEnologique de Bourgogne.
12 Bvd Bretonnière; 21200 Beaune (Côte d'Or).

FR.01.10.12 Laboratoire de Recherches de la Chaire de Science Economiques.

FR.01.10.13 Laboratoire de Recherches de la Chaire de Zootechnie.

FR.01.10.14 Phytotechnie

FR.01.11.00 **Groupe des Laboratoires et de Service – La Minière. 78280 Guyancourt (Yvelines).**

FR.01.11.01 Station de Zoologie et de Biocenotiques Forestières.

FR.01.11.02 Station de Rechercheste de Lutte Biologique.

FR.01.11.03 Groupe d'Etude et de Contrôle des Variétés et des Semences.

FR.01.11.04 Service d Expérimentation

FR.01.11.05 Service de Production de Semences.

FR.01.11.06 Domaine Expérimental de La Minière.

FR.01.12.00 **Centre de Recherches de Montpellier, Ecole Nationale Supérieure Agronomique 9 Place Viala; 34060 Montpellier Cedex**

FR.01.12.01 Station d'Amélioration des Plantes.

FR.01.12.02 Station de Technologie Végétale.

FR.01.12.03 Laboratoire de Physiologie Animale.

FR.01.12.04 Station de Recherches Viticoles – Domaine du Chapitre.

FR.01.12.05 Station d'Economie Rurale.

FR.01.12.06 Laboratoire de Recherches de la Chaire de Botanique et de Pathologie Végétale de l'ENSA

FR.01.12.07 Laboratoire de Recherches de la Chaire d'Arboriculture fruitière de l'ENSA

FR.01.12.08 Laboratoire de Recherches de la Chaire de Génétique de l'ENSA

FR.01.12.09 Laboratoire de Recherches de la Chaire de Biochimie et Physiologie Végétale de l'ENSA.

FR.01.12.10 Laboratoire de Recherches de la Chaire de Zoologie de l'ENSA.

FR.01.12.11 Laboratoire de Recherches de la Chaire de Géologie pédologie de l'ENSA et service d'Etude des Sols.

FR.01.12.12 Laboratoire de Recherches de la Chaire de Physiologie générale de la Faculté des Sciences de Marseille. Centre St.Térome – Traverse de la Barasse – 13013 Marseille

FR.01.12.13 Domaine Expérimental du Chapitre. 34750 Villeneuve des Maguelonne.

FR.01.12.14 Laboratoire de Technologie des Blés Durs et du Riz.

FR.01.13.00 Centre de Recherches de Rennes Ecole Nationale Supérieure Agronomique 65 Rue de Saint–Brieuc; 35042 Rennes (Ille et Vilaine).

FR.01.13.01 Station d'Amélioration des Plantes.

FR.01.13.02 Station d'Economie et de Sociologie Rurales

FR.01.13.03 Station de Recherches Cidricoles.

FR.01.13.04 Station de Recherches Zootechniques.

FR.01.13.05 Laboratoire de Recherches de Technologie Laitière.

FR.01.13.06 Laboratoire de Recherches de la Chaire de Zoologie de L'ENSA

FR.01.13.07 Laboratoire de Recherches de la Chaire de Botanique et Pathologie Végétale de l'ENSA.

FR.01.13.08 Laboratoire de Recherches de Science du Sol de la Chaire de Géologie Pedologie de l'ENSA.

FR.01.13.09 Station de Testage des Porcs 35650 Le Rheu (Iue–Et–Vilaine).

FR.01.13.10 Laboratoire de Recherches sur la Traite.

FR.01.14.00 Centre de recherches de Toulouse Chemin de Borde–Rouge, Auzeville Bp No. 12 31320 Castanet–Tolosan.

FR.01.14.01 Station d'Agronomie

FR.01.14.02 Laboratoire de Technologie des Produits végétaux

FR.01.14.03 Station d'œ et de Technologie Végétale de Narbonne. 11, Boulevard du Général de Gaulle; 11104 Narbonne (aude).

FR.01.14.04 Laboratoire de Méthodologie Génétique

FR.01.14.05 Laboratoire de Génétique des Petits Ruminants

FR.01.14.06 Laboratoire de Génétique Cellulaire.

FR.01.14.07 Station de Pharmacologie–Toxicologie 180 Chemin de Tournefeuille; 31300 Toulouse (Haute–Garonne)

FR.01.14.08 Laboratoire d'Economie Rurale–Faculté de Droit et des Sciences Economiques. Place Anatole France – 31000 Toulouse.

FR.01.14.09 Domaine Expérimental d' Auzeville

FR.01.14.10 Domaine Expérimental de Langlade. Pompertuzat 31450 Montgiscard.

FR.01.14.11 Station de Testage des Lapins.

FR.01.14.12 Laboratoire de Recherches sur les Additifs Alimentaires 180, Chemin Tournefeuille – 31300 Toulouse.

FR.01.15.00 Centre de Recherches de Tours Domaine de L'Orfrasière; Bp No 1 – Nouzilly, 37380 Monnaie

FR.01.15.01 Station de Recherches sur la Physiologie de la Reproduction. La Reproduction.

FR.01.15.02 Laboratoire de Neuroendocrinologie sexuelle

FR.01.15.03 Station de Pathologie de la Reproduction.

FR.01.15.04 Station de Pathologie Aviaire.

FR.01.15.05 Laboratoire de Parasitologie.

FR.01.15.06 Station Expérimentale d'Artiguères Benquet – BP No 68 – 40002 Mont–de–Marsan.

FR.01.15.07 Domaine Expérimental de l'Orfrasière

FR.01.15.08 Station Expérimentale d' Insémination Artificielle. 86480 Rouillé (Vienne).

FR.01.15.09 Station Expérimentale d'Aviculture du Magneraud – St.–Georges–du–Bois 17700 – Surgères Bp No 52 (Charente–Maritime).

FR.01.16.00 Centre de Recherches Agronomiques des Antilles et de la Guyane Domaine "Duclos" 97170 Petit–Bourg (Guadeloupe)

FR.01.16.01 Station d' Amélioration des Plantes

FR.01.16.02 Station d' Agronomie.

FR.01.16.03 Station de Bioclimatologie.

FR.01.16.04 Station de Pathologie Végétale

FR.01.16.05 Station de Zoologie et de Lutte Biologique.

FR.01.16.06 Station de Recherches Zootechniques.

FR.01.16.07 Station de Technologie.

FR.01.16.08 Domaine Expérimental de Gardel.

97160 Le Moule

FR.01.16.09 Domaine Expérimental de Saint–Francois 97118 Saint–François.

FR.01.17.00 Institut National Agronomique Paris – Grignon. 16 Rue Claude Bernard; 75231 Paris Cédex 05.

FR.01.17.01 Laboratoire de Recherches de la Chaire d'Agriculture.

FR.01.17.02 Laboratoire de Recherches de la Chaire d'Anatomie et Physiologie Animale.

FR.01.17.03 Laboratoire de Recherches de la Chaire de Botanique et Pathologie Végétale.

FR.01.17.04 Laboratoire de Recherches de la Chaire de Chimie

FR.01.17.05 Laboratoire de Recherches de la Chaire d'Ecologie et Bioclimatologie

FR.01.17.06 Laboratoire de Recherches de la Chaire d'Economie Planification et Sociologie Rurales

FR.01.17.07 Laboratoire de Recherches de la Chaire de Génétique

FR.01.17.08 Laboratoire de Recherches de la Chaire de Géologie

FR.01.17.09 Laboratoire de Recherches de la Chaire d'Industries Agricoles.

FR.01.17.10 Laboratoire de Recherches de la Chaire de Microbiologie

FR.01.17.11 Laboratoire de Recherches de la Chaire de Zoologie.

FR.01.17.12 Laboratoire de Recherches de la Chaire de Zootechnie.

FR.01.17.13 Laboratoire de Recherches de la Chaire de Microbiologie des Sols

FR.01.17.14 Laboratoire de Recherches de la Chaire de Microbiologie Technologie.

FR.01.18.00 Institut Natinal Agronomique Paris – Grignon 78850 – Thiverval – Grignon (Yvelines).

FR.01.18.01 Laboratoire de Recherches de la Chaire d'Agriculture.

FR.01.18.02 Laboratoire de Recherches de la Chaire de Botanique et Pathologie Végétale.

FR.01.18.03 Laboratoire de Chimie Biologique et de Photophysiologie

FR.01.18.04 Laboratoire de Recherches de la Chaire de Géologie

FR.01.18.05 Laboratoire de Recherches de la Chaire de

Physico– Chimie et Science du Sol.

FR.01.18.06 Laboratoire de Recherches de la Chaire de Technologie et Microbiologie

FR.01.18.07 Laboratoire de Recherches de la Chaire de Zootechnie.

FR.01.18.08 Laboratoire d'Economie Rurale.

FR.01.18.09 Laboratoire d'Acridologie Appliquée

FR.01.19.00 Ecole Nationale Vétérinaire d'Alfort. 7 Avenue du Général de Gaule; 94701 Maisons–Alfort (Val–de–Marne)

FR.01.19.01 Laboratoire de Recherches de la Chaire de Physique et Chimie Biologiques et Médicales

FR.01.19.02 Laboratoire de Recherches de la Chaire de Pharmacie Toxicologie.

FR.01.19.03 Laboratoire de Recherches de la Chaire d'Anatomie

FR.01.19.04 Laboratoire de Recherches de la Chaire de Physiologie Thérapeutique et Pharmacodynamie

FR.01.19.05 Laboratoire de Recherches de la Chaire d'Histologie et Anatomie Pathologique.

FR.01.19.06 Laboratoire de Recherches de la Chaire d'Hygiène et Industrie des Aliments d'Origine Animale

FR.01.19.07 Laboratoire de Recherches de la Chaire de Parasitologie

FR.01.19.08 Laboratoire de Recherches de la Chaire de Pathologie Médicale des Equidés, Carnivores et Volailles.

FR.01.19.09 Laboratoire de Recherches de la Chaire de Pathologie Médicale duBétail et des Animaux de Basse–Cour.

FR.01.19.10 Laboratoire de Recherches de la Chaire de Pathologie Chirurgicale.

FR.01.19.11 Laboratoire de Recherches de la Chaire de Pathologie de la Reproduction.

FR.01.19.12 Laboratoire de Recherches de la Chaire de Maladies Contagieuses Zoonoses et Législation Sanitaire .

FR.01.19.13 Laboratoire de Recherches de la Chaire de Pathologie Générale, Microbiologie, Immunologie.

FR.01.19.14 Laboratoire de Recherches de la Chaire de Zootechnie et Economie Rurale

FR.01.19.15 Laboratoire de Recherches de la Chaire de Nutrition et Alimentation.

FR.01.19.16 Laboratoire de Génétique.

FR.01.19.17 Laboratoire de Synthèse Organique

FR.01.19.18 Laboratoire de Biologie Physico–Chimique.

FR.01.19.19 Laboratoire de Chirurgie Expérimentale.

FR.01.20.00 Ecole Nationale Vétérinaire de Lyon Marcy l'Etoile – 69260 Charbonnieres.

FR.01.20.01 Laboratoire de Recherches de la Chaire de Physique et Chimie biologiques et Médicales.

FR.01.20.02 Laboratoire de Recherches de la Chaire de Pharmacie Toxicologie.

FR.01.20.03 Laboratoire de Recherches de la Chaire d.Anatomie

FR.01.20.04 Laboratoire de Recherches de la Chaire de Physiologie, Thérapeutique, Pharmacodynamie.

FR.01.20.05 Laboratoire de Recherches de la Chaire d'Histologie et Anatomie Pathologique.

FR.01.20.06 Laboratoire de Recherches de la Chaire d'Hygiène et Industrie des Aliments d'Origine Animale

FR.01.20.07 Laboratoire de Recherches de la Chaire de Parasitologie

FR.01.20.08 Laboratoire de Recherches de la Chaire de Pathologie Médicale Des Equidés et des Carnivores

FR.01.20.09 Laboratoire de Recherches de la Chaire de Pathologie Médicale duBétajl et des Animaux de Basse–cour

FR.01.20.10 Laboratoire de la Chaire de Pathologie Chirurgicale

FR.01.20.11 Laboratoire de Recherches de la Chaire de Pathologie de la Reproduction.

FR.01.20.12 Laboratoire de Recherches de la Chaire de Maladies Contagieuses,Zoonoses et Législation Sanitaire

FR.01.20.13 Laboratoire de Recherches de la Chaire de Pathologie Générale, Microbiologie et Immunologie.

FR.01.20.14 Laboratoire de Recherches de la Chaire de Zootechnie Economie Rurale.

FR.01.20.15 Laboratoire de Recherches de la Chaire d'Alimentation.

FR.01.21.00 Ecole Nationale Vétérinaire de Toulouse Chemin des Capelles 31076 – Toulouse (Haute–Garonne).

FR.01.21.01 Laboratoire de Recherches de la Chaire de Physique et Chimie Biologiques et Médicales

FR.01.21.02 Laboratoire de Recherches de la Chaire de Pharmacie toxicologie.

FR.01.21.03 Laboratoire de Recherches de la Chaire d'Anatomie

FR.01.21.04 Laboratoire de Recherches de la Chaire de Physiologie Thérapeutique, Pharmacodynamie

FR.01.21.05 Laboratoire de Recherches de la Chaire

d'Histologie et Anatomie Pathologique.

FR.01.21.06 Laboratoire de Recherches de la Chaire d'Hygiène et Industrie des Aliments d'Origine Animale

FR.01.21.07 Laboratoire de Recherches de la Chaire de Parasitologie

FR.01.21.08 Laboratoire de Recherches de la Chaire de Pathologie Médicale des Equidés et Carnivores.

FR.01.21.09 Laboratoire de Recherches de la Chaire de Pathologie Médicale du Bétail et des Animaux de Basse–cour.

FR.01.21.10 Laboratoire de Recherches de la Chaire de Pathologie Chirurgicale.

FR.01.21.11 Laboratoire de Recherches de la Chaire de Pathologie de la Reproduction.

FR.01.21.12 Laboratoire de Recherches de la Chaire de Maladies Contagieuses, Zoonoses et Législation Sanitaire

FR.01.21.13 Laboratoire de Recherches de la Chaire de Pathologie Générale, Microbiologie, Immunologie.

FR.01.21.14 Laboratoire de Recherches de la Chaire de Zootechnie et Economie Rurale

FR.01.21.15 Laboratoire de Recherches de la Chaire d'Alimentation

FR.01.22.00 Services Isolés.

FR.01.22.01 Charente–Maritime (17).
Domaine Expérimental de Saint–Laurent–de–la–Prée – 17450 Fouras.

FR.01.22.02 Corse (20) Station de Recherches Agronomiques de Corse
San Giuliano – 20230 San–Nicolao

FR.01.22.03 Essonne (91) Laboratoire de Technologie Alimentaire, Ecole Nationale Supérieure des Industries Agricoles et Alimentaires, au Cerdia.
Le Noyer–Lambert – 91300 Massy

FR.01.22.05 Laboratoire de Recherches et d'Etudes sur l'Economie des Industries Agricoles et Alimentaires.
3 rue du Caducée – BP 333 – 94153 MIN RUNGIS.

FR.01.22.06 Eure–Et–Loir (28): Station Agronomique de Chartres.
Rue du Maréchal Leclerc – 28110 Luce.

FR.01.22.07 Finistère (29), Station d'Amélioration de la Pomme de Terre et des Plantes á Bulbes
Domaine BP no. 5 – 29208 Landerneau Expérimental de Keraïber – Ploudaniel – 29260 Lesneven

FR.01.22.08 Finistère (29) Station d'Agronomie.
rue de Stang Vihan – 29000 Quimper.

FR.01.22.09 Indre (36) Station d'Agronomie
30 rue Vieille Prison – 36000 Chateauroux

FR.01.22.10 Marne (51) Station de Science du Sol.
Route de Montmirail, Fagnières – 51000 Châlons–sur–Marne.

FR.01.22.11 Nord (59): Laboratoire de Recherches de la
Chaire de Malteire. Brasserie. Eaux et Boissons Gazeuses
Supèrieure des Industries Agricoles et Alimentaires, Centre de
Douai.
105 Rue de L'Université – 59500 Douai

FR.01.22.12 Paris (75) Station d'Economie et de Sociologie
Ruales.
6 Passage Tenaille – 75014 Paris.

FR.01.22.13 Laboratoire de Recherches sur la Qualité des
Blés, Ecole Française de Meunerie.
16 Rue Nicolas Fortin, 75013 Paris

FR.01.22.14 Laboratoire de Contróle et de Technologie des
Blés, Ecole Française de Meunerie
16 Rue Nicolas Fortin, 75013 Paris

FR.01.22.15 Laboratoire de Synthése de Produits Naturels.
[Laboratoire Associe] Ecole Nationale Supérieure de Chimie.
11 Rue Pierre Curie 75005 Paris.

FR.01.22.16 Pas–de–Calais (62): Station de Science du Sol,
Laboratoire d'Analyse des Sols
273 Rue de Cambrai – 62000 Arras.

FR.01.22.17 Rhone (69), Laboratoire de Biologie,
(Laboratoire associe) Institut National des Sciences
Appliquées
20 Avenue Albert Einstein – 69100 Villeurbanne

FR.01.22.18 Savoie (Haute) (74) Station d'Hydrobiologie
Lacustre
74203 Thonon–les–Bains

FR.01.22.19 Seine Maritime (76): Station Agronomique.
1 Rue Dufay – 76100 Rouen.

FR.01.22.20 Laboratoire de Zoologie
16 Rue Dufay – 76100 Rouen.

FR.01.22.21 Section Régionale de Rouen du Service,
d'Expérimentation et d'Information
16 Rue Dufay – 76100 Rouen.

FR.01.22.22 Somme (80), Station d'Agronomie.
23 Rue Debray – 80000 Amiens.

FR.01.22.23 Vienne (86), Station d'Amélioration des Plantes
Fourragéres.
86600 Lusignan.

FR.01.22.24 Vosges (88), Domaine Expérimental de
Mirecourt Domaine du Joly
88500 Mirecourt

FR.01.22.25 Yveline (78), Station de Recherches de
Virologie et d'Immunologie, Laboratoire d'Ichtyopathologie.

FR.01.22.26 Stations Travaillant en Liaison avec I.N.R.A.

FR.01.22.27 Station d'Agronomie de L'Yonne.
89000 Auxerre.

FR.01.22.28 Domaine de Saint Pol de Leon.

FR.01.23.01 Centre de Recherches. Forestieres d'Orleans.

FR.01.23.03 Station de Recherches sur la Forct et
l'Environnement

**FR.01.24.00 Centre de Recherches de Nantes Chemin de la
Geraudiere; 44072 Nantes Cedex.**

FR.01.25.02 Unité d'Économie des Industries Agricoles et
Alimentaires.

FR.01.25.03 Station d'Agronomi de e'Aisne.
Rue Ferdinand Christ; 02000 Laon

**FR.01.26.00 Laboratoire de Recherches et d'Études
Économiques, Université de Sciences Sociales de Grenoble
INRA–IREP Boite Postale 47x, Centre de TRI; 38040 Grenoble
Cedex**

UNITED KINGDOM

**GB.01.00.00 Agricultural Research Council.
160, Great Portland Street, London W1N 6DT.**

GB.01.01.00 Animal Breeding Research Organisation. West Mains Road, Edinburgh, EH9 3JQ.

GB.01.01.01 Applied Genetics.

GB.01.01.02 Disease Studies.

GB.01.01.03 Immunology.

GB.01.01.04 Growth and Efficiency.

GB.01.01.05 Statistics and Computing.

GB.01.01.06 Physiological Genetics.

GB.01.01.07 Experiments Division.

GB.01.02.00 Food Research Institute. Colney Lane, Norwich, NR4 7UA.

GB.01.02.02 Interdivisional Projects.

GB.01.02.03 Biochemistry.

GB.01.02.04 Microbiology.

GB.01.02.05 Science Services and Development.

GB.01.02.06 Nutrition.

GB.01.03.00 Institute of Animal Physiology. Babraham Cambridge, CB2 4AT.

GB.01.03.01 Applied Biology.

GB.01.03.02 Biochemistry.

GB.01.03.03 Experimental Pathology and Immunology.

GB.01.03.04 Directors Group.

GB.01.03.05 Physiology.

GB.01.03.07 Biophysics.

GB.01.03.08 Cell Biology.

GB.01.03.09 Animal Research Station.

GB.01.03.10 Unit of Reproductive Physiology and Biochemistry.

GB.01.04.00 Institute for Research on Animal Diseases. Compton, Near Newbury, Berkshire, RG16 ONN.

GB.01.04.01 Director.

GB.01.04.02 Biochemistry.

GB.01.04.03 Cellular Pathology.

GB.01.04.04 Functional Pathology.

GB.01.04.05 Microbiology.

GB.01.04.06 Parasitology.

GB.01.04.08 Statistics.

GB.01.04.10 Immunochemistry.

GB.01.05.00 Letcombe Laboratory. Wantage, Oxfordshire, OX12 9JT.

GB.01.05.01 Chemistry and Electronics

GB.01.05.02 Physiology.

GB.01.05.03 Field Studies.

GB.01.05.04 Joint Chemistry and Physiology.

GB.01.05.05 Environmental Radioactivity.

GB.01.06.00 Meat Research Institute. Langford, Bristol, BS18 7DY.

GB.01.06.01 Meat Production.

GB.01.06.02 Carcase Handling and Characterisation.

GB.01.06.03 Carcase Hygiene.

GB.01.06.04 Meat Distribution.

GB.01.06.05 Meat Quality and Safety Improvement

GB.01.06.06 Meat Quality Assessment.

GB.01.06.07 Meat Animal By–Products

GB.01.06.09 Information Service

GB.01.07.00 Poultry Research Centre. King's Buildings, West Mains Road, Edinburgh EH93JS.

GB.01.07.01 Anatomy.

GB.01.07.02 Biochemistry and Data Processing Section.

GB.01.07.03 Environmental Physiology.

GB.01.07.04 Ethology

GB.01.07.07 Nutrition.

GB.01.07.08 Reproductve Physiology.

GB.01.07.09 Statistics

GB.01.08.00 Weed Research Organisation Begbroke Hill, Yarnton, Oxford, OX5 1PF.

GB.01.08.01 Herbicide Evaluation.

GB.01.08.02 Chemistry.

GB.01.08.03 Microbiology

GB.01.08.04 Botany

GB.01.08.06 Perennials

GB.01.08.07 Aquatic Weeds

GB.01.08.09 Tropical Weeds.

GB.01.08.10 Information.

GB.01.08.11 Animals.

GB.01.08.12 Grass and Fodder.

GB.01.08.13 Weed Control.

GB.01.08.14 Unit of Developmental Botany

GB.01.09.00 Animal Virus Research Institute Pirbright, Woking, Surrey, GU24 ONF.

GB.01.09.01 Biochemistry

GB.01.09.02 Epidemiology

GB.01.09.03 Disease Security.

GB.01.09.04 Experimental Pathology.

GB.01.09.05 General Virology.

GB.01.09.06 Genetics

GB.01.09.07 Laboratory Animal Science.

GB.01.09.08 Vaccine Research.

GB.01.10.00 East Malling Research Station East Malling, Maidstone Kent, ME19 6BJ

GB.01.10.01 Zoology.

GB.01.10.02 Fruit Breeding.

GB.01.10.03 Fruit Nutrition

GB.01.10.04 Fruit Storage

GB.01.10.05 Plant Pathology

GB.01.10.06 Plant Physiology.

GB.01.10.07 Plant Protective Chemistry

GB.01.10.08 Pomology

GB.01.10.09 Statistics.

GB.01.10.10 Plant Propagation.

GB.01.11.00 Glasshouse Crops Research Institute. Worthing Road, Rustington, Littlehampton, Sussex, BN16 3PU.

GB.01.11.01 Plant Breeding.

GB.01.11.02 Plant Physiology.

GB.01.11.03 Biochemistry and Crop Protection.

GB.01.11.05 Plant Nutrition and Soil Chemistry.

GB.01.11.06 Crop Science Department.

GB.01.11.07 Biometrics.

GB.01.11.08 Entomology.

GB.01.11.09 Microbiology.

GB.01.11.10 Virology.

GB.01.11.12 Technical Services.

GB.01.12.00 Grassland Research Institute. Hurley, Maidenhead, Berkshire, SL6 5LR.

GB.01.12.01 Soils and Plant Nutrition.

GB.01.12.02 Agronomy.

GB.01.12.03 Plant and Crop Physiology.

GB.01.12.05 Animal Nutrition and Production.

GB.01.12.06 Animal Husbandry and Production.

GB.01.12.07 Biomathematics

GB.01.12.08 Ecology.

GB.01.12.09 Grassland Intelligence Survey Team.

GB.01.12.10 Permanent Pasture Group.

GB.01.12.11 Veterinary Research Group.

GB.01.13.00 Houghton Poultry Research Station. Houghton, Huntingdon, Cambridgeshire PE17 2DA.

GB.01.13.01 Directors Group.

GB.01.13.03 Experimental Husbandry.

GB.01.13.04 Leukosis Experimental Unit.

GB.01.13.05 Parasitology.

GB.01.13.07 Physiology and Biochemistry.

GB.01.13.08 Microbiology.

GB.01.13.09 Development and Services.

GB.01.14.00 John Innes Institute. Colney Lane, Norwich, NR4 7UH.

GB.01.14.01 Applied Genetics.

GB.01.14.02 Genetics.

GB.01.14.03 Ultrastructural Studies.

GB.01.14.04 Virus Research.

GB.01.15.00 Long Ashton Research Station. Long Ashton
Bristol BS18 9AF.

GB.01.15.02 Cider and Fruit Juices.

GB.01.15.03 Microclimatology.

GB.01.15.04 Organic Chemistry.

GB.01.15.05 Physical Chemistry.

GB.01.15.07 Plant Nutrition.

GB.01.15.08 Plant Pathology.

GB.01.15.09 Plant Physiology and Biochemistry.

GB.01.15.10 Pomology and Plant Breeding.

GB.01.15.11 Spray Applications.

GB.01.15.14 Zoology.

GB.01.16.00 National Institute of Agricultural Engineering.
Wrest Park, Silsoe, Bedford, MK45 4HS.

GB.01.16.01 Cultivation and Traction.

GB.01.16.02 Tractor and Ergonomics.

GB.01.16.03 Machinery.

GB.01.16.04 Control and Instrumentation

GB.01.16.05 Engineering.

GB.01.16.06 Rowcrop.

GB.01.16.07 Harvesting and Conservation

GB.01.16.08 Material Handling

GB.01.16.09 Spraying

GB.01.16.10 Farm Buildings.

GB.01.16.11 Systems.

GB.01.17.00 National Insitute for Research in Dairying.
Shinfield, Reading, RG2 9AT.

GB.01.17.01 Bacteriology

GB.01.17.03 Chemistry.

GB.01.17.04 Dairy Husbandry

GB.01.17.05 Process Engineering

GB.01.17.07 Feeding and Metabolism.

GB.01.17.08 Nutrition.

GB.01.17.09 Physics.

GB.01.17.10 Physiology

GB.01.17.11 Pig Husbandry

GB.01.17.12 Biochemistry

GB.01.18.00 National Vegetable Research Station
Wellesbourne, Warwick, CV35 9EF.

GB.01.18.01 Plant Breeding.

GB.01.18.02 Chemistry.

GB.01.18.03 Biochemistry

GB.01.18.04 Plant Physiology.

GB.01.18.06 Entomology

GB.01.18.07 Plant Pathology

GB.01.18.08 Weeds.

GB.01.18.09 Statistics

GB.01.18.11 Nematology

GB.01.19.00 Plant Breeding Institute Maris Lane,
Trumpington Cambridge CB2 2LQ.

GB.01.19.01 Cereals.

GB.01.19.02 Forage and Grasses.

GB.01.19.04 Sugar Beet.

GB.01.19.05 Cytogenetics

GB.01.19.06 Pathology and Entomology.

GB.01.19.07 Physiology

GB.01.19.08 Chemistry.

GB.01.19.09 Statistics

GB.01.19.10 Forage, Oil and Potatoes

GB.01.20.00 Rothamsted Experimental Station. Harpenden,
Herts, AL5 2JQ.

GB.01.20.01 Biochemistry

GB.01.20.02 Botany

GB.01.20.03 Chemistry.

GB.01.20.04 Entomology

GB.01.20.05 Insecticides

GB.01.20.06 Nematology

GB.01.20.07 Pedology

GB.01.20.08 Physics.

GB.01.20.09 Plant Pathology

GB.01.20.10 Soil Microbiology

GB.01.20.11 Molecular Structures.

GB.01.20.14 Statistics

GB.01.21.00 Welsh Plant Breeding Station. Plas Gogerddan,
Aberystwyth, Dyfed, SY233EB.

GB.01.21.01 Arable Crop Breeding

GB.01.21.02 Chemistry.

GB.01.21.03 Cytology

GB.01.21.04 Developmental Genetics.

GB.01.21.05 Grassland Agronomy.

GB.01.21.06 Herbage Breeding.

GB.01.21.07 Plant Pathology

GB.01.21.09 Seed Multiplication and Herbage Seed
Research.

GB.01.21.10 Plant Biochemistry.

GB.01.22.00 Hop Research Department. Wye College,
University of London, Ashford, Kent

GB.01.22.01 Hop Research Department.
Delete.

GB.01.23.00 Broom's Barn Experimental Station Higham,
Bury St. Edmunds, Suffolk

GB.01.23.01 Broom's Barn Experimental Station
Delete.

GB.01.24.00 Soil Survey of England and Wales. R.E.S.,
Harpenden, Herts.

GB.01.24.01 Soil Survey of England and Wales.
Delete.

GB.01.25.00 Unit of Animal Genetics. Institute of Animal
Genetics, University of Edinburgh, West Main Road,
Edinburgh EH9 3JN.

GB.01.25.01 Group One

GB.01.25.02 Group Two

GB.01.25.04 Group Four.

GB.01.25.05 Group Five.

GB.01.26.00 Unit of Developmental Botany 181A, Huntingdon
Road, Lambridge

GB.01.26.01 Unit of Developmental Botany
181a Huntingdon Road, Cambridge, CB3 0DY

GB.01.27.00 Unit of Invertebrate Chemistry and Physiology
Department of Applied Biology, University of Cambridge,
Pembroke Street, Cambridge

GB.01.27.01 Unit of Invertebrate Chemistry Physiology.
Cambridge University of Cambridge
Delete.

GB.01.28.00 Unit of Invertebrate Chemistry and Physiology
University of Sussex, Falmer, Brighton, Sussex.

GB.01.28.01 Unit of Invertebrate Chemistry Physiology.
Brighton University of Sussex
Delete.

GB.01.29.00 Unit of Muscle Mechanisms. Department of
Zoology, University of Oxford, South Parks Road, Oxford.

GB.01.29.01 Unit of Muscle Mechanisms.
Delete.

GB.01.30.00 Unit of Nitrogen Fixation. The Chemical
Laboratory, University of Sussex, Brighton, Sussex, BN1 9QT.

GB.01.30.01 Unit of Nitrogen Fixation.
Delete.

GB.01.32.01 Unit of Soil Physics
219c Huntingdon Road, Cambridge, CB3 0DL

GB.01.33.01 Unit of Statistics. University of Edinburgh
University of Edinburgh, James Clerk Maxwell Building, The
King's Buildings Mayfield Road, Edinburgh, EH9 3JZ.

GB.01.34.00 Unit of Statistics Group, Cambridge Department
of Applied Biology, Pembroke Street, Cambridge.

GB.01.34.01 Unit of Statistics Group Cambridge.
Delete.

GB.01.35.01 Unit of Systemic Fungicides.
Wye College, Ashford, Kent, TN25 5AH

GB.02.00.00 Agricultural Research Council Grants.
160, Great Portland Street, London W1N 6DT

GB.02.01.00 University of Aberdeen.

GB.02.01.01 Agriculture.

GB.02.01.02 Chemistry.

GB.02.01.03 Soil Science.

GB.02.01.04 Biochemistry

GB.02.01.05 Botany

GB.02.02.00 Animal Health Trust.

GB.02.02.01 Small Animal Centre.

GB.02.03.00 **University of Bath.**

GB.02.03.01 Biological Science.

GB.02.03.02 School of Pharmacy and Pharmacology.

GB.02.04.00 **Queen's University of Belfast.**

GB.02.04.01 Biochemistry

GB.02.04.02 Botany

GB.02.05.00 **University of Birmingham.**

GB.02.05.01 Biochemistry

GB.02.05.02 Chemistry.

GB.02.05.03 Genetics

GB.02.05.04 Zoology and Comparative Physiology

GB.02.05.05 Virology

GB.02.05.06 Microbiology

GB.02.06.00 **University of Birmingham.**

GB.02.06.01 Anatomy.

GB.02.06.02 Animal Husbandry.

GB.02.06.03 Organic Chemistry

GB.02.06.04 Physiology

GB.02.06.05 Veterinary Medicine

GB.02.06.06 Zoology.

GB.02.07.00 **University of Cambridge**

GB.02.07.01 Anatomy.

GB.02.07.02 Applied Biology

GB.02.07.03 Biochemistry

GB.02.07.04 Botany

GB.02.07.05 Genetics

GB.02.07.06 Parasitology

GB.02.07.07 Pathology.

GB.02.07.09 Physiologcal Laboratory

GB.02.07.10 Veterinary Medicine

GB.02.07.11 Zoology.

GB.02.08.00 **University of Dundee.**

GB.02.08.01 Biological Science.

GB.02.09.00 **University of East Anglia.**

GB.02.09.01 Biological Science.

GB.02.10.00 **University of Edinburgh**

GB.02.10.01 Botany

GB.02.10.02 Veterinary Pathology.

GB.02.10.03 Veterinary Pharmacology

GB.02.10.04 Veterinary Physiology

GB.02.10.05 Animal Health

GB.02.10.07 Genetics

GB.02.11.00 **Edinburgh School of Agriculture.**

GB.02.11.01 Animal Production

GB.02.11.02 Forestry and Natural Resources

GB.02.11.03 Soil Science.

GB.02.12.00 **East of Scotland College of Agriculture**

GB.02.12.01 Agriculture.

GB.02.12.02 Agricultural Biochemistry

GB.02.12.03 Agricultural Biochemistry and Animal Nutrition.

GB.02.12.04 Animal Production

GB.02.12.05 Crop Production

GB.02.12.06 Microbiology

GB.02.12.07 Soil Science.

GB.02.12.08 Zoology.

GB.02.13.00 **Writtle Institute of Agriculture, Essex**

GB.02.13.01 Joint Study

GB.02.15.00 **University of Glasgow**

GB.02.15.01 Botany

GB.02.15.02 Botany and Chemistry.

GB.02.15.03 Parasitology

GB.02.15.04 Surgery.

GB.02.15.05 Veterinary Parasitology

GB.02.15.06 Veterinary Pathology.

GB.02.15.07 Veterinary Pharmacology

GB.02.15.08 Animal Husbandry.

GB.02.15.09   Veterinary Medicine

GB.02.15.11   Agriculture Chemistry

GB.02.16.00   Hatfield Polytechnic.

GB.02.16.01   Biological Science.

GB.02.17.00   University of Hull.

GB.02.17.01   Plant Biology

GB.02.17.02   Plant Pathology

GB.02.17.03   Biochemistry

GB.02.18.00   University of Kent at Canterbury

GB.02.18.01   Biology.

GB.02.18.02   Chemistry.

GB.02.19.00   University of Lancaster

GB.02.19.01   Biological Science.

GB.02.20.00   University of Leed.

GB.02.20.01   Animal Physiology and Nutrition.

GB.02.20.02   Biophysics

GB.02.20.03   Plant Science

GB.02.20.04   Pure and Applied Zoology

GB.02.20.05   Dental School

GB.02.21.00   University of Leicester

GB.02.21.01   Genetics

GB.02.22.00   University of Liverpool

GB.02.22.01   Avian Medicine.

GB.02.22.02   Biochemistry

GB.02.22.03   Botany

GB.02.22.04   Veterinary Clinical Studies.

GB.02.22.05   Veterinary Pathology.

GB.02.22.06   Animal Husbanry

GB.02.22.07   Veterinary Preventive Medicine

GB.02.22.08   Veterinary Anatomy.

GB.02.23.00   Bedford College

GB.02.23.01   Botany

GB.02.24.00   Charing Cross Hospital Medical School.

GB.02.24.01   Biochemistry

GB.02.25.00   Guy's Hospital Medical School

GB.02.25.01   Biochemistry and Chemistry.

GB.02.25.02   Biochemistry, Chemistry and Biochemistry
Purine Laboratory

GB.02.26.00   Imperial College of Science and Technology.

GB.02.26.01   Botany

GB.02.26.02   Geology.

GB.02.26.03   Zoology and Applied Entomology

GB.02.26.04   Electric and Electronic

GB.02.27.00   Queen Elizabeth College.

GB.02.27.01   Biology.

GB.02.27.02   Biochemistry

GB.02.27.03   Chemistry.

GB.02.27.04   Food Science.

GB.02.27.05   Physiology

GB.02.28.00   Queen Mary College

GB.02.28.01   Plant Biology and Microbiology

GB.02.28.02   Electronic Engineering.

GB.02.29.00   Royal Veterinary College

GB.02.29.01   Animal Husbandry.

GB.02.29.02   Medicine

GB.02.29.03   Pathlogy

GB.02.30.00   St. Thomas's Hospital Medical School

GB.02.30.01   Gynaecology.

GB.02.31.00   Westfield College

GB.02.31.01   Botany and Biochemistry

GB.02.32.00   Wye College

GB.02.32.01   Agriculture.

GB.02.32.02   Biological Science.

GB.02.32.03   Horticulture

GB.02.32.04   Plant Sciences.

GB.02.32.05   Poultry Reseach

GB.02.32.06   Physical Sciences

GB.02.33.00   Victoria University of Manchester.

GB.02.33.01   Botany

GB.02.33.02   Corrosion and Protection Centre.

GB.02.33.03   Cryptogamic Botany.

GB.02.34.00   National Cranfield Institute of Technology.

GB.02.34.01   Department not known

GB.02.35.00   University of Newcastle Upon Tyne

GB.02.35.01   Agriculture.

GB.02.35.02   Agricultural Biology.

GB.02.35.03   Agricultural Biochemistry

GB.02.35.04   Agricultural Engineering.

GB.02.35.05   Agricultural Zoology.

GB.02.35.06   Soil Science.

GB.02.36.00   North East London Polytechnic

GB.02.36.01   Biology.

GB.02.37.00   University of Nottingham.

GB.02.37.01   Agriculture and Horticulture.

GB.02.37.02   Applied Biochemistry and Nutrition

GB.02.37.03   Biochemistry

GB.02.37.04   Botany

GB.02.37.05   Pharmacy

GB.02.37.06   Physiology and Environment Studies

GB.02.37.07   Zoology.

GB.02.38.00   University of Oxford.

GB.02.38.01   Agricultural Science.

GB.02.38.02   Botany School

GB.02.38.03   Human Anatomy

GB.02.38.04   Forestry

GB.02.38.05   Microbiology

GB.02.39.00   University of Reading

GB.02.39.01   Agriculture and Horticulture.

GB.02.39.02   Agricultural Botany

GB.02.39.03   Botany

GB.02.39.04   Food Science.

GB.02.39.05   Microbiology

GB.02.39.06   Physiology and Biochemistry

GE.02.39.07   Soil Science.

GB.02.39.08   Zoology.

GB.02.40.00   National College of Food Technology University of Reading

GB.02.40.01   Department not known

GB.02.41.00   University of Sheffield

GB.02.41.01   Biochemistry

GB.02.41.02   Botany Botany

GB.02.41.03   Chemistry.

GB.02.41.04   Physiology

GB.02.41.05   Electrical and Electronic Engineering.

GB.02.42.00   University of Southampton

GB.02.42.01   Biology.

GB.02.42.02   Physiology and Biochemistry

GB.02.43.00   University of Stirling.

GB.02.43.01   Biochemistry

GB.02.43.02   Biology.

GB.02.44.00   University of Sussex.

GB.02.44.01   Biochemistry

GB.02.44.02   Biology.

GB.02.44.03   Molecular Science

GB.02.45.00   University College of Wales, Aberystwyth.

GB.02.45.01   Agricultural Botany

GB.02.45.02   Biochemistry and Agricultural Biochemistry

GB.02.45.03   Botany and Microbiology

GB.02.45.04   Agriculture.

GB.02.45.05   Botany, Microbiology and Agriculture

GB.02.45.06   Zoology.

GB.02.46.00   University College of North Wales, Bangor.

GB.02.46.01   Applied Zoology

GB.02.46.02 Biochemistry and Soil Science.

GB.02.46.03 School of Plant Biology.

GB.02.46.04 Zoology.

GB.02.47.00 University College, Cardiff.

GB.02.47.01 Biochemistry

GB.02.47.02 Botany

GB.02.47.03 Zoology.

GB.02.48.00 University College of Swansea.

GB.02.48.01 Botany and Microbiology Biological Science.

GB.02.48.02 Chemical Engineering.

GB.02.49.00 University of Warwick

GB.02.49.01 Biological Science.

GB.02.49.02 Molecular Science

GB.02.49.03 Engineering.

GB.02.50.00 University of York.

GB.02.50.01 Biology.

GB.02.51.00 University of St. Andrews.

GB.02.51.01 Plant Physiology.

GB.02.52.00 University of Durham.

GB.02.52.01 Botany

GB.02.53.00 Royal Free Hospital, London

GB.02.54.00 University College, London

GB.02.55.00 Open University

GB.02.55.01 Faculty of Science.

GB.02.56.00 Brewing Research Foundation.

GB.02.57.00 Polytechnic of Leicester.

GB.02.57.01 Department not known

GB.02.58.00 Chelsea College, London.

GB.02.58.01 Department not known

GB.02.59.00 Seale – Hayne Agricultural College.

GB.02.59.01 Department not known

GB.02.60.00 Surrey

GB.02.60.01 Biochemistry

GB.02.61.00 The West of Scotland Agricultural College.

GB.02.61.01 Microbiology

GB.02.62.00 University of Salford

GB.02.62.01 Biology.

GB.02.63.00 Charing Cross Hospital Medical School.

GB.02.63.01 Department not known

GB.03.00.00 Department of Agriculture and Fisheries for Scotland. Chesser House, 500 Gorgie Road, Edinburgh EH 11 3AW

GB.03.01.00 Animal Diseases Research Association. Moredun Institute, 408 Gilmerton Road, Edinburgh, EH17 7JH.

GB.03.01.01 Biochemistry

GB.03.01.02 Physical Chemistry.

GB.03.01.03 Electron Microscope

GB.03.01.04 Parasitology

GB.03.01.05 Microbiology

GB.03.01.06 Pathology.

GB.03.01.09 Directors Unit.

GB.03.01.10 Clinical Research

GB.03.01.11 Physiology

GB.03.01.12 Experimental Surgery.

GB.03.02.00 Hannah Research Institute. Ayr, Scotland KA6 5HL.

GB.03.02.01 Applied Studies

GB.03.02.02 Biochemistry

GB.03.02.03 Chemistry.

GB.03.02.04 Physiology

GB.03.03.00 Hill Farming Research Organisation. Bush Estate, Penicuik, Midlothian, EH26 OPH

GB.03.03.01 Reproduction

GB.03.03.02 Nutrition and Metabolism.

GB.03.03.03 Agronomy and Systems.

GB.03.03.04 Plants and Soils.

GB.03.03.05 Red Deer.

GB.03.04.00 Macaulay Institute for Soil Research.

Criagiebuckler, Aberdeen, AB9 2QJ.

GB.03.04.01 Pedology

GB.03.04.02 Spectrochemistry

GB.03.04.03 Soil Organic Chemistry

GB.03.04.04 Plant Physiology.

GB.03.04.05 Microbiology

GB.03.04.06 Soil Fertility.

GB.03.04.07 Statistics

GB.03.04.08 Soil Survey

GB.03.05.00 Scottish Institute of Agricultural Engineering.
Bush Estate, Penicuik, Midlothian, EH26 OPH

GB.03.05.01 Primary Cultivations.

GB.03.05.02 Farm Transport.

GB.03.05.03 Soil Structure.

GB.03.05.04 Sloping Ground: Implements and Machines

GB.03.05.05 Fertilizer Distributing Mechanisms.

GB.03.05.06 Crop Drying.

GB.03.05.07 Potato Crop Mechanisation.

GB.03.05.08 Potato Grading.

GB.03.05.09 Machinery Controls.

GB.03.05.10 Raspberry Harvesting.

GB.03.05.11 Liaison.

GB.03.05.14 Fermentation Equipment.

GB.03.05.15 Application of Physical Analysis

GB.03.06.00 Rowett Research Institute. Greenburn Road,
Bucksburn, Aberdeen, AB2 95B.

GB.03.06.01 Protein Biochemistry.

GB.03.06.02 Carbohydrate Metabolism

GB.03.06.03 Lipid Biochemistry.

GB.03.06.04 Nutritional Biochemistry.

GB.03.06.05 Microbiology

GB.03.06.06 Physiology (Calorimetry).

GB.03.06.07 Veterinary Pathology.

GB.03.06.08 Experimental Pathology.

GB.03.06.10 Chemical and Physical Analysis

GB.03.06.11 Applied Nutrition

GB.03.06.12 Physiology (Digestion).

GB.03.06.13 Directors Unit.

GB.03.07.00 Scottish Horticultural Research Institute
Invergowrie, Dundee, DD2 5DA.

GB.03.07.01 Crop Research

GB.03.07.02 Mycology

GB.03.07.03 Plant Breeding.

GB.03.07.04 Virology

GB.03.07.05 Zoology.

GB.03.08.00 Scottish Plant Breeding Station Pentlandfield,
Roslin, Midlothian, EH25 9RF

GB.03.08.03 Brassicas.

GB.03.08.04 Cereals.

GB.03.08.05 Potatoes

GB.03.08.06 Agronomy

GB.03.09.00 Agricultural Scientific Services

GB.03.09.01 Plant Varieties and Seeds Division.

GB.03.09.02 Potatoes and Plant Health Division.

GB.03.09.03 Pest Control and Pesticides Division.

GB.04.00.00 **Department of Agriculture for Northern
Ireland.**
**Dundonald House, Upper Newtownards Road,
Belfast BT4 3SB**

GB.04.01.00 Agricultural and Food Bacteriology Division

GB.04.01.01 Milk and Water.

GB.04.01.02 Poultry, Meat, Eggs and Animal Foods

GB.04.01.03 Fish and Bakery Products

GB.04.01.04 Horticultural Products.

GB.04.01.05 Soil, Animal Waste and Effluent

GB.04.02.00 Biometrics Division

GB.04.02.01 Biometrics Division

GB.04.03.00 Field Botany Research Division.

GB.04.03.01 Agronomy and Physiology

GB.04.03.02 Plant Breeding.

GB.04.03.03   Plant Testing

GB.04.04.00 **Agricultural and Food Chemistry Division.**

GB.04.04.01   Animal Nutrition.

GB.04.04.02   Soils and Plant Nutrition.

GB.04.04.03   Biochemistry

GB.04.04.04   Food Chemistry.

GB.04.05.00 **Enniskillen College of Agriculture**

GB.04.05.01   Grassland Experimental Centre.

GB.04.06.00 **Agricultural Economics and Statistics Division.**

GB.04.06.01   Agricultural Economics and Statistics Division.

GB.04.06.02   Agricultural Economics and Statistics Division.

GB.04.07.00 **Agricultural Entomology Division**

GB.04.07.01   Nematology

GB.04.07.02   Entomology

GB.04.08.00 **Fresh Water Biological Investigation Unit.**

GB.04.08.01   Freshwater Biological Investigation Unit.

GB.04.09.00 **Fisheries Research Laboratory.**

GB.04.09.01   Freshwater Section.

GB.04.09.02   Marine Section.

GB.04.10.00 **Forestry Division**

GB.04.10.01   Forestry Division

GB.04.11.00 **Greenmount Agricultural and Horticultural College**

GB.04.11.01   Experimental Section.

GB.04.11.02   Hill Farm

GB.04.12.00 **Horticultural Centre.**

GB.04.12.01   Glasshouses.

GB.04.12.02   Mushrooms.

GB.04.12.03   Nursery Stock

GB.04.12.04   Soft Fruits

GB.04.12.05   Top Fruit

GB.04.12.06   Vegetables

GB.04.13.00 **Agricultural Research Institute and Crop and Animal Husbandry Research Division**

GB.04.13.01

GB.04.13.02   Dairy.

GB.04.13.03   Beef Cattle

GB.04.13.04   Buildings.

GB.04.13.05   Poultry.

GB.04.13.06   Sheep.

GB.04.13.07   Sucklers

GB.04.13.08   Crops.

GB.04.13.09   Grass.

GB.04.14.00 **Loughry College of Agricultural and Food Technology.**

GB.04.14.01   Agriculture.

GB.04.14.02   Communications

GB.04.14.03   Agricultural Engineering.

GB.04.14.04   Food and Dairy Technology.

GB.04.14.05   Poultry.

GB.04.14.06   Agricultural Science.

GB.04.15.00 **Plant Pathology Research Division**

GB.04.15.01   Cereals and Grass

GB.04.15.02   Forestry

GB.04.15.03   Potatoes (Excluding Virus Diseases)

GB.04.15.04   Fruit and Horticulture.

GB.04.15.05   Virology

GB.04.16.00 **Veterinary Research Laboratories**

GB.04.16.01   Virology

GB.04.16.02   Pathology.

GB.04.16.03   Bacteriology

GB.04.16.04   Biochemistry

GB.04.16.05   Parasitology

GB.04.16.06   Immunology

GB.04.16.07   Mastitis

GB.04.16.08   Reproductive Physiology

GB.04.16.09   Omagh Laboratory.

GB.05.00.00 Ministry of Agriculture, Fisheries and Food
Great Westminster House, Horseferry Road, London, SW1P 2AE.

GB.05.01.00 Agricultural Development and Advisory Service

GB.05.01.01 Fruit (EHS)

GB.05.01.02 Glasshouse Products (EHS).

GB.05.01.03 Vegetables (EHS).

GB.05.01.04 Bulbs (EHS)

GB.05.01.05 Flowers (EHS)

GB.05.01.06 Hops (EHS).

GB.05.01.21 Cereals (EHF)

GB.05.01.22 Pigs (EHF).

GB.05.01.23 Beef Production (EHF).

GB.05.01.24 Dairy Cattle (EHF)

GB.05.01.25 Soil Management (EHF).

GB.05.01.26 Farm Waste (EHF)

GB.05.01.27 Arable Crops (EHF)

GB.05.01.28 Sheep Production (EHF)

GB.05.01.29 Poultry (EHF)

GB.05.01.30 Grassland and Forage Crops (EHF).

GB.05.01.41 Fruit (REC).

GB.05.01.42 Glasshouse Products (REC).

GB.05.01.43 Vegetables (REC).

GB.05.01.44 Bulbs (REC).

GB.05.01.45 Flowers (REC).

GB.05.01.46 Hops (REC).

GB.05.01.61 Cereals (REC).

GB.05.01.62 Pigs (REC).

GB.05.01.63 Beef Production (REC).

GB.05.01.64 Dairy Cattle (REC).

GB.05.01.65 Soil Management (REC).

GB.05.01.66 Farm Waste (REC).

GB.05.01.67 Arable Crops (REC).

GB.05.01.68 Sheep Production (REC).

GB.05.01.69 Poultry (REC).

GB.05.01.70 Grassland and Forage Crops (REC).

GB.05.02.00 Agricultural Scientific Services, Slough.

GB.05.02.01 Pest Control Chemistry

GB.05.02.02 Rodent Pests.

GB.05.02.03 Storage Pests

GB.05.02.04 Land Pests and Birds

GB.05.03.00 Agricultural Scientific Services, Harpenden

GB.05.03.01 Chemistry Department.

GB.05.03.09 Chemistry Department.

GB.05.03.10 Entomology (Plant) Health.

GB.05.03.11 Entomology Taxonomy.

GB.05.03.12 Entomology Pest (Assessment).

GB.05.03.13 Entomology Nematology.

GB.05.03.14 Entomology (Pesticide) Surveys.

GB.05.03.20 (Plant) Pathology Disease (Assessment).

GB.05.03.21 (Plant) Pathology Legislation.

GB.05.03.22 (Plant) Pathology Bacteriology.

GB.05.03.23 (Plant) Pathology Seed (Pathology).

GB.05.03.24 (Plant) Pathology Certification (Schemes).

GB.05.03.25 Plant Pathology.

GB.05.04.00 Veterinary Investigation Centres.

GB.05.04.04 Thirsk.

GB.05.04.11 Newcastle.

GB.05.04.22 Leeds.

GB.05.04.32 Sutton Bonington Loughborough.

GB.05.04.34 Moulton Northampton.

GB.05.04.35 Lincoln.

GB.05.04.41 Wolverhampton.

GB.05.04.43 Worcester.

GB.05.04.54 Cambridge.

GB.05.04.55 Norwich.

GB.05.04.61 Weybridge.

GB.05.04.62 Wye.

GB.05.04.63 Reading.

GB.05.04.64 Winchester.

GB.05.04.71 Starcross.

GB.05.04.73 Langford Bristol.

GB.05.04.74 Polwhele Truro.

GB.05.04.81 Bangor.

GB.05.04.83 Aberystwyth.

GB.05.04.84 Carmarthen.

GB.05.05.00 **Central Veterinary Laboratory.**

GB.05.05.01 Bacteriology.

GB.05.05.02 Biochemistry.

GB.05.05.03 Biological Products Standards.

GB.05.05.04 Diseases of Breeding Department.

GB.05.05.05 Parasitology.

GB.05.05.06 Pathology Department.

GB.05.05.07 Poultry Department.

GB.05.05.08 Virology

GB.05.05.09 Animal Production and Maintenance Department.

GB.05.05.10 Epidemiology.

GB.05.05.11 Reading Cattle Breeding Centre.

GB.05.05.12 Clinical Pathology Lasswade.

GB.05.06.00 **British Food Manufacturing Industries Research Association.**

GB.05.06.01 Analytical.

GB.05.06.02 Biochemistry.

GB.05.06.03 Chemistry.

GB.05.06.04 Confectionery.

GB.05.06.05 Fruit and Vegetables.

GB.05.06.06 Gels and Protein.

GB.05.06.07 In–Line Process Control.

GB.05.06.08 Microbiology.

GB.05.06.09 Microscopy.

GB.05.06.10 Oils and Fats.

GB.05.06.11 Food Engineering.

GB.05.06.12 Meat and Fish.

GB.05.06.19 Inter–Departmental Work.

GB.05.07.00 **British Industrial Biological Research Association.**

GB.05.07.01 British Industrial Biological Research Association, Department of.

GB.05.07.02 British Industrial Biological Research Association, Department of.

GB.05.08.00 **Campden Food Preservation Research Association.**

GB.05.08.01 Agricultural and Quality.

GB.05.08.02 Analytical Chemistry.

GB.05.08.03 Biochemistry.

GB.05.08.04 Product Investigations.

GB.05.08.05 Process Development.

GB.05.08.06 Inter–Departmental Projects.

GB.05.08.07 Inter–Departmental Projects.

GB.05.08.09 Inter–Departmental Projects.

GB.05.09.00 **Flour Milling and Baking Research Association.**

GB.05.09.01 Applied Biochemistry.

GB.05.09.02 Baking Technology.

GB.05.09.03 Fundamental and Process Control.

GB.05.09.04 Milling Technology.

GB.05.09.05 Technological Services.

GB.05.09.09 Inter–Departmental Projects.

GB.05.10.00 **Torrey Research Station.**

GB.05.10.01 Protein Chemistry.

GB.05.10.02 Flavours, Products Innovation and Pollution.

GB.05.10.03 Fish Bacteriology and Bacteriology Services.

GB.05.10.04 Biotechnology.

GB.05.10.05 Physics, Computing and Statistics.

GB.05.10.06 Engineering.

GB.05.10.07 Processing Technology and Product Quality.

GB.05.10.09   Fishery By–Products.

GB.05.10.10   Physical Chemistry, Shellfish and Quality Assessment.

GB.05.10.11   Bacterial Biochemistry.

GB.05.11.00   **Food Science Division.**

GB.05.11.01   Food Science Division.

GB.06.00.00   **Scottish Agricultural Colleges. Chesser House, 500 Gorgie Road, Edinburgh, EH11 3AW.**

GB.06.01.00   **East of Scotland College of Agriculture. The Kings Buildings, West Mains Rd, Edinburgh EH9 3DG.**

GB.06.01.01   Agricultural Biochemistry.

GB.06.01.02   Animal Production.

GB.06.01.03   Animal Nutrition.

GB.06.01.04   Beekeeping.

GB.06.01.05   Biochemistry.

GB.06.01.06   Crop Production.

GB.06.01.07   Crop Protection.

GB.06.01.08   Economics.

GB.06.01.09   Engineering and Farm Mechanisation.

GB.06.01.10   Farm Buildings.

GB.06.01.11   Farm Business Management.

GB.06.01.12   Horticulture.

GB.06.01.13   Microbiology.

GB.06.01.14   Soil Science.

GB.06.01.15   Spectrochemistry.

GB.06.01.16   Veterinary Investigation.

GB.06.01.17   Advisory.

GB.06.02.00   **North of Scotland College of Agriculture. School of Agriculture, 581 King Street, Aberdeen, AB9 1UD.**

GB.06.02.01   Craibstone.

GB.06.02.02   Balnastraid.

GB.06.02.03   Clashnoir.

GB.06.02.04   Weyland.

GB.06.02.05   Aldroughty.

GB.06.02.06   Achany.

GB.06.02.07   Tillycorthie.

GB.06.02.08   Animal Husbandry.

GB.06.02.09   Poultry Husbandry.

GB.06.02.10   Zoology.

GB.06.02.11   Veterinary Hygiene and Animal Physiology.

GB.06.02.12   Veterinary Investigation Service.

GB.06.02.13   Crop Husbandry.

GB.06.02.14   Grassland Husbandry.

GB.06.02.15   Horticulture.

GB.06.02.16   Botany.

GB.06.02.17   Plant Pathology.

GB.06.02.18   Entomology, Nematology, Beekeeping.

GB.06.02.19   Chemistry and Biochemistry.

GB.06.02.20   Bacteriology.

GB.06.02.21   Engineering.

GB.06.02.22   Farm Buildings.

GB.06.02.23   Agricultural Economics.

GB.06.02.24   Scottish Farm Buildings Investigation Unit.

GB.06.02.25   Regional Advisory Service, Aberdeen Region.

GB.06.02.26   Regional Advisory Service, Highland Region.

GB.06.03.00   **West of Scotland College of Agriculture. Auchincruive, Ayr, KA6 5HW.**

GB.06.03.01   Agriculture.

GB.06.03.02   Agricultural Engineering.

GB.06.03.03   Dairy Technology.

GB.06.03.04   Horticulture and Beekeeping.

GB.06.03.05   Poultry Husbandry.

GB.06.03.06   Agricultural Chemistry.

GB.06.03.07   Botany.

GB.06.03.08   Microbiology.

GB.06.03.09   Zoology.

GB.06.03.10   Plant Pathology.

GB.06.03.11   Agricultural Economics.

GB.06.03.12 Veterinary Medicine.

GB.06.03.13 Agronomy.

GB.06.03.14 Crop Husbandry.

GB.06.03.15 Animal Husbandry.

GB.06.03.16 West of Scotland College of Agriculture, Department of.

GB.06.03.17 Farm Buildings.

GB.06.03.18 Glasshouse Investigation Unit.

GB.06.03.19 West of Scotland College of Agriculture, Department of.

GB.06.03.20 West of Scotland College of Agriculture, Department of.

GB.06.03.22 West of Scotland College of Agriculture, Department of.

GB.06.03.24 West of Scotland College of Agriculture, Department of.

GB.06.03.25 West of Scotland College of Agriculture, Department of.

IRELAND

IRELAND

IE.01.00.00 **Department of Agriculture**
Kildare Street; Dublin 2.

IE.01.01.01 Group G – Cereal Breeding and Seed Improvement.
Backweston Farm; Leixlip; Co. Dublin.

IE.01.01.02 Cereal Station.
Ballinacurra; Midleton; Co. Cork.

IE.01.02.00 **Veterinary Research Laboratory Abbotstown;**
Co. Dublin.

IE.03.00.00 **Department of Tourism and Transport.**

IE.03.01.00 **Agricultural Meteorology Unit, Meteorological**
Service. Glasnevin Heights, Dublin 9.

IE.04.00.00 **Department of Industry, Commerce and**
Energy
Kildare Street; Dublin 2.

IE.04.01.00 **Mines and Minerals Division.**

IE.04.01.01 Geological Survey of Ireland
Hume Street; Dublin 2.

IE.05.00.00 **Department of Fisheries**
22 Upper Merrion Street; Dublin 2.

IE.05.01.00 **Forest and Wildlife Service.**

IE.06.00.00 **An Foras Taluntais**
19 Sandymount Avenue; Dublin 4.

IE.06.01.00 **An Foras Taluntais Dunsinea Research Centre**
Castleknock; Co. Dublin.

IE.06.01.03 An Foras Taluntais Grange Research Centre
Dunsany; Co. Meath.

IE.06.02.00 **An Foras Taluntais Johnstown Castle Research**
Centre. Co. Wexford.

IE.06.02.01 An Foras Taluntais Lullymore Research Centre.
Rathangan; Co. Kildare.

IE.06.03.00 **An Foras Taluntais Kinsealy Research Centre**
Malahide Road; Dublin 5.

IE.06.03.01 An Foras Taluntais Clonroche Research Centre.
Co. Wexford.

IE.06.03.02 An Foras Taluntais Ballygagin Research Centre
Dungarvan; Co. Waterford.

IE.06.04.00 **An Foras Taluntais Moorepark Research Centre.**
Fermoy; Co. Cork.

IE.06.05.00 **An Foras Taluntais Oakpark Research Centre.**

IE.06.05.02 An Foras Taluntais Nematology Laboratory
Irish Sugar Company; Co. Carlow.

IE.06.06.00 **An Foras Taluntais Economics and Rural Welfare**
Research Centre. 19 Sandymount Avenue; Dublin 4.

IE.06.07.00 **An Foras Taluntais Western Research Centre.**
Creagh; Ballinrobe; Co. Mayo.

IE.06.07.01 An Foras Taluntais Ballinamore Research
Centre.
Co. Leitrim.

IE.06.07.02 An Foras Taluntais Belclare Research Centre
Tuam; Co. Galway.

IE.06.07.06 An Foras Taluntais Glenamoy Research Centre
Co. Mayo.

IE.06.07.07 An Foras Taluntais Maam Research Centre
Co. Galway.

IE.07.00.00 **Institute for Industrial Research and**
Standards
Ballymun Road; Dublin 9.

IE.07.02.00 **Forest Products Department**

IE.07.03.00 **Environmental Technology Department.**

IE.07.04.00 **Food Technology Department**

IE.08.00.00 **An Foras Forbartha (Institute for**
Physical Planning and Construction Research).
St. Martin's House; Waterloo Road; Dublin 4.

IE.08.02.00 **Planning Division**

IE.11.00.00 **University College.**

IE.11.01.00 **Faculty of Dairy Science**

IE.11.01.01 Department of Dairy and Food Chemistry.

IE.11.01.02 Department of Dairy and Food Microbiology

IE.11.01.04 Department of Agriculture

IE.11.01.05 Department of Dairy and Food Economics.

IE.11.01.06 Department of Dairy and Food Engineering.

IE.11.02.00 **Faculty of Science.**

IE.11.02.02 Department of Botany.

IE.11.02.03 Department of Biochemistry.

IE.11.03.00 **Faculty of Arts**

IE.11.03.01 Department of Social Theory and Institutions.

IE.11.03.02 Department of Economics

IE.12.00.00 **University College.**

IE.12.01.00 **Faculty of General Agriculture**

IE.12.01.01 Department of Agricultural Biology

IE.12.01.02   Department of Agricultural Chemistry

IE.12.01.04   Department of Applied Agricultural Economics.

IE.12.01.05   Department of Agriculture
Lyons Estate; Newcastle P.O.; Co. Dublin.

IE.12.01.06   Department of Plant Pathology.

IE.12.01.07   Department of Forestry.

IE.12.01.08   Department of Horticulture.

IE.12.01.09   Department of Agricultural Extension

IE.12.02.00 **Faculty of Veterinary Medicine**

IE.12.02.01   Department of Veterinary Anatomy

IE.12.02.02   Department of Veterinary Hygiene and Animal
Husbandry.
Ballsbridge; Dublin 4.

IE.12.02.03   Department of Veterinary Biochemistry and
Physiology.

IE.12.02.04   Department of Veterinary Surgery, Obstetrics
and Radiology

IE.12.02.05   Department of Veterinary Pathology and
Microbiology.

IE.12.02.06   Department of Veterinary Preventive Medicine.

IE.12.02.07   Department of Veterinary Parasitology.

IE.12.02.08   Department of Veterinary Medicine and
Pharmacology.

IE.12.03.00 **Faculty of Engineering and Architecture. Upper
Merrion Street; Dublin 2.**

IE.12.03.01   Department of Agricultural Engineering

IE.12.04.00 **Faculty of Science.**

IE.12.04.01   Department of Botany.

IE.12.04.03   Department of Zoology

IE.12.05.00 **Faculty of Arts**

IE.12.05.01   Department of Political Economy and National
Economics of Ireland.

IE.12.05.02   Department of Geography

IE.13.00.00 **University College.**

IE.13.01.00 **Faculty of Science.**

IE.13.01.01   Department of Microbiology.

IE.13.01.02   Department of Biochemistry.

IE.13.01.03   Department of Oceanography.

IE.13.02.00 **Faculty of Commerce**

IE.13.02.01   Department of Economics

IE.14.00.00 **Trinity College (University of Dublin).
Dublin 2.**

IE.14.02.00 **Faculty of Science.**

IE.14.02.01   Department of Botany.

IE.14.02.02   Department of Zoology

IE.14.02.03   Department of Biochemistry.

IE.14.02.05   Department of Genetics.

IE.14.04.00 **Faculty of Medical and Dental Sciences.**

IE.14.04.01   Department of Community Health

IE.15.00.00 **Economic and Social Research Institute.
4 Burlington Road; Dublin 4.**

IE.18.01.01   Inland Fisheries Trust Incorporated

IE.18.01.02   Salmon Research Trust of Ireland Incorporated.

IE.18.01.11   Bord lascaigh Mhara (Sea Fisheries
Development Board)

IE.18.01.18   National Economic and Social Council.
Oisin House; Pearse Street; Dublin 2.

IE.18.01.19   Electricity Supply Board
Lower Fitzwilliam Street; Dublin 2.

ITALY

ITALY

IT.01.00.00 **Enti Autonomi (Pubblici e Privati).**

IT.01.01.00 Istituto Nazionale di Economia Agraria. Via Barberini 36; 00100 Roma.

IT.01.01.01 Osservatorio di Economia Agraria per Il Piemonte, la Valle D' Aosta e la Liguria
Via Michelangelo 32; 10126 Torino

IT.01.01.02 Osservatorio di Economia Agraria per la Lombardia
Via Celoria 2; 20133 Milano.

IT.01.01.03 Osservatorio di Economia Agraria per il Veneto il Trentino– Alto Adige ed il Friuli– Venezia Giulia
Via Gradenigo 6/A; 35100 Padova.

IT.01.01.04 Osservatorio di Economia Agraria per L' Emilia Romagna.
Via Filippo Re 10; 40126 Bologna.

IT.01.01.05 Osservatorio di Economia Agraria per la Toscana
Piazzale Delle Cascine; 50144 Firenze.

IT.01.01.06 Osservatoiro di Economia Agraria per l' Umbria e le Marche, Facoltà di Agraria S Pietro.
Borgo S. Pietro – 06100 Perugia.

IT.01.01.07 Osservatorio di Economia Agraria per il Lazio e l' Abruzzo
Via del Castro Laurenziano 9; 00161 Roma

IT.01.01.08 Osservatorio di Economia Agraria per la Campania, la Calabria e il Molise facoltà di agraria.
80055 Portici Napoli

IT.01.01.09 Osservatorio di Economia Agraria per la Puglia e la Lucania.
Via Amendola 165/A; 70126 Bari

IT.01.01.10 Osservatorio di Economia Agraria per la Sicilia
Parco Orleans; 90128 Palermo.

IT.01.01.11 Osservatorio di Economia Agraria per la Sardegna.
Via E. de Nicola; 07100 Sassari.

IT.01.01.12 Osservatorio di Economia Agraria per l' Europa.
Via Boncompagni 16; 00187 Roma.

IT.01.02.00 **Istituto di Tecnica e Propaganda Agraria** Via Caio Mario 37; 00100 Roma.

IT.01.03.00 **Fondazione Problemi Montani Arco Alpino.** 20100 Milano.

IT.01.05.00 **Osservatorio di Genetica Animale** Via Pastrengo 20, 10100 Torino.

IT.01.09.00 **Centro lattiero casearo di assistenza e sperimentazione "Antonio Bizzozero".** Via Torelli 17; 43100 Parma.

IT.01.13.00 **Istituto Nazionale Della Nutrizione** Via Lancisi, 29 – 00100 Roma.

IT.01.14.00 **Ente Nazionale Risi Centro di Ricerche** Piazza Martini delle Liberta 31.27100 Mortava Pavia.

IT.01.16.00 **Irvam Istituto per le Ricerche e le Informazioni di Mercato e la Valovizzazione della Produzione Agricola.** Via Castelfidardo 43; 00185 Roma.

IT.01.17.00 **Istituto Nazionale di Coniglicoltura** Strada Provinciale Pavia 4; 15100 Alessandria.

IT.01.18.00 **Ente Nazionale per la Cellulosa e per la Carta.** Via Regina Margherita 270; 00198 Roma.

IT.01.18.01 Centro di Sperimentazione Agricola e Forestale.
Via Casalotti 300; 00166 Roma

IT.01.18.02 Istituto di Sperimentazione per la Pioppicoltura
Casale Monferrato; 15100 Alessandria Cas. Post. 24; Tel 4654

IT.01.18.03 Centro di Sperimentazione Cartaria
Via Assisi 163; Roma.

IT.01.18.04 Centro di Sperimentazione Grafica.
Via Assisi 163; 00181 Roma.

IT.01.18.05 Laboratorio di Cartotecnica Speciale
Locate Triulzi; 20085 Milano.

IT.01.19.00 **Centro di Specializzazione e Ricerche Economico Agrarie Per il Mezzogiorno, Universita di Napoli, Facoltà di Agraria 80055 Portici Napoli**

IT.01.20.01 Accademia Economico Agraria dei Georgofili.
Logge degli Uffizi; 50100 Firenze

IT.01.20.02 Accademia Italiana di Scienze Forestali
P.zza Edison, 11; 50100 Firenze

IT.01.20.03 Accademia Nazionale di Agricoltura
Via Farini, 14; 40100 Bologna

IT.01.21.00 **Centro lombardo per l'orto–floro–frutticoltura.** Minoprio – 22070 Como.

IT.01.22.00 **Ist. Scientifico di Chimica e Biochimica. "G. Ronzoni."** Via C. Colombo, 81 – 20090 Milano.

IT.01.23.00 **Ist. Ricerche Farmacologiche – "Mario Negri.".** Via Eritrea, 62 – 20090 Milano.

IT.01.24.00 **Associazione Italiana Allevatori.** Via Tomassetti, 9. 00100 Roma

IT.01.25.00 **Istituto Internazionale di Studi Ligure.** 17024 Borgo Finale Ligure – (Savona)

IT.01.26.00 **Accademia Italiana della vite e del vino.** Via Volta dei Mercanti, 1; FIRENZE.

IT.01.27.00 **Istituto Nazionale di Sociologia Rurale.** Via Boncompagni, 8; 00187 ROMA.

IT.01.28.00 **CSATA– Centro Studi ed Applicazioni delle**

ITALY

Tecnologie Avanzate Via Amendola, 173 – 70100 Bari.

IT.02.00.00 Ministero dell' Agricoltura e delle Foreste.

IT.02.01.00 Istituto Sper. per lo Studio e la Difesa del Suolo P. Zza M. D' Azeglio; 50121 Firenze

IT.02.02.00 Istituto Sper. per la Nutrizione delle Piante Via Della Navicella 2; 00184 Roma.

IT.02.03.00 Istituto Sper. per la Patologia Vegetale. Via Casal de' Pazzi 250; 00156 Roma

IT.02.04.00 Istituto Sper. per la Zoologia Agraria. Via Lanciola – Cascine del Riccio ; 50125 Firenze.

IT.02.05.00 Istituto Sper. Agronomico. Via C. Ulpiani, 70125 Bari.

IT.02.06.00 Istituto Sper. per la Meccanizzaione Agricola Via Xx Settembre 98/ E; 00187 Roma.

IT.02.07.00 Istituto Sper. per la Zootecnia. Via O. Panvinio 11; 00162 Roma.

IT.02.08.00 Istituto Sper. per la Cerealicoltura Via Cassia 176; 00191 Roma.

IT.02.09.00 Istituto Sper. per le Colture Foraggere V. le Piacenza 25; 20075 Lodi (Milano).

IT.02.10.00 Istituto Sper. per l'Orticoltura Via F.Conforti 11; 84100 Salerno

IT.02.11.00 Istituto Sper. per le Colture Industriali Via Corticella 133 40129 Bologna.

IT.02.12.00 Istituto Sper. per la Floricoltura Corso Degli Inglesi 362; 18038 Sanremo (Imperia).

IT.02.13.00 Istituto Sper. per la Viticoltura. Via xxviii Aprile 1; 31015 Conegliano Veneto (Treviso).

IT.02.14.00 Istituto Sper. per la Olivicoltura Viale Delle Medaglie d'Oro 74; 87100 Cosenza.

IT.02.15.00 Istituto Sper. per la Frutticoltura. Via Fioranello, 52 Ciampino Aeroporto 00040 Roma.

IT.02.16.00 Istituto Sper. per L' Agrumicoltura Corso Savoia 166; 95024 Acireale Catania

IT.02.17.00 Istituto Sper. per la Selvicoltura Viale S. Margherita 80/82 52100 Arezzo.

IT.02.18.00 Istituto Sper. per L' Assestamento forestale e l' Alpilcoltura Villa S. Carlo–Villazzano; 38050 Trento.

IT.02.19.00 Istituto Sper. per la Valorizzazione Tecnologica dei Prodotti Agricoli. Via Venezian 26 20133 Milano.

IT.02.20.00 Istituto Sper. per l' Enologia Via Micca 35; 14100 Asti.

IT.02.21.00 Istituto Sper. per l' Elaiotecnica Via C. Battisti

198; 65100 Pescara.

IT.02.22.00 Istituto Sper. Lattiero– Caseario Via C. Besana 8; 20075 Lodi (Milano).

IT.02.23.00 Istituto Sperimentale per il Tabacco Via Vitiello 66; 84018 Scafati (Salerno).

IT.02.24.00 Laboratorio Centrale di Idrobiologia Via Brisse 27; 00149 Roma

IT.02.25.00 Ufficio Centrale di Ecologia Agraria. Via Del Caravita 7; 00186 Roma

IT.02.26.00 Stabilimento Ittiogenico. Largo Torrelunga 7; 25100 Brescia

IT.02.27.00 Stabilimento Ittiogenico. Via della Stazione Tiburtina 11; 00182 Roma.

IT.02.28.00 Istituto Nazionale di Biologia e della Selvaggina. Via Stradelli Guelfi 23/A; 40064 Ozzano Emilia Bologna.

IT.02.29.00 Istituto Sperimentale Talassografico Via R Gessi 2; 34100 Trieste

IT.02.30.00 Istituto Sperimentale Talassografico Via Roma 3; 74100 Taranto

IT.02.31.00 Istituto Sperimentale Talassografico Spianata S Ranieri;98100 Messina

IT.02.32.00 Osservatorio per le Malattie delle Piante. Via Martinez 8/ B; 95024 Acireale Catania.

IT.02.33.00 Osservatorio per le Malattie delle Piante. Lungomare N. Sauro Palazzo Dell' Agricoltura; 70100 Bari.

IT.02.34.00 Osservatorio per le Malattie delle Piante. Sezione Entomologia Piazza della Costituzione 8; 40128 Bologna.

IT.02.35.00 Osservatorio per le Malattie delle Piante. Piazza Costituzione 8; 40126 Bologna.

IT.02.36.00 Osservatorio per le Malattie delle Piante. Corso Liberto Palazzo Plossliner; 39100 Bolzano

IT.02.37.00 Osservatorio Regionale per le Malattie delle Piante Viale Trento 50; 09100 Cagliari.

IT.02.38.00 Osservatorio per le Malattie delle Piante. Via Dei Normanni; 88100 Catanzaro.

IT.02.39.00 Osservatorio per le Malattie delle Piante. Via Bolognesi 163; 50139 Firenze.

IT.02.40.00 Osservatorio per le malattie delle piante. Via Nino Bixio 6; 16128 Genova.

IT.02.41.00 Osservatorio per le Malattie delle Piante. Corso Italia 25; 34170 Gorizia.

IT.02.42.00 Osservatorio per le malattie delle piante. Via Moretto da Brescia 7; Milano.

IT.02.44.00 Osservatorio per le Malattie delle Piante. Via G.

ITALY

Cavedoni 10; 41100 Modena

IT.02.45.00 Osservatorio per le Malattie delle Piante. Via
Nicolò Garzilli – 80128 Palermo

IT.02.46.00 Osservatorio per le Malattie delle Piante. Via S.
Epifanio 4; 27100 Pavia

IT.02.47.00 Osservatorio per le Malattie delle Piante, Sezione
Entomologia Via Solatia 1/1; 06100 Perugia.

IT.02.48.00 Osservatorio per le Malattie delle Piante, Sezione
Patologia Vegetale Via Assisana 31; 06100 Perugia.

IT.02.49.00 Osservatorio per le Malattie delle piante Viale
Riviera 81; 61100 Pescara

IT.02.50.00 Osservatorio per le malattie delle piante. Via
Reggia di Portici 69; 80146 Napoli

IT.02.52.00 Osservatorio per le malattie delle piante. Viale G.
Pisano 18; 56100 Pisa

IT.02.54.00 Osservatorio per le Malattie delle Piante. Via
Tevere 5/ B; 00198 Roma.

IT.02.55.00 Osservatorio per le Malattie delle Piante
Osservatorio per le Malattie delle Piante. C.so Cavallotti 51;
18038 Sanremo Imperia.

IT.02.56.00 Osservatorio per le Malattie delle Piante. Via S.
Secondo 39; 10128 Torino.

IT.02.57.00 Osservatorio per le Malattie delle Piante. Via G.
Murat 1; 34100 Trieste.

IT.02.58.00 Osservatorio per le Malattie delle Piante.
Lungadige dei Capuleti 1; 37100 Verona.

IT.02.59.00 Osservatorio per le malattie delle piante. Via
Gazzoletti; TRENTO.

IT.03.00.00 Ministero degli Affari Esteri.

IT.03.01.00 Istituto Agronomico per l'oltremare. V. Cocchi 4;
50131 Firenze.

IT.03.02.00 Istituto Agronomico Mediterrano. C. P. 135;
70100 Bari

IT.04.00.00 Ministero della Pubblica Istruzione.

IT.04.01.00 Universitata' Degli Studi di Bari – Facolta di
Agraria.

IT.04.01.01 Istituto di Agronomia e Coltivazioni Erbacee.
V. Amendola. 165/ A; 70100 Bari.

IT.04.01.02 Istituto di Economia e Politica Agraria.
V. Amendola. 165/ A; 70100 Bari.

IT.04.01.03 Istituto di Meccanica Agraria.
Via Amendola 165/A; 70100 Bari

IT.04.01.04 Istituto di Chimica Agraria.
V. Amendola 165/ A; 70100 Bari

IT.04.01.05 Istituto di Zootecnia
V. Amendola 165/ A; 70100 Bari

IT.04.01.06 Istituto di Patologia Vegetale
V. Amendola 165/ A; 70100 Bari

IT.04.01.07 Istituto di Industrie Agrarie.
V. Amendola 165/ A; 70100 Bari

IT.04.01.08 Istituto di Estimo Rurale e Contabilita'
V. Amendola 165/ A; 70100 Bari

IT.04.01.09 Istituto di Microbiologia Agraria.
V. Amendola 165/ A; 70100 Bari

IT.04.01.10 Istituto di Costruzioni Rurali
Via Amendola 165/A; 70100 Bari

IT.04.01.11 Istituto d Anatomia e Fisiologia degli Animali
Domestici
V. Amendola 165/ A; 70100 Bari

IT.04.01.12 Istituto di Coltivazioni Arboree
V. Amendola 165/ A; 70100 Bari

IT.04.01.13 Istituto di Entomologia Agraria.
V. Amendola 165/ A; 70100 Bari

IT.04.01.15 Istituto di Miglioramento Genetico delle Piante
Agrarie.
Via Amendola 165/A, 70100 Bari

IT.04.01.16 Istituto di Silvicoltura.
70100 Bari.

IT.04.01.17 Istituto di Legislazione Forestale
Via Amendola 165/a; 70100 Bari.

IT.04.01.18 Istituto di Economia e Finanza

IT.04.01.19 Isituto di Botanica, Facolta di Scienze
Viale Japigia – 70100 Bari.

IT.04.01.20 Cattedra di Istologia, Facolta di Veterinaria
Viale Japigia – 70100 Bari.

IT.04.01.21 Cattedra di Zoologia, Facolta di Veterinaria.
Viale Japigia – 70100 Bari.

IT.04.01.22 Cattedra di Chimica, Facoltà di Veterinaria.
Viale Japigia – 70100 Bari.

IT.04.01.23 Cattedra di Anatomia Veterinaria, Facoltà Di
Veterinaria
Viale Japigia – 70100 Bari.

IT.04.01.24 Cattedra Alimentazione e Nutrizione Animale.
Via Japigia 88/1 – 70100 Bari

IT.04.02.00 Universita' Degli Studi di Bologna – Facolta' di
Agraria.

IT.04.02.01 Istituto Agronomia Generale e Coltivazioni
Erbacee.
V. Filippo Re 6; 40100 Bologna

ITALY

IT.04.02.02 Istituto di Chimica Agraria.
V. S. Giacomo 7; 40100 Bologna

IT.04.02.03 Istituto di Coltivazioni Arboree.
V. Filippo Re 6; 40100 Bologna

IT.04.02.04 Istituto di Economia e Politica Agraria.
V. Filippo Re 10; 40100 Bologna.

IT.04.02.05 Istituto di Entomologia Agraria.
V. Filippo Re 6; 40100 Bologna

IT.04.02.06 Istituto di Estimo Rurale e Contabilita'.
V. Filippo Re 10; 40100 Bologna.

IT.04.02.07 Istituto diIdraulica Agraria
V.Filippo Re 4; 40100 Bologna.

IT.04.02.08 Istituto di Industrie Agrarie.
V.S.Giacomo 7; 40100 Bologna

IT.04.02.09 Istituto di Meccanica Agraria.
V. Filippo Re 4; 40100 Bologna

IT.04.02.10 Istituto di Microbiologia Agraria e Tecnica
V. Filippo Re 6; 40100 Bologna

IT.04.02.11 Istituto di Patologia Vegetale
V.Filippo Re 8; 40100 Bologna.

IT.04.02.12 Istituto de Zoocolture.
V.S.Giacomo 9; 40100 Bologna

IT.04.02.13 Istituto di Costruzioni Rurali e Topografia
V. Filippo Re 4; 40100 Bologna

IT.04.02.14 Istituto di Miglioramento Genetico delle Piante
Agrarie.
V. Filippo Re 6; 40100 Bologna

IT.04.02.15 Scuola di Specializzazione in Frutticoltura
Industriale
40100 Bologna

IT.04.02.16 Centro Sperimentale trasform.conserv.
ortofrutticoli.
Via Filippo Re 8; 40100 Bologna.

IT.04.02.17 Istituto di Tecnologie Chimiche Speciali.
Viale Risorgimento 4; 41100 Bologna

IT.04.02.18 Istituto di Alimentazione Animale.
40100 Bologna

IT.04.02.19 Centro sperimentale Cadriano.
41100 Bologna

IT.04.02.20 Istituto Genio Rurale.
Via Filippo Re, 6 – 40100 Bologna.

IT.04.02.21 Istituto di Mineralogia e Geologia
40100 Bologna

IT.04.02.22 Istituto di Orticoltura e Floricoltura
40100 Bologna

IT.04.02.23 Istituto di Allevamenti Zootecnici
40100 Bologna

IT.04.02.24 Istituto di Zooeconomia
40100 Bologna

IT.04.02.25 Scuola di Specializzazione in Fitopatologia.
Via Filippo Re 6; 40100 Bologna.

IT.04.02.26 Istituto di Botanica, Facoltà di Scienze
Via Irnerio, 42 – 40100 Bologna

IT.04.02.27 Istituto di Anatomia Animali Domestici con
Istol. Emb. Biochimica, Facoltà di Veterinaria
Via Belmeloro 8; 40100 Bologna.

IT.04.02.28 Istituto di Fisiologia Generale e Speciale
Animali Domestici e Chimica Biologica, Facoltà di Medicina
Veterinaria
40100 Bologna

IT.04.02.29 Istituto di Malattie Infettive Profil. e Polizia
Veterinaria, Facoltà di Medicina Veterinaria
Via S. Giacomo 9; 40100 Bologna.

IT.04.02.30 Istituto di Patologia Generale e Anatomia
Patologica, Facoltà di Medicina Veterinaria
Via Belmeloro 10; 40100 Bologna

IT.04.02.31 Istituto di Biochimica, Facoltà di Medicina
Veterinaria
Via Belmeloro 8, 40100 Bologna.

IT.04.02.32 Istituto di Patologia Speciale e Clinica Chirurgica
Veterinaria – Facoltà di Medicina Veterinaria
Viale Filopanti 9; 40100 Bologna.

IT.04.02.33 Istituto di Patologia e Clinica Medica, Facoltà di
Medicina Istituto di Patologia e Clinica Medica, Facolta di
Medicina Veterinaria.
Viale Filopanti 5; 40100 Bologna.

IT.04.02.34 Istituto di Patologia Aviaria, Facoltà di Medicina
Veterinaria
Via Belmeloro 10; 40100 Bologna

IT.04.02.35 Istituto di Ostetricia e Ginecologia Veterinaria –
Facoltà di Medicina Veterinaria.
Viale Filopanti 7; 40100 Bologna.

IT.04.02.36 Istituto di zootecnia e nutrizione
animale–Facoltà di medicina veterinaria.
Via S. Giacomo 11; 40126 Bologna.

IT.04.02.37 Istituto di Farmacologia Veterinaria, Facoltà di
Medicina Veterinaria.
Strada Maggiore 45; 40100 Bologna

IT.04.02.38 Istituto di Selvicoltura.
Facoltà di Agraria; 40100 Bologna

IT.04.02.39 Istituto di Zoologia.
40100 Bologna

IT.04.02.40 Istituto di Microbiologia e Industrie Agrarie

Facoltà di Agraria; Reggio Emilia.

IT.04.02.41 Istituto di Edilizia Zootecnica.
Via F.lli Rosselli, 107 – 42100 Caviolo (Reggio Emilia).

IT.04.02.42 Ist. Produzioni Foraggere.
Via F. Crispi, 3 – 42100 Reggio Emilia.

IT.04.02.43 Ist. Chimica – G. Ciamician. Fac. Chimica.
Via Scheni, 2 – 10012 Bologna.

IT.04.02.44 Ist. Chimica Organica.
Via Risorgimento, 4 – 10012 Bologna.

IT.04.02.45 Istituto Allevamenti Zootecnici.
42100 Reggio Emilia.

IT.04.02.46 Istituto di Farmacologia.
10012 Bologna

IT.04.02.47 Centro Fitofarmaci, c/o Ist. Patologia Vegetale.
Via Filippo Re, 8 – 10012 Bologna.

IT.04.02.48 Istituto di Geologia e Paleontologia.
Via Zamboni, 63/67 40100 Bologna.

IT.04.03.00 Università degli Studi di Catania–Facoltà di Agraria. Via Valdisavoia , 5 – 95100 Catania

IT.04.03.01 Istituto di Entomologia Agraria.
Via Valdisavoia, 5 – 95100 Catania.

IT.04.03.02 Istituto di Patologia Vegetale
V. Valdisavoia 1; 95100 Catania

IT.04.03.03 Istituto di Zootecnica Generale.
V. Valdisavoia 5; 95100 Catania

IT.04.03.04 Istituto di Chimica Agraria.
V. Valdisavoia 5; 95100 Catania

IT.04.03.05 Istituto di Agronomia Generale e Coltivazioni
Erbacee.
V. Valdisavoia 5; 95100 Catania

IT.04.03.06 Istituto di Economia e Politica Agraria
V. Valdisavoia 5; 95100 Catania

IT.04.03.07 Istituto di Coltivazioni Arboree
Via Valdisavoia 5 95100 Catania.

IT.04.03.08 Istituto di Industrie Agrarie.
V. Valdisavoia 5; 95100 Catania

IT.04.03.09 Istituto di Topografia e Costruzioni Rurali
V. Valdisavoia 1; 95100 Catania

IT.04.03.10 Azienda Agraria Didattico Sperimentale.
Pantano d' Arci; 84— Catania.

IT.04.03.11 istituto di Idraulica Agraria
V. Valdisavoia 1; 95100 Catania

IT.04.03.12 Istituto di Meccanica Agraria.
V. Valdisavoia 5; 95100 Catania

IT.04.03.13 Istituto di Orticoltura e Floricoltura
V. Valdisavoia 5; 95100 Catania

IT.04.03.14 Istituto di Estimo Rurale e Contabilita'.
V. Valdisavoia 5; 95100 Catania

IT.04.03.15 Istituto di Produzione Animale
95100 Catania

IT.04.03.16 Istituto di Microbiologia
95100 Catania

IT.04.03.17 Istituto di Botanica Facoltà di Scienze.
95100 Catania

IT.04.03.18 Istituto di Topografia Facoltà di Scienze.
95100 Catania

IT.04.03.19 Istituto di Zootecnia Facoltà di Scienze
95100 Catania

IT.04.03.20 Policattedra di Biologia Animale.
Via Androne, 81 – 95124 Catania.

IT.04.04.00 Universita' Cattolica del S. Cuore Facolta' Agraria

IT.04.04.01 Istituto di Agronomia Generale e Coltivazioni
Erbacee.
V. Emilia Parmense 84; 29100 Piacenza.

IT.04.04.02 Istituto di Entomologia Agraria.
V. Emilia Parmense 84; 29100 Piacenza.

IT.04.04.03 Istituto di Coltivazioni Arboree
V. Emilia Parmense 84; 29100 Piacenza

IT.04.04.04 Istituto di Patologia Vegetale
V. Emilia Parmense 84; 29100 Piacenza.

IT.04.04.05 Istituto di Botanica e Genetica Vegetale.
Via E. Parmense, 84 – 29100 Piacenza

IT.04.04.06 Istituto di Chimica
V. Emilia Parmense 84; 29100 Piacenza.

IT.04.04.07 Istituto di Industrie Agrarie.
V. Emilia Parmense 84; 29100 Piacenza.

IT.04.04.08 Istituto di Microbiologia e Tecnica.
V. Emilia Parmense 84; 29100 Piacenza.

IT.04.04.09 Istituto di Microbioligia Lattiero– Casearia.
V. Emilia Parmense 84; 29100 Piacenza.

IT.04.04.10 Istituto di Economia e Politica Agraria Estimo
Rurale e Contabilita'.
V. Emilia Parmense 84; 29100 Piacenza.

IT.04.04.11 Istituto di Zootecnica Generale.
V. Emilia Parmense 84; 29100 Piacenza.

IT.04.04.12 Istituto della Scienza della Nutrizione.
V. Emilia Parmense 84; 29100 Piacenza.

IT.04.04.13 Istituto di Meccanica Agraria.
V. Emilia Parmense 84; 29100 Piacenza.

IT.04.04.14 Istituto di Topografia e Costruzioni Rurali
V. Emilia Parmense 84; 29100 Piacenza.

IT.04.04.15 Istituto di Genetica Vegetale.
V. Emilia Parmense 84; 20100 Piacenza.

IT.04.05.00 Universita' degli Studi di Firenze, Facoltà di Agraria.

IT.04.05.01 Istituto di Agricoltura Montana e Alpicoltura
P. Zzale delle Cascine 18; 50100 Firenze

IT.04.05.02 Istituto di Argronomia Generale e Coltivazioni
Erbacee
P. Zzale delle Cascine 18; 50100 Firenze

IT.04.05.03 Istituto di Anatomia degli Animali Domestici.
P. Zzale delle Cascine 28; 50100 Firenze

IT.04.05.04 Istituto di Assestamento Forestale
P. Zzale delle Cascine 18; 50100 Firenze

IT.04.05.05 Istituto di Botanica Agraria e Forestale.
P. Zzale delle Cascine 18; 50100 Firenze

IT.04.05.06 Istituto di Chimica Agraria e Forestale
P. Zzale delle Cascine 18; 50100 Firenze

IT.04.05.07 Istituto di Coltivazioni Arboree
P. Zzale delle Cascine 18; 50100 Firenze

IT.04.05.08 Istituto di Economia e Estimo Forestale
P. Zzale delle Cascine 18; 50100 Firenze

IT.04.05.09 Istituto di Economia e Politica Agraria
P. Zzale delle Cascine 18; 50100 Firenze

IT.04.05.10 Istituto di Geologia Applicata
P. Zzale delle Cascine 18; 50100 Firenze

IT.04.05.11 Istituto di Industrie Agrarie.
P. Zzale delle Cascine 18; 50100 Firenze

IT.04.05.12 Istituto di Meccanica Agraria.
P. Zzale delle Cascine 18; 50100 Firenze

IT.04.05.13 Istituto di Microbiologia Agraria e Tecnica
P.zzale delle Cascine 27 50100 Firenze.

IT.04.05.14 Istituto di Mineralogia e Geologia Agraria.
P. Zzale delle Cascine 18; 50100 Firenze

IT.04.05.15 Istituto di Patologia Agraria e Forestale
P. Zzale delle Cascine 28; 50100 Firenze

IT.04.05.16 Istituto di Selvicoltura e Assestamento
Forestale.
P. Zzale delle Cascine 28; 50100 Firenze

IT.04.05.17 Istituto di Sistemazioni Idraulico Forestali.
P. Zzale delle Cascine 18; 50100 Firenze

IT.04.05.18 Istituto di Zoognostica
P. Zzale delle Cascine 18; 50100 Firenze

IT.04.05.19 Istituto di Zoologia Forestale
P. Zzale delle Cascine 18; 50100 Firenze

IT.04.05.20 Istituto di Zootecnica.
P. Zzale delle Cascine 18; 50100 Firenze

IT.04.05.22 Istituto di Esercitazioni Pratiche in Campagna.
50100 Firenze

IT.04.05.23 Istituto di Zootecnica Speciale.
50100 Firenze

IT.04.05.25 Istituto di Idronomia Montana e Idraulica
50100 Firenze

IT.04.05.26 Istituto di Estimo rurale e Contabilità.
P.le delle Cascine; 50100 Firenze

IT.04.05.27 Ist. Tecnologie e Utilizzazioni Forestali.
Via S. Bonaventura – 50145 Firenze.

IT.04.05.28 Ist. Anatomia Comparata. Fac Scienze Mat. Fis.
e Nat.
Via Romana, 17 – 50125 Firenze.

IT.04.05.29 Istituto di Costruzioni Rurali e Forestali.
Via delle Cascine; FIRENZE.

IT.04.06.00 Università degli Studi di Milano –Facoltà di Agraria–. Via Celoria, 2 – 20100 Milano

IT.04.06.01 Istituto di Agronomia e Coltivazioni Erbacee.
Via Celoria, 2 20100 Milano.

IT.04.06.02 Istituto di Anatomia e Fisiologia degli Animali
Domestici.
V. Celoria 2; 20100 Milano.

IT.04.06.03 Istituto di Chimica Agraria.
V. Celoria 2; 20100 Milano.

IT.04.06.04 Istituto di Chimica Organica
V. Celoria 2; 20100 Milano.

IT.04.06.05 Istituto di Coltivazioni Arboree
V. Celoria 2; 20100 Milano.

IT.04.06.06 Istituto di Economia e Politica Agraria
V. Celoria 2; 20100 Milano.

IT.04.06.07 Istituto di Entomologia Agraria.
V. Celoria 2; 20100 Milano.

IT.04.06.08 Istituto di Idraulica Agraria.
V. Celoria 2; 20100 Milano.

IT.04.06.09 Istituto di Industrie Agrarie (Enologia Caseificio
Oleificio)
V. Celoria 2; 20100 Milano.

IT.04.06.10 Istituto di Ingegneria Agraria.
V. Celoria 2; 20100 Milano.

IT.04.06.11 Istituto di Microbiologia Agraria e Tecnica
V. Celoria 2; 20100 Milano.

IT.04.06.12 Istituto di Patologia Vegetale
V. Celoria 2; 20100 Milano.

IT.04.06.13 Istituto di Zootecnia Generale
V. Celoria 2; 20100 Milano.

IT.04.06.14 Istituto di Zoognostica
20100 Milano.

IT.04.06.15 Istituto di Genetica.
20100 Milano.

IT.04.06.16 Istituto di Microbiologia lattiero– casearia.
20100 Milano.

IT.04.06.17 Istituto di Tecnologie Alimentari.
20100 Milano.

IT.04.06.18 Istituto di Scienze Botaniche.
20100 Milano.

IT.04.06.19 Istituto di Agronomia Chimica e Agronomia Zootecnica
20100 Milano.

IT.04.06.20 Istituto di Biochimica Generale.
Via Celoria 2, 20133 Milano

IT.04.06.21 Azienda della Facoltà di Agraria di Milano con sede in Pavia.

IT.04.06.22 Istituto di Microbiologia Industriale.
20100 Milano.

IT.04.06.23 Istituto di Mineralogia e Geologia (Agraria).
20100 Milano.

IT.04.06.25 Istituto di Topografia e Costruzioni Rurali
20100 Milano.

IT.04.06.26 Istituto di Anatomia degli Aninali Domestici
Facoltà di Veterinaria.
Via Celoria 10; 20100 Milano.

IT.04.06.27 Istituto di Anatomia Patologica
Veterinaria,Facolta di Veterinaria.
Via Celoria 10; 20100 Milano.

IT.04.06.28 Istituto di Approvigionamenti Annonari,Facolta di Veterinaria.
20100 Milano.

IT.04.06.29 Istituto di Clinica Chirurgica Veterinaria,Facolta di Veterinaria
20100Milano

IT.04.06.30 Istituto di Farmacologia Veterinaria,Facoltà di Veterinaria
Via Celoria 10,20100 Milano

IT.04.06.31 Istituto di Fisiologia degli Animali
Domestici,Facoltà di Veterinaria.
Via Celoria 10; 20100 Milano.

IT.04.06.33 Istituto di Ispezione degli Alimenti di Origine Animale, Facoltà di Veterinaria

Via Celoria 10; 20100 Milano.

IT.04.06.34 Istituto di Medicina Legale Veterinaria, Facoltà di Veterinaria.
20100 Milano.

IT.04.06.35 Istituto di Microbiologia e Immunologia Veterinaria Facoltà di Veterinaria
20100 Milano.

IT.04.06.36 Istituto di Clinica Medica Veterinaria, Facoltà di Veterinaria
20100 Milano.

IT.04.06.37 Istituto di Malattie Infettive, Facoltà di Veterinaria.
Via Celoria 10; 20100 Milano.

IT.04.06.38 Istituto di Zootecnia Generale – Falcoltà di Veterinaria.
Via Celoria 10; 20100 Milano.

IT.04.06.39 Istituto di Tecnica Conserviera degli Alimenti di Origine Animale – Facoltà di Veterinaria
20100 Milano.

IT.04.06.40 Istituto di Radiologia Veterinaria, Facolta di Veterinaria
20100 Milano.

IT.04.06.41 Istituto Policattedra di Patologia Generale Veterinaria, Facoltà di Veterinaria.
20100 Milano.

IT.04.06.42 Istituto di Patologia Aviare, Facoltà di Veterinaria.
Via Celoria 10, 20100 Milano.

IT.04.06.43 Istituto di Parassitologia,Facoltà di Veterinaria.
Via Celoria 10; 20100 Milano.

IT.04.06.44 Istituto di Clinica Ostetrica Vetenaria, Facoltà di Veterinaria.
Via Celoria 10; 20100 Milano.

IT.04.06.45 Ist. Agronomia Generale e Coltivazioni Erbacee.
Via Celoria, 2 – 20133 Milano.

IT.04.06.46 Istituto di Alimentazione Animale.
Via Celoria, 2 20100 Milano

IT.04.06.47 Istituto di Tossicologia.
Via Vanvitelli, 32; MILANO.

IT.04.07.00 Universita' degli Studi di Napoli – Facolta' di Agraria.

IT.04.07.01 Istituto di Agronomia Generale e Coltivationi Erbacee.
V. Dell' Universita' 100; 80055 Napoli Portici.

IT.04.07.02 Istituto di Coltivazioni Arboree
V. Dell' Università ' 100; 80055 Napoli Portici.

IT.04.07.03 Istituto di Entomologia Agraria.

V. Dell ' Università ' 100; 80055 Napoli Partici

IT.04.07.04   Istituto di Botanica Generale.
V. Dell ' Universita ' 100; 80055 Napoli Portici.

IT.04.07.05   Istituto di Chimica Agraria.
V. Dell ' Universita ' 100; 80055 Napoli Portici.

IT.04.07.06   Istituto di Economia e Politica Agraria
V. Dell ' Universita ' 100;80055 Napoli Partici

IT.04.07.07   Istituto di Patologia Vegetale.
V. Dell ' Universita ' 100; 80055 Napoli Portici.

IT.04.07.08   Istituto di Estimo Rurale e Contabilita '
V. Dell ' Universita ' 100; 80055 Napoli Portici.

IT.04.07.09   Istituto di Zootecnica Generale.
V. Dell ' Universita ' 100 ; 80055 Napoli Portici

IT.04.07.10   Istituto di Idraulica Agraria Topografia e
Costruzioni Rurali
V. Parco Gussone; 80055 Napoli Portici.

IT.04.07.11   Istituto di Industrie Agrarie.
V. Parco Gussone; 80055 Napoli Portici.

IT.04.07.13   Istituto di Microbiologia Agraria.
V. Dell ' Universita ' 100; 80055 Napoli Portici,

IT.04.07.14   Istituto di Microbiologia Lattiero–Casearea
V. Dell 'Universita ' 100; 80055 Napoli Portici

IT.04.07.17   Istituto di Economia Agraria
V. Dell ' Universita ' 100; 80055 Napoli Portici.

IT.04.07.19   Istituto di Malattie delle Piante.
V. Liberta' 183; 80055 Napoli Portici.

IT.04.07.20   Istituto di Mineralogia e Geologia Agraria.
V. Dell 'Universita ' 100; 80055 Napoli Portici

IT.04.07.22   Istituto Di Produzione Animale.
V. Dell 'Universita' 100; 80055 Napoli Portici.

IT.04.07.23   Stazione di Microbiologia Industriale.
Parco Gussone – 80055 Portici Napoli

IT.04.07.24   Cattedra di Zoognostica
80055 Napoli Portici

IT.04.07.25   Istituto di fisica e meccanica agraria.
80055 Napoli Portici

IT.04.07.26   Istituto di produzione animale.
80055 PORTIEI (Napoli).

IT.04.07.27   Istituto di Mineralogia
80055 Napoli Portici

IT.04.07.28   Istituto di Anatomia Animali Domestici.
80055 Napoli Portici

IT.04.07.29   Istituto di Diritto Agrario.
80055 Napoli Portici

IT.04.07.30   Istituto di Fisiopatologia Vegetale.
80055 Napoli Portici

IT.04.07.32   Istituto di Economia e Politica Agraria.
Via Partenope 36; 80100 Napoli.

IT.04.07.33   Istituto di Anatomia Normale, Facoltà di
Veterinaria.
80100 Napoli.

IT.04.07.34   Istituto di Malattie Infettive, Facolta di
Veterinaria
80100 Napoli.

IT.04.07.35   Istituto di Anatomia Patologica, ;facolta di
Veterinaria.
80100 Napoli.

IT.04.07.36   Istituto di Chirurgia Veterinaria, Facolta di
Veterinaria.
80100 Napoli.

IT.04.07.37   Istituto di Clinica Medica Facolta di Veterinaria.
80100 Napoli.

IT.04.07.38   Istituto di Zootecnica Generale, Facolta di
Veterinaria.
80100 Napoli.

IT.04.07.39   Istituto di Botanica, Facolta di Veterinaria.
Via Federico Delpino 1, 80137 Napoli

IT.04.07.40   Istituto di Ispezioni Alimenti, Facolta di
Veterinaria
80100 Napoli.

IT.04.07.41   Istituto di Istologia ed Embriologia, Facolta di
Veterinaria.
80100 Napoli.

IT.04.07.42   Istituto di Zoognostica, Facolta di Veterinaria
80100 Napoli.

IT.04.07.43   Istituto di Patologia Aviare, Facolta di
Veterinaria
80100 Napoli.

IT.04.07.44   Istituto di Biologia Generale e Genetica, Facolta
di Scienze.
80100 Napoli.

IT.04.07.45   Istituto diChimica Biologica
Facoltà di Medicina e Chirurgia; 80100 Napoli

IT.04.07.46   Ist. Chimica Organica. Fac. Scienze Naturali.
Via Mezzocannone, 16 – 80134 Napoli.

IT.04.08.00 Universita ' degli Studi di Padova – Facolta di
Agraria.

IT.04.08.01   Istituto di Agronomia Generale e Coltivazioni
Erbacee.
V. Gradenigo 6; 35100 Padova.

IT.04.08.02   Istituto di Chimica Agraria ed Industrie Agrarie
V. Gradenigo 6; 35100 Padova.

IT.04.08.03 Istituto di Coltivazioni Arboree
v. Gradenigo 6; 35100 Padova

IT.04.08.04 Istituto di Economia e Politica Afraria
V. Gradenigo 6; 35100 Padova.

IT.04.08.05 Istituto di Entomologia Agraria.
V. Gradenigo 6; 35100 Padova.

IT.04.08.06 Istituto di Estimo Rurale e Contabilita'.
V. Gradenigo 6; 35100 Padova.

IT.04.08.07 Istituto di Meccanica Agraria.
V. Gradenigo 6; 35100 Padova.

IT.04.08.08 Istituto di Patologia Vegetale
V. Gradenigo 6; 35100 Padova.

IT.04.08.09 Istituto di Selvicoltura.
V. Gradenigo 6; 35100 Padova.

IT.04.08.10 Istituto di Zootecnica.
V. Gradenigo 6; 35100 Padova.

IT.04.08.12 Azienda Agraria Sperimentale
Via Gradenigo 6, 35100 Padova.

IT.04.08.13 Istituto di Microbiologia Agraria.
35100 Padova.

IT.04.08.14 Istituto di Economia e Politica Agraria, Facolta
di Economia e Commercio.
Via dell' Artigliere 8; 35100 Padova.

IT.04.08.15 Istituto di Botanica e Fisiologia Vegetale.
Via Gradenigo – 35100 PADOVA

**IT.04.09.00 Universita' degli Studi di Palermo – Facolta' di
Agraria.**

IT.04.09.01 Istituto di Chimica Agraria.
V. Le delle Scienze; 90100 Palermo.

IT.04.09.02 Istituto di Botanica Generale.
V. Archirafi; 90100 Palermo.

IT.04.09.03 Istituto di Entomologia Agraria.
V. Le delle Scienze; 90100 Palermo.

IT.04.09.04 Istituto di Agronomia Generale e Coltivazioni
Erbacee.
V. Le delle; Scienze 90100 Palermo

IT.04.09.05 Istituto di Economia e Politica Agraria.
V. Le delle Scienze; 90100 Palermo

IT.04.09.06 Istituto di Microbiologia Agraria.
V. Le delle Scienze; 90100 Palermo

IT.04.09.07 Istituto di Zootecnica Generale.
V. Le delle Scienze; 90100 Palermo

IT.04.09.08 Istituto di Coltivazioni Arboree
V. Le delle Scienze; 90100 Palermo

IT.04.09.09 Istituto di Estimo Rurale e Contabilita'.
V. Le delle Scienze; 90100 Palermo

IT.04.09.10 Istituto di Industrie Agrarie.
V. Le delle Scienze; 90100 Palermo.

IT.04.09.11 Istituto di Principi di Economia Politica e
Statistica
V. Le delle Scienze; 90100 Palermo

IT.04.09.12 Istituto di Economia e Politica Agraria
V. Le delle Scienze; 90100 Palermo

IT.04.09.13 Istituto di Meccanica Agraria con Applicazioni
di Disegno.
V. Le delle Scienze; 90100 Palermo

IT.04.09.14 Istituto di Orticoltura e Floricoltura
90100 Palermo

IT.04.09.15 Istituto di Idraulica Agraria.
90100 Palermo

**IT.04.10.00 Universita ' degli Studi di Perugia–Facolta di
Agraria.**

IT.04.10.01 Istituto di Agronomia generale e Coltivazioni
Erbacee
V. San Pietro; 06100 Perugia

IT.04.10.02 Istituto di Allevamento Vegetale
Borgo XX Giugno; 06100 Perugia.

IT.04.10.03 Istituto di Botanica Generale.
V. San Pietro; 06100 Perugia

IT.04.10.04 Istituto di Chimica Agraria.
V. San Pietro; 06100 Perugia

IT.04.10.05 Istituto di Coltivazioni Arboree
V. San Pietro; 06100 Perugia

IT.04.10.06 Istituto di Ecologia Agraria
V. San Pietro; 06100 Perugia

IT.04.10.07 Istituto di Economia e Politica Agraria
V. San Pietro; 06100 Perugia

IT.04.10.08 Istituto di Entomologia Agraria.
V. San Pietro; 06100 Perugia

IT.04.10.09 Istituto di Estimo Rurale e Contabilita'.
V. San Pietro; 06100 Perugia

IT.04.10.10 Istituto di Idraulica Agraria.
V. San Pietro; 06100 Perugia

IT.04.10.11 Istituto di Industrie Agrarie.
V. San Pietro; 06100 Perugia

IT.04.10.12 Istituto di Meccanica Agraria.
V. San Pietro; 06100 Perugia

IT.04.10.13 Istituto di Microbiologia Agraria e Tecnica
V. San Pietro; 06100 Perugia

IT.04.10.14 Istituto di Mineralogia e Geologia
Borgo XX Giugno 74 06100 Perugia. Borgo XX Giugno; 06100
Perugia.

IT.04.10.15 Istituto di Patologia Vegetale
V.SanPietro; 06100 Perugia

IT.04.10.16 Istituto di Topografia e Costruzione Rurali
Borgo XX Giugno 74 06100 Perugia.

IT.04.10.17 Istituto di Zoocolture.
V.San Pietro; 06100 Perugia.

IT.04.10.18 Istituto di Zootecnica Generale e Speciale.
V.San Pietro; 06100 Perugia.

IT.04.10.19 Cattedra di Orticoltura e Frutticoltura.
06100 Perugia

IT.04.10.20 Cattedra di Zoognostica
06100 Perugia

IT.04.10.21 Cattedra di Zoologia.
06100 Perugia

IT.04.10.22 Cattedra di Genetica.
06100 Perugia

IT.04.10.23 Istituto di Microbiologia Lattiero–casearia.
Via San Costanzo; 06100 Perugia.

IT.04.10.24 Cattedra di Coltivazioni Tropicali e Subtropicali
06100 Perugia

IT.04.10.25 Istituto di Fisiopatologia.
06100 Perugia

IT.04.10.26 Istituto di Micologia
Via S. Pietro 06100 Perugia.

IT.04.10.27 Ist. Scienze dell'Alimentazione, Fac. Farmacia.
Via S. Costanzo – 06100 Perugia.

IT.04.10.28 Istituto di Fisiologia Veterinaria e Chimica
Biologica.
Via S. Costanzo, 4 06100 Perugia

IT.04.10.29 Istituto Malattie Infettive e Polizia Veterinaria.
Via S. Costanzo, 4 – 06100 Perugia

IT.04.11.00 Universita' degli Studi di Pisa – Facolta' di
Agraria.

IT.04.11.01 Istituto di Agronomia
V.S.Michele degli Scalzi 4; 56100 Pisa.

IT.04.11.02 Istituto di Anatomia e Fisiologia degli Animali
Domestici.
V.Del Borghetto 80; 56100 Pisa

IT.04.11.03 Istituto di Chimica Agraria.
Via S. Michele degli Scalzi, 2 – 56100 Pisa

IT.04.I1.04 Istituto di Coltivazioni Arboree
V. del Borghetto 80; 56100 Pisa

IT.04.11.05 Istituto di Economia Agraria ed Estimo.
V. del Borghetto 80; 56100 Pisa

IT.04.11.06 Istituto di Entomologia Agraria.
Via S. Michele degli Scalzi 4 56100 Pisa.

IT.04.11.07 Istituto di Idraulica Agraria.
V. del Borghetto 80; 56100 Pisa

IT.04.11.08 Istituto di Industrie Agrarie.
V.S.Michele degli Scalzi 2; 56100 Pisa.

IT.04.11.09 Istituto di Mineralogia e Geologia
V. del Borghetto 80; 56100 Pisa

IT.04.11.10 Istituto di Patologia Vegetale e Microbiologia
Agraria
V.S.Michele degli Scalzi 6; 56100 Pisa.

IT.04.11.11 Istituto di Topografia e Costruzioni Rurali
Via Del Borghetto 80 56100 Pisa.

IT.04.11.12 Istituto di Meccanica Agraria.
Via del Borghetto, 80.

IT.04.11.13 Istituto di Zootecnica Speciale.
V.Del Borghetto 80; 56100 Pisa

IT.04.11.14 Istituto di Genetica.
V. del Borghetto 80; 56100 Pisa

IT.04.11.15 Istituto di Orticoltura e Floricoltura
V.Le B.Buozzi 29; 56100 Pisa.

IT.04.11.16 Istituto di Zoocolture.
V.S.Picro A Grado; 56100 Pisa.

IT.04.11.17 Podere Sperimentale
V.Matteotti 1; 56100 Pisa

IT.04.11.18 Istituto di Zootecnica Generale.
56100 Pisa.

IT.04.11.19 Istituto di Microbiologia Agraria e Tecnica.
Via del Borghetto 80 ; 56100 Pisa.

IT.04.11.20 Istituto di Anatomia ed Istologia Animali
Domestici, Facolta di Medicina Veterinaria.
Viale B. Buozzi 14; 56100 Pisa

IT.04.11.21 Istituto di Fisiologia Generale e Chimica
Biologica, Facolta' di Medicina Veterinaria.
Viale Buozzi 14; 56100 Pisa

IT.04.11.22 Istituto di Parassitologia, Facolta' di Medicina
Veterinaria.
viale B Buozzi 14; 56100 Pisa

IT.04.11.23 Istituto di Patologia Generale e Anatomia
Patologica, Facolta' di Medicina Veterinaria
Viale delle Piagge 2 ; 56100 Pisa.

IT.04.11.24 Istituto di Malattie Infettive, Facolta' di
Medicina Veterinaria
Viale B Buozzi 14; 56100 Pisa.

IT.04.11.25   Istituto di Patologia Speciale e Clinica Chirurgica
Facolta' di Medicina Veterinaria
Viale B Buozzi 14; 56100 Pisa.

IT.04.11.26   Istituto di Patologia Speciale e Clinica Medica
Facolta' di Medicina Veterinaria
Viale B Buozzi 14; 56100 Pisa.

IT.04.11.27   Istituto di Zootecnica e Zoognostica, Facolta' di
Medicina Veterinaria.
Viale delle Pioggie 2; 56100 Pisa.

IT.04.11.28   Ist. Zoologia, Lab. di Biochimica.
Via delle Piagge, 2 – 56100 Pisa.

IT.04.11.29   Scuola Superiore di studi universitari e
perfezionamento.
Via Carducci, 40; PISA.

IT.04.11.30   Istituto di Chimica Organica.
Via Risorgimento, 35; PISA.

IT.04.11.31   Istituto di Botanica.
Via Luca Ghini, 5; PISA.

**IT.04.12.00   Universita' degli Studi di Torino – Facolta' di
Agraria.**

IT.04.12.01   Istituto di Agronomia Generale e Coltivazioni
Erbacee.
V.Michelangelo 32; 10100 Torino

IT.04.12.02   Istituto di Chimica Agraria.
V.P.Giuria 15; 10100 Torino.

IT.04.12.03   Istituto di Coltivazioni Arboree
V.P.Giuria 15; 10100 Torino.

IT.04.12.04   Istituto di Apicoltura e Bachicoltura.
V.Oremea 99; 10100 Torino

IT.04.12.05   Istituto di Economia e Politica Agraria
V.Michelangelo 32; 10100 Torino

IT.04.12.06   Istituto di Entomologia Agraria e Apicoltura.
V.P.Giuria 15; 10100 Torino.

IT.04.12.07   Istituto di Idraulica Agraria.
Corso Raffaello 8; 10100 Torino

IT.04.12.09   Istituto di Industrie Agrarie.
V.P.Giuria 15; 10100 Torino.

IT.04.12.10   Istituto di Meccanica Agraria.
V.Michelangelo 32; 10100 Torino

IT.04.12.11   Istituto di Microbiologia Agraria e Tecnica
V.Pietro Giuria 15; 10100 Torino

IT.04.12.12   Istituto di Produzione delle Sementi.
V.P.Giuria 15; 10100 Torino.

IT.04.12.14   Istituto di Orticoltura e Floricoltura
V.P.Giuria 15; 10100 Torino.

IT.04.12.15   Istituto di Patologia Vegetale

V.P.Giura 15; 10100 Torino

IT.04.12.16   Istituto di Zootecnica Generale.
V.Genova 6; 10100 Torino.

IT.04.12.17   Istituto di Zootecnica Speciale.
Via Valperga Caluso 21 ; 10125 Torino.

IT.04.12.18   Istituto di frutticoltura industriale.
Via Ormea 99; 10126 Torino.

IT.04.12.19   Istituto di Botanica Generale.
10100 Torino.

IT.04.12.21   Istituto di Anatomia Animali Domestici,
Facolta' di Medicina Veterinaria.
Via Nizza 52; 10100 Torino.

IT.04.12.22   Istituto di Patologia Generale ed Anatomia
Patologica, Facolta' di Medicina Veterinaria.
Via Nizza 52; 10100 Torino.

IT.04.12.23   Istituto di Patologia Aviare, Facolta' di Medicina
Veterinaria
Via Nizza 52; 10100 Torino.

IT.04.12.24   istituto delle Malattie Infettive Profilassi e
Polizia Veterinaria, Facolta' di Medicina Veterinaria
Via Nizza 52; 10100 Torino.

IT.04.12.25   Istituto Patologia Speciale e Clinica Chirurgica,
Facolta' di Medicina Veterinaria
Via Nizza 52; 10100 Torino.

IT.04.12.26   Istituto Patologia Speciale e Clinica Medica,
Facolta' di Medicina Veterinaria
Via Nizza 52; 10100 Torino.

IT.04.12.27   Istituto di Zootecnica Generale Veterinaria,
Facolta' di Medicina Veterinaria
Via Nizza 52; 10100 Torino.

IT.04.12.28   Istituto di Zootecnica Speciale, Facolta' di
Medicina Veterinaria.
Via Nizza 52; 10100 Torino.

IT.04.12.29   Istituto di Clinica Ostetricia Veterinaria,
Facolta' di Medicina Veterinaria.
Via Nizza 52; 10100 Torino.

IT.04.12.30   Istituto di Microscopio Elettronico, Facolta' di
Medicina Veterinaria.
Via Nizza 52; 10100 Torino.

IT.04.12.31   Istituto di Fisiologia e Chimica Biologica,
Facolta' di Medicina Veterinaria.
Via Campana 16; 10126 Torino.

IT.04.12.32   Istituto di Istologia ed Embriologia, Facolta' di
Medicina Veterinaria
Via Nizza 52; 10100 Torino.

IT.04.12.33   Istituto di Botanica Speciale Veterinaria,
Facolta' di Medicina Veterinaria
Via Mattioli 25; 10126 Torino

ITALY

IT.04.12.34 Ist. Chimica Organica. Fac. Scienze Mat. Fis. e Nat.
Via Bidone, 36 – 10100 Torino

IT.04.12.35 Istituto Produzione delle Sementi.
10100 Torino.

IT.04.12.36 Istituto di Ispezione degli alimenti di origine animale.
Via Nizza, 52 10100 Torino.

IT.04.13.00 Università degli Studi di Sassari–Facoltà di Agraria. V.Enrico De Nicola; 07100 Sassari

IT.04.13.01 Istituto di Entomologie Agraria.
Via E. De Nicole; 07100 SASSARI.

IT.04.13.02 Istituto di Agronomia Generale Coltivazioni
V.Enrico De Nicola; 07100 Sassari.

IT.04.13.03 Istituto di Economia Agraria Estimo e Contabilita'
V. Enrico de Nicola; 07100 Sassari

IT.04.13.04 Istituto di Topografia e Costruzioni Rurali
V. Enrico de Nicola; 07100 Sassari

IT.04.13.05 Istituto di Chimica Agraria.
V. Enrico de Nicola; 07100 Sassari

IT.04.13.06 Istituto di Meccanica Agraria.
V. Enrico de Nicola; 07100 Sassari

IT.04.13.07 Istituto di Patologia Vegetale
V. Enrico de Nicola; 07100 Sassari

IT.04.13.08 Istituto di Idraulica Agraria.
V. Enrico de Nicola; 07100 Sassari

IT.04.13.09 Istituto di Industrie Agrarie.
Viale Italia ; 07100 Sassari.

IT.04.13.10 Istituto di Coltivazioni Arboree
V. Enrico de Nicola; 07100 Sassari

IT.04.13.11 Istituto di Mineralogia e Geologia
V. Enrico de Nicola; 07100 Sassari

IT.04.13.12 Istituto di Microbiologia Agraria e Tecnica
V. Enrico de Nicola; 07100 Sassari

IT.04.13.13 Istituto di Zootecnica.
V. Enrico de Nicola; 07100 Sassari

IT.04.13.14 Istituto Entomologia Agraria.
Via E. De Nicola – 07100 Sassari.

IT.04.13.15 Istituto di Fisiologia Generale e Speciale degli Animali domestici.
Via Vienna, 2 07100 Sassari

IT.04.14.00 Università degli Studi di Siena, Facolta' di Scienze

IT.04.14.01 Istituto di Zoologia.
Via P. A. Mattioli 4; 53100 Siena

IT.04.14.02 Istituto di Botanica.
Via p.A.Mattioli 4;53100 Siena

IT.04.15.00 Universita' degli Studi di Parma, Facolta' di Medicina Veterinaria.

IT.04.15.01 Istituto di Zootecnica Generale.
Quartiere del Cornocchio; 43100 Parma

IT.04.15.02 Istituto di Economia Rurale e di Zooeconomia.
Quartiere del Cornocchio; 43100 Parma

IT.04.15.03 Istituto di Anatomia Animali Domestici.
Quartiere del Cornocchio; 43100 Parma

IT.04.15.04 Istituto di Clinica Medica Veterinaria.
Quartiere del Cornocchio; 43100 Parma

IT.04.15.05 Istituto di Malattie Infettive Veterinarie.
Quartiere del Cornocchio – 43100 Parma.

IT.04.15.06 Istituto di Biochimica Veterinaria
Quartiere del Cornocchio; 43100 Parma

IT.04.15.07 Istituto di Microbiologia Veterinaria.
Quartiere del Cornocchio; 43100 Parma

IT.04.15.08 Istituto di Anatomia Patologica Veterinaria
Quartiere del Cornocchio; 43100 Parma

IT.04.15.09 Istituto di Clinica Ostetrica e Ginecologia Veterinaria.
Via del Taglio, 43100 Parma.

IT.04.15.10 Istituto di Clinica Chirurgica Veterinaria.
Quartiere del Cornocchio; 43100 Parma

IT.04.15.11 Istituto di Ispezione Alimenti Origine Animale
Quartiere del Cornocchio; 43100 Parma

IT.04.15.12 Cattedra di Radiologia Veterinaria
Quartiere del Cornocchio; 43100 Parma

IT.04.15.13 Cattedra di Botanica Veterinaria
Quartiere del Cornocchio; 43100 Parma

IT.04.15.14 Istituto di Patologia Generale e Anatomia Patologica
Via Carissimi 11; 43100 Parma

IT.04.15.15 Istituto di Patologia Speciale e Clinica Medica Veterinaria.
Quartiere del Cornocchio; 43100 Parma

IT.04.15.16 Istituto Policattedra di Ostetricia e Ginecologia e Patologia Speciale e Clinica Chirurgica Veterinaria.
Via Carissimi 11; 43100 Parma

IT.04.15.17 Istituto di Ricerche Economico Agrarie e Forestali, Facolta' di Economia e Commercio
Via Kennedy; 43100 Parma.

IT.04.15.18 Orto Botanico – Facoltà di Scienze
Via Ferini 70; 43100 Parma.

IT.04.15.19 Istituto di Zoologia, Facolta di Scienze.

Via Universita 12; 43100 Parma.

IT.04.15.20   Istituto di Genetica, Facolta di Scienze.
borgo Carissimi 11; 43100 Parma.

IT.04.15.21   Cattedra di Alimentazione e Nutrizione
Facolta' di Medicina Veterinaria; 43100 Parma.

**IT.04.16.00 Universita' degli studi di Roma –Facolta'di
Economia e Comm. Istituto Di Economia e Politica agraria.
00100 Roma.**

IT.04.16.02   Istituto di Botanica, Facolta' di Scienze.
Città Universitaria; 00100 Roma.

IT.04.16.03   Istituto di Allevamento Vegetale, Facoltà di
Scienze.
00100 Roma.

IT.04.16.04   Istituto di Genetica, Facoltà di Scienze
Citta Universitaria; 00100 Roma

IT.04.16.05   Istituto di Biochimica, Facoltà di Scienze
00100 Roma.

IT.04.16.06   Istituto di Zoologia, Facoltà di Scienze
00100 Roma.

IT.04.16.07   Orto Botanico – Facoltà di Scienze
Città Universitaria – 00100 Roma

IT.04.16.08   Istituto di Istologia ed Embriologia, Facoltà di
Scienze.
00100 Roma.

IT.04.16.09   Ist. Chimica Biologica. Fac. Scienze Mat. Fis. e
Nat.
Piazzale Aldo Moro – 00100 Roma.

**IT.04.17.00 Universita' degli Studi di Messina, Facoltà di
Medicina Veterinaria.**

IT.04.17.01   Istituto di Anatomia Animali Domestici.
Via S. Cecilia 30; 98100 Messina

IT.04.17.03   Istituto di Patologia Generale e Anatomia
Patologica.
Via S.Cecilia 30; 98100 Messina.

IT.04.17.04   Istituto delle Malattie Infettive e Polizia
Veterinaria.
Via S. Cecilia 30; 98100 Messina

IT.04.17.05   Istituto di Ostetricia e Ginecologia Veterinaria.
Via S. Cecilia 30; 98100 Messina

IT.04.17.06   Istituto Patologia Speciale e Clinica Medica
Veterinaria.
Via S. Cecilia 30; 98100 Messina

IT.04.17.07   Istituto di Patologia Speciale e Clinica Chirurgica
Veterinaria
Via S. Cecilia 30; 98100 Messina

IT.04.17.08   istituto di Fisiologia Generale
Via S. Cecilia 30; 98100 Messina.

IT.04.17.09   Istituto Ispezione Alimenti Origine Animale.
Via S. Cecilia 30; 98100 Messina

IT.04.17.10   Istituto di Zootecnica.
Via S. Cecilia 30; 98100 Messina

IT.04.17.11   Istituto di Biochimica.
Via S. Cecilia 30; 98100 Messina

IT.04.17.12   Istituto di Anatomia Topografica e Clinica
Operativa.
Via S. Cecilia 30; 98100 Messina

IT.04.17.13   Istituto di Medicina Operativa Veterinaria.
via S. Cecilia 30; 98100 Messina

IT.04.17.14   Ist. Idrobilogia e Pescicoltura. Fac. Scienze Mat.
Fis. e Nat.
Via Dei Verdi, 75 – 98100 Messina.

IT.04.17.15   Istituto di Chimica Organica.
Via Dei Verdi; MESSINA.

**IT.04.18.00 Universita' degli Studi di Pavia – Facolta' di
Economia e Comm.**

IT.04.18.01   Istituto di Economia e Politica Agraria
Strada Nuova 65; 27100 Pavia.

IT.04.18.02   Istituto di Zoologia.
Facoltà di Scienze Matematiche e Fisiche; 27100 Pavia

IT.04.18.03   Centro Micologia Medica. Fac. Medicina.
27100 Pavia.

IT.04.18.04   Istituto di Chimica Organica.
Via Bassi, 4 27100 Pavia.

IT.04.18.05   Istituto Mineralogia, Petrografia e Geochimica.
Via Bassi, 4 27100 Pavia.

**IT.04.19.00 Universita' degli Studi di Venezia – Facolta' di
Economia e Comm.**

IT.04.19.01   Laboratorio di Economia e Politica Agraria.
Via Cà Foscari 30100 Venezia.

**IT.04.20.00 Università degli Studi di Ferrara, Facoltà di
Scienze.**

IT.04.20.01   Istituto di Chimica
44100 Ferrara

IT.04.20.02   Ist. di Chimica Farmaceutica e Tossicologica.
Fac. Farmacia
Via Scandiana, 21 – 44100 Ferrara

IT.04.20.03   Istituto di Zoologia.
44100 Ferrara

**IT.04.21.00 Ministero della Pubblica Istruzione (Istituti
Extra–Universitari)**

IT.04.21.01   Istituto Nazionale di Apicoltura
Via S.Giacomo 9; 40126 Bologna

IT.04.21.02 Erbario Tropicale di Firenze
Via Lamarmora 4; 50121 Firenze.

IT.04.21.03 Giardino Coloniale Borzi
Via Archirafi 38; 90123 Palermo

IT.04.21.04 Stazione Zoologica e Acquario.
Villa Comunale; 80125 Napoli.

IT.04.21.05 Istituto Nazionale di Entomologia.
Via Catone 34; 00192 Roma

IT.04.22.00 Università Libera Dell'Aquila. 67100 L'Aquila.

IT.04.22.01 Ist. di Botanica.
Piazza Annunziata, 1 – 67100 L'Aquila.

IT.04.23.00 Università Studi di Modena. Via Campi, 183 –
41100 Modena.

IT.04.23.01 Ist. Chimica Organica, Catted. di Chimica
Sostanze Coloranti.
Via Campi, 183 – 41100 Modena.

IT.04.24.00 Università degli Studi di Cagliari. V.le Università
– 09100 Cagliari

IT.04.24.01 Istituto di Geologia, Paleontologia e Geografia.
Località Sà Duchessa – 09100 Cagliari.

IT.04.25.00 Università degli Studi di Trieste. V.le Europa
34100 Trieste

IT.04.25.01 Istituto di Geologia e Paleontologia.
Viale Europa, 1 34127 Trieste

IT.05.00.00 Ministero della Sanita.

IT.05.01.00 Istituto Sperimentale Italiano "Lazzaro
Spallanzani" per la Fecondazione Artificiale Via Monte
Ortigara 35; 20137 Milano.

IT.05.02.00 Istituto Sperimentale per l'Igiene ed I1 Controllo
Veterinario della pesca. Via Paolucci; 65100 Pescara

IT.05.03.00 Istituto Zooprofilattico Sperimentale della
Lombardia e dell'Emilia. Via Cremona 282; 25100 Brescia.

IT.05.04.00 Istituto Zooprofilattico Sperimentale delle Venezie
Via Orus 2; 35100 Padova.

IT.05.05.00 Istituto Zooprofilattico Sperimentale del Piemonte
e della Liguria Via Bologna 148; 10154 Torino

IT.05.06.00 Istituto Zooprofilattico Sperimentale dell'Umbria
e delle Marche Viale G. Salvemini 1 ; 06100 Perugia.

IT.05.07.00 Istituto Zooprofilattico Sperimentale del Lazio e
della Toscana. Via Appia Nuova; 1411 Capannelle Roma

IT.05.08.00 Istituto Zooprofilattico Sperimentale dell' Abruzzo
Via Campo Boario – Teramo; 64100 Teramo

IT.05.09.00 Istituto Zooprofilattico Sperimentale della Puglia.
Via Manfredonia – Foggia; 71100 Foggia

IT.05.10.00 Istituto Zooprofilattico Sperimentale del
Mezzogiorno Via Salute 3; 80055 Napoli Portici

IT.05.11.00 Istituto Zooprofilattico Sperimentale della
Sardegna. Via Duca degli Abruzzi 8; 07100 Sassari.

IT.05.12.00 Istituto Zooprofilattico Sperimentale della Sicilia
Via Generale Turba, 2; 90100 Palermo.

IT.05.13.00 Istituto Superiore di Sanità. V.le Regina Elena,
299; 00100 Roma

IT.06.00.00 Consiglio Nazionale delle Ricerche
Piazzale Aldo Moro, 7 – 00185 Roma Tel 49931
pref. 06

IT.06.01.00 Istituto di Fitovirologia Applicata. Tel. 011/
341017 – 346654

IT.06.02.00 Istituto di Radiobiochimica ed Ecofisiologia
Vegetale Via Salaria Km 29 300 – Cas. Post. 10 – 00016
Monterotondo – Staz. Roma Tel. 06/9005123

IT.06.03.00 Istituto per 10 Studio delle Biosintesi Vegetali nelle
Piante d'interesse Agrarie Tel. 02/292170 – 230985

IT.06.04.00 Istituto per la Chimica del Terreno. Tel.
050/48337.

IT.06.05.00 Istituto del Germoplasma. Via Amendola, 165/A –
70126 Bari Tel. 080/5834 63 –4.

IT.06.06.00 Istituto di Nematologia Agraria Applicata ai
Vegetali. Tel. 080/583377

IT.06.07.00 Istituto per 10 Studio dei Problemi Agronomici
dell'Irrigazione nel Mezzogiorno Via dell'Argine, 1085– 80147
Ponticelli – Napoli Tel. 081/7563557

IT.06.08.00 Istituto di Ricerche sull'Adattamento dei Bovini e
dei Bufalini all'Ambiente del Mezzogiorno Via dell'Argine,
1085 – 80147 Ponticelli – Napoli Tel. 081/753557

IT.06.09.00 Centro di Studio per il Miglioramento Genetico
degli Agrumi. C/O Istituto Coltivazioni Arboree Univ; Viale
Delle Scienze; 903128 Palermo.

IT.06.10.00 Centro di Studio Sulle Rilevazioni Contabili
Aziendali. C/O Istituto di Estimo Rurale e Contabilita
Universita; Via Filippo Re 10; 40126 Bologna.

IT.06.11.00 Centro di Studio per la Patologia delle Specie
Legnose Montane. Piazzale delle Cascine, 28 –50144 Firenze

IT.06.12.00 Centro di Studio per i Colloidi del Suolo C/O
Istituto Chimica Agraria e Forestale Universita; Piazzale delle
Cascine 28; 50144 Firenze.

IT.06.13.00 Centro di Studio Sui Microorganismi Autotrofi.
C/O Istituto Microbiologia Agraria e Tecnica Università;
Piazzale delle Cascine 27; 50144 Firenze

IT.06.14.00 Centro di Studio Sulla Micologia del Terreno C/O
Istituto ed Orto Botanico Università; Viale P Mattioli 25; 10125
Torino

IT.06.15.00 Centro di Studio per la Propagazione dell'Olivo. C/O Istituto Coltivazioni Arboree Università; San Pietro; 06100 Perugia.

IT.06.16.00 Centro di Studio Sulla Chimica degli Antiparassitari C/O Istituto di Chimica Agraria Univ; Borgo XX Giugno; 06100 Perugia.

IT.06.17.00 Centro di Studio Sulle Tossine e Parassiti Sistemici dei Vegetali. C/O Istituto Patologia Vegetale Univ; via Amendola 165/ A. 701126 Bari.

IT.06.18.00 Centro di Studi per gli Antiparassitari. C/O Istituto Patologia Vegetale Univ; Via Filippo Re 8; 40126 Bologna.

IT.06.19.00 Centro di Studio per la Microbiologia del Suolo. C/O Istituto Patologia Vegetale e Microbiologia Generale e Agraria; Via Del Borghetto 80; 56100 Pisa

IT.06.20.00 Centro di Studio per Il Miglioramento Genetico delle Piante Foraggere. C/O Istituto Allevamento Vegetale Univ; Borgo XX Giugno; 06100 Perugia.

IT.06.21.00 Centro di Studio per il Miglioramento Genetico della Vite. C/O Istituto Coltivazioni Arboree–Univ; Via Pietro Giuria 15; 10126 Torino

IT.06.22.00 Centro di Studio per le Malattie della Barbabietola da Zucchero. C/O Istituto Patologia Vegetale–Univ; Via Gradenigo 6; 35100 Padova

IT.06.23.00 Centro di Studio per la Genesi Classificazione e Cartografia del Suolo C/O Istituto Geologia Applicata Fac Agraria–Università; Piazzale delle Cascine; 50144 Firenze.

IT.06.24.00 Centro di Studio per la Conservazione dei Foraggi C/O Istituto Agronomia Generale e Coltivazioni Erbacee–Univ; Via Filippo Re 8; 40126 Bologna.

IT.06.25.00 Centro di Studio per L'Alimentazione degli Animali in Produzione Zootecnica. C/O Istituto Zootecnica Generale–Univ; Via Nizza 52; 10126 Torino

IT.06.26.00 Centro per lo Studio Tecnologico Bromatologico e Microbiologico del Latte C/O Istituto Industrie Agrarie Univ; Via Celoria 2; 20100 Milano.

IT.06.27.00 Centro di studio sull' Orticoltura Industriale C/O Istituto agronomia–Università;Via Amendola 165/A; 70126 Bari

IT.06.28.00 Centro di Studio sulla Propagazione delle Specie Legnose C/O Istituto Coltivazioni Arboree–Univ; Via Donizzetti 6; 50100 Firenze

IT.06.29.00 Centro per lo Studio dei Diserbanti. C/O Istituto Agronomia Generale–Univ; Via Gradenigo 6; 35100 Padova

IT.06.30.00 Centro di Studio per la Tecnica Frutticola. C/O Istituto Coltivazioni Arboree–Univ; Via Filippo Re 6; 40126 Bologna.

IT.06.31.00 Centro di Studio sulle Colture Precoci Ortive in Sicilia. C/O Istituto Agronomia Generale e Coltivazioni Erbacee – Università; Via Valdisavoia 5; 95123 Catania.

IT.06.32.00 Gruppo di Ricerca per I Virus e le Virosi delle Piante.

IT.06.33.00 Istituto per la Meccanizzazione Agricola Via Onorato Vigliani, 104; 10138 Torino.

IT.06.34.00 Ist. Italiano di Idrobiologia. Largo Vittorio Tonolli, 50–52 – Verbania Pallanza (Novara)

IT.06.35.00 Istituto Elaborazione della Informazione. Via S. Maria, 46 – 56100 Pisa.

IT.06.36.00 Istituto per la Protezione Idrogeologica del bacino padano Tel. 011/779045 – 779613.

IT.06.37.00 Istituto di Geologia Applicata alla Pianificazione Viaria e all'Uso del Sottosuolo Tel. 049/760933–.

IT.06.38.00 Centro di Studio sulla Farmacologia delle Infrastrutture Cellulari c/o Ist. Farmacologia. Via Vanvitelli, 32 – 20100 Milano

IT.06.39.00 Istituto per lo Studio delle Proprietà Fisiche di Biomolecole e Cellule. Via Buonarrotti, 9 56100 Pisa

IT.06.40.00 Istituto per la Tecnica del Freddo Istituto per la Tecnica del Freddo Tel. 049/760933 Tel. 049/760933

IT.06.41.00 Istituto per lo Studio dello Sfruttamento Biologico delle Lagune Istituto per lo Studio dello Sfruttamento Biologico delle Lagune Via Fraccacreta, 1 Lesina– Foggia.

IT.06.42.00 Centro di Studio sulla Biologia molecolare e cellulare delle piante. Istituto di Scienze Botaniche. Via G. Colombo, 60 – 20133 Milano.

IT.06.43.00 Istituto di Mutagenesi e Differenziamento. Via Cisanello, 147/b; PISA.

IT.07.00.00 Ministero dell'Industria e del Commercio

IT.07.01.00 Comitato Nazionale Energia Nucleare Viale Regina Margherita 125; 00100 Roma.

IT.07.01.01 Laboratorio per le Applicazioni in Agricoltura S.P. Anguillarese Km. 1.300; 00600 Roma Casaccia.

IT.07.02.00 Istituto Sperimentale per le Industrie degli Olii e dei Grassi Via G. Colombo 79; 20133 Milano.

IT.07.03.00 Istituto Sperimentale per l' Ind. delle Pelli e delle Materie Concianti Via Poggio Reale 39; 80143 Napoli.

IT.07.04.00 Stazione Sperimentale per la Cellulosa Carta e Fibre Tessili Vegetali e Artificiali p. Zza Leonardo Da Vinci 26;20133 Milano.

IT.07.05.00 Stazione Sperimentale per l' Ind. delle Conserve Alimentari. Viale Faustino Tanara 33; 43100 Parma.

IT.07.06.00 Stazione Sperimentale per l' Ind. delle Essenze e dei Derivati degli Agrumi C. So V. Emanuele 131; 89100 Reggio Calabria

IT.07.07.00 Stazione Sperimentale per la Seta. Via G. Colombo, 81 20133 Milano.

IT.08.00.00 Cassa per il Mezzogiorno.

IT.09.00.00 Amministrazioni Regionali

IT.09.03.00 Lombardia.

IT.09.03.01 Istituto Superiore Lattiero Caseario. Via L. Pilla 25; Bis 46100 Mantova.

IT.09.07.00 Friuli – Venezia Giulia.

IT.09.07.01 Centro Regionale per la Sperimentazione Agricola per il Friuli – Venezia Giulia. Via Marangoni 97, 33100 Udine

IT.09.09.00 Toscana.

IT.09.09.01 Istituto Regionale di Cerealicoltura Via S.Michele degli Scalzi, 2; 56100Pisa.

IT.09.19.00 Sicilia.

IT.09.19.01 Cantina Sperimentale della Regione Siciliana. Casella postale 153, 98057 Milazzo (Messina).

IT.09.19.02 Cantina Sperimentale Noto. Largo Pantheon 7; 96010 Siracusa.

IT.09.19.03 Regione Sicilia Assessorato all'Agricoltura. Viale della Regione Siciliana. – 90100 Palermo.

IT.09.19.04 Ente Siciliano per la Promozione Industriale – ESPI–. Via G. Garibaldi, 36 – 98100 Messina.

IT.09.20.00 Sardegna

IT.09.20.01 Stazione Sperimentale del Sughero. Via Oschifi, 9; 07029 Tempio Pausania (Sassari)

IT.09.20.02 Centro Regionale Agrario Sperimentale. Via Alberti, 22 – 09100 Cagliari.

IT.10.00.00 Amministrazioni provinciali.

IT.10.21.01 Stazione sperimentale agraria e forestale. San Michele all'Adige (Trento).

IT.10.28.01 Istituto sperimentale di frutticoltura della provincia di Verona. Via S. Giacomo 25 ; 37100 Verona.

IT.12.00.00 Enti a scopo di Lucro

IT.12.01.00 Istituto Nazionale per le Piante da Legno C.so Casale, 476; Torino.

IT.12.02.00 Consorzio Sementiero Appulo–Lucano. Via Matteotti, 3 – Consalvo Bari.

IT.12.03.00 Centro Miglioramento Genetico Vegetale. Via Strampelli, 13 – 00040 Ardea (Roma).

IT.12.04.00 Azienda Cerealicola Sementiera. 70024 Gravina di Puglia.

IT.12.05.00 Società Michaelles. Via della Zecca, 1 10012 Bologna.

IT.12.06.00 U.N.I.D.A.L. S. p. A. Via Corsica, 21 – 20100 Milano.

IT.12.07.00 A.N.I.C. S. p. A. Piazza Baldrini, 1–20097 San Donato Milanese.

IT.12.08.00 Plasmon Dietetici S. p. A. Via Migliana, 45 – 04013 LATINA – (Lazio).

IT.12.09.00 Società AMMINEX. Via Travicella, 55 – 00100 Roma.

IT.12.10.00 I.B.P. Industrie Buitoni Perugia. Via Cortanese, 4 – 06100 Perugia.

IT.12.11.00 Centro Ricerche SIPCAM S.p.A. Via G. Colasso, 3 – 20100 Milano.

IT.12.12.00 Società SIAPA. Galliera – 40126 Bologna

IT.12.13.00 Montedison S. p. A. Via Bonfandini, 148 – 20100 Milano.

IT.12.14.00 Rumianca S. p. A. Corso Montevecchio, 37/39 – 10100 Torino.

IT.12.15.00 S.T.I.M.A.T. S. p. A. Via Pasteur, 78 – 00100 Roma.

IT.12.16.00 Società S.I.V.A.L.C.O. Via Mazzini, 200 – 44022 Comacchio–Ferrara.

IT.12.17.00 Centro Ittiologico Valli Venete. Cà Pirodi– Cà Venier – 45100 Rovigo.

IT.12.18.00 C.I.R.M. S.r.1. Via Mancini, 1 – 10100 Milano

IT.12.19.00 GEOTECNECO S.p.A. Via S. Lorenzo in Campo – 61047 Pesaro.

THE NETHERLANDS

NL.01.00.00 Onderzoekinstellingen onder het Ministerie van Landbouw en Visserij [Research Institutes under the Ministry of Agriculture and Fisheries.]

NL.01.01.00 Directie Landbouwkundig Onderzoek (Division for Agricultural Research). Bezuidenhoutseweg 73, P.O.Box 20401, 2500 EK 's-Gravenhage.

NL.01.01.01 BGD–Stichting Bureau voor Gemeenschappelijke Diensten (Officeof Joint Services). Bornsesteeg 53; P. O. Box 33; 6700 AA Wageningen.

NL.01.01.02 CABO – Centrum voor Agrobiologisch Onderzoek (Centre for Agrobiological Research). Bornsesteeg 65; P.O.Box 14; 6700 AA Wageningen.

NL.01.01.03 IB – Instituut voor Bodemvruchtbaarheid (Institute for Soil Fertility). Oosterweg 92; P. O. Box 30003; 9750 RA Haren(Gr.).

NL.01.01.04 CABO – Centrum voor Agrobiologisch Onderzoek (Centre for Agrobiological Research). Bornsesteeg 65; P.O.Box 14; 6700 AA Wageningen.

NL.01.01.05 IBVL – Instituut voor Bewaring en Verwerking van Landbouwprodukten (Institute for Research on Storage and Processing of Agricultural Produce). Bornsesteeg 59; P.O.Box 18; 6700 AA Wageningen.

NL.01.01.06 IMAG – Instituut voor Mechanisatie, Arbeid en Gebouwen (Institute of Agricultural Engineering). Mansholtlaan 10–12; P.O.Box 43; 6700 AA Wageningen.

NL.01.01.08 IPO – Instituut voor Plantenziektenkundig Onderzoek (Research Institute for Plant Protection). Binnenhaven 12; P.O.Box 42; 6700 AA Wageningen.

NL.01.01.09 IPS – Instituut voor Pluimvee–onderzoek "Het Spelderholt" (Spelderholt Institute for Poultry Research). Het Spelderholt 9; 7361 DA Beekbergen.

NL.01.01.10 ITAL – Instituut voor Toepassing van Atoomenergie in de Landbouw (Institute for Atomic Sciences in Agriculture). Keijenbergseweg 6; P.O.Box 48; 6700 AA Wageningen.

NL.01.01.12 IVO – Instituut voor Veeteeltkundig Onderzoek "Schoonoord" (Research Institute for Animal Husbandry "Schoonoord"). Driebergseweg 10D; P.O.Box 501; 3700 AM Zeist.

NL.01.01.13 RIVRO – Rijksinstituut voor het Rassenonderzoek van Cultuurgewassen (Government Institute for Research on Varieties of Cultivated Plants). Nieuwe Wageningseweg 1, P.O.Box 32, 6700 AA Wageningen.

NL.01.01.14 IVT – Instituut voor de Veredeling van Tuinbouwgewassen (Institute for Horticultural Plant Breeding). Mansholtlaan 15; P.O.Box 16; 6700 AA Wageningen.

NL.01.01.15 IVVO–Instituut voor Veevoedingsonderzoek

"Hoorn" (Institute for Livestock Feeding and Nutrition Research "Hoorn"). Runderweg 2; P.O.Box 160; 8200 AD Lelystad.

NL.01.01.16 LEI – Landbouw–Economisch Instituut (Agricultural Economics Research Institute). Conradkade 175; P.O.Box 29703; 2502 LS Den Haag.

NL.01.01.17 PUDOC – Centrum voor Landbouwpublikaties en Landbouwdocumentatie (Centre for Agricultural Publishing and Documentation). Prinses Marijkeweg 17; P.O.Box 4; 6700 AA Wageningen.

NL.01.01.18 SI – Sprenger Instituut (Institute for Research on Storage and Processing of Horticultural Produce). Haagsteeg 6, P.O.Box 17; 6700 AA Wageningen.

NL.01.01.19 Stiboka – Stichting voor Bodemkartering [Soil Survey Institute.]. Prinses Marijkeweg 11; P.O.Box 98; 6700 AB Wageningen.

NL.01.01.20 SVP – Stichting voor Plantenveredeling (Foundation for Agricultural Plant Breeding). Droevendaalsesteeg 1; P.O.Box 117; 6700 AC Wageningen.

NL.01.01.21 TDFL–Stichting Technische en Fysische Dienst voor de Landbouw (Technical and Physical Engineering Research Service). Mansholtlaan 12; P.O.Box 356; 6700 AJ Wageningen.

NL.01.02.00 Directies Akkerbouw en Tuinbouw, Veehouderij en Zuivel en Verwerking en Afzet van Agrarische Producten (Divisions Arable Farming and Horticulture, Farm Husbandry and Dairy–produce and Processing and Marketing of Agricultural Products).

NL.01.02.01 Aalsmeer – Proefstation voor de Bloemisterij (Research Station for Floriculture). Linnaeuslaan 2a; 1431 JV Aalsmeer.

NL.01.02.03 Boskoop–Proefstation voor de Boomkwekerij (Research Station for Arboriculture). Valkenburgerlaan 3; P.O.Box 118; 2770 AC Boskoop.

NL.01.02.04 Horst – Proefstation voor de Champignoncultuur (Mushroom Experimental Station). Peelheideweg 1; P.O.Box 6042; 5960 AA Horst.

NL.01.02.05 Lisse – Laboratorium voor Bloembollenonderzoek (Bulb Research Centre). Vennestraat 22; P.O.Box 85; 2160 AB Lisse.

NL.01.02.06 Naaldwijk – Proefstation voor de Groenten– en Fruitteelt onder Glas (Glasshouse Crops Research and Experiment Station). Zuidweg 38; P.O.Box 8; 2670 AA Naaldwijk.

NL.01.02.07 PAGV – Proefstation voor de Akkerbouw en de Groenteteelt in de Vollegrond (Research Station for Arable Farming and Field Production of Vegetables). Edelhertweg 1; P. O. Box 430; 8200 AK Lelystad.

NL.01.02.08 PR – Proefstation voor de Rundveehouderij (Research and Advisory Station for Cattle Husbandry). Runderweg 6; 8219 PK Lelystad.

NL.01.02.09 RLPS – Rijkslandbouwproefstation voor Meststoffen– en Veevoederonderzoek (Government Agricultural Experiment Station for analysis of fertilizers and feedingstuffs).
Kruiseherengang 21; 6211 NW Maastricht.

NL.01.02.10 RPvZ–Rijksproefstation voor Zaadcontrole (Government Seed Testing Station).
Binnenhaven 1; P. O. Box 9104; 6700 HE Wageningen.

NL.01.02.11 RZS – Leiden – Rijks Zuivel station [Government Dairy Station.]
Vreewijkstraat 12b; 2311 XH Leiden.

NL.01.02.12 Wilhelminadorp – Proefstation voor de Fruitteelt (Research Station for Fruit Growing).
Brugstraat 51; 4475 AN Wilhelminadorp.

NL.01.02.13 MOC – Melkhygienisch Onderzoek Centrum [Milk Hygiene Research Centre.]
Duivendaal 6; P.O.Box 343; 6700 AH Wageningen.

NL.01.03.00 Plantenziektenkundige Dienst (Plant Protection Service).

NL.01.03.01 LIO – Laboratorium voor Insekticidenonderzoek [Laboratory for Research on Insecticides.]
Prinses Marijkeweg 22; 6709 PG Wageningen.

NL.01.03.02 PD – Plantenziektenkundige Dienst [Plant Protection Service.].
Geertjesweg 15; P.O.Box 9102; 6700 HC Wageningen.

NL.01.04.00 Veterinaire Dienst (Veterinary Division).

NL.01.04.01 CDI – Centraal Diergeneeskundig Instituut – Afdeling Biologie (Central Veterinary Institute – Biological Department).
Prof. Poelslaan 35; P.O.Box 6007; 3002 AA Rotterdam.

NL.01.04.02 CDI – Centraal Diergeneeskundig Instituut – Afdeling Virologie (Central Veterinary Institute – Virology Department).
Houtribweg 39; 8221 RA Lelystad.

NL.01.05.00 Landinrichtingsdienst (Government Service for Land and Water Use).

NL.01.05.01 ICW – Instituut voor Cultuurtechniek en Waterhuishouding [Institute for Land and Water Management Research.].
Prinses Marijkeweg 11; P.O.Box 35; 6700 AA Wageningen.

NL.01.05.02 ILRI – International Institute for Land Reclamation and Improvement.
Prinses Marijkeweg 11; P.O.Box 45; 6700 AA Wageningen.

NL.01.06.00 Staatsbosbeheer (State Forest Service in the Netherlands).

NL.01.06.01 DORSCHKAMP – Rijksinstituut voor Onderzoek in de Bos– en Landschapsbouw "De Dorschkamp" (Dorschkamp Research Institute for Forestry and Landscape Planning).
Bosrandweg 20; P.O.Box 23; 6700 AA Wageningen.

NL.01.06.02 RIN – Rijksinstituut voor Natuurbeheer (Research Institute forNature Management).
Kemperbergerweg 67; 6816 RM Arnhem.

NL.01.07.00 Directie van de Visserijen [Government Division for Fisheries.]

NL.01.07.02 RIVO – Rijksinstituut voor Visserij–onderzoek (Netherlands Institute for Fishery Investigations).
Haringkade 1; P.O.Box 68; 1970 AB IJmuiden.

NL.02.00.00 Landbouwhogeschool [Agricultural University.]

NL.02.00.01 Vakgroep Agrarische Bedrijfseconomie (Department of Farm Management).
Hollandseweg 1; P.O.Box 8130; 6700 EW Wageningen.

NL.02.00.02 Vakgroep Agrarische Geschiedenis (Department of Rural History).
Hollandseweg 1; P.O.Box 8130; 6700 EW Wageningen.

NL.02.00.03 Vakgroep Agrarische Sociologie van de Niet – Westerse Gebieden (Department of Rural Sociology of the Tropics and Sub–tropics).
Hollandseweg 1; P.O.Box 8130; 6700 EW Wageningen.

NL.02.00.04 Vakgroep Algemene Agrarische Economie [Department of Agricultural Economics.]
Hollandseweg 1; P.O.Box 8130; 6700 EW Wageningen.

NL.02.00.05 Vakgroep Algemene en Regionale Landbouwkunde (Department of General Agriculture – Agricultural Geography).
Hollandseweg 1; P.O.Box 8130; 6700 EW Wageningen.

NL.02.00.06 Vakgroep Biochemie (Department of Biochemistry).
De Dreijen 11; P.O.Box 8128; 6700 ET Wageningen.

NL.02.00.07 Vakgroep Bodemkunde en Bemestingsleer [Department of Soils and Fertilizers.]
De Dreijen 3; P.O.Box 8005; 6700 EC Wageningen.

NL.02.00.08 Vakgroep Bodemkunde en Geologie [Department of Soil Science and Geology.].
Duivendaal 10; P.O.Box 37; 6700 AA Wageningen.

NL.02.00.09 Vakgroep Boshuishoudkunde (Department of Forest Management).
Generaal Foulkesweg 64; P.O.Box 342; 6700 AH Wageningen.

NL.02.00.10 Vakgroep Bosbouwtechniek (Department of Forest Technique).
Generaal Foulkesweg 64; P.O.Box 342; 6700 AH Wageningen.

NL.02.00.12 Vakgroep Cultuurtechniek [Department of Land and Water Use.]
Nieuwe Kanaal 11; P.O.Box 9101; 6700 HB Wageningen.

NL.02.00.13 Vakgroep Dierkunde [Department of Zoology.]
Prof. Ritzema Bosweg 32A; P.O.Box 61; 6700 AB Wageningen.

NL.02.00.14 Vakgroep Entomologie [Department of

Entomology.]
Binnenhaven 7; P.O.Box 8031; 6700 EH Wageningen.

NL.02.00.15 Vakgroep Erfelijkheidsleer [Department of Genetics.]
Generaal Foulkesweg 53; 6703 BM Wageningen.

NL.02.00.16 Vakgroep Fysische en Kolloidchemie (Department of Physical and Colloid Chemistry).
De Dreijen 6; 6700 EK Wageningen.

NL.02.00.17 Vakgroep Dierfysiologie (Department of Animal Physiology).
Haarweg 10; 6709 PJ Wageningen.

NL.02.00.18 Vakgroep Fytopathologie [Department of Phytopathology.].
Binnenhaven 9; P.O.Box 8025; 6700 EE Wageningen.

NL.02.00.19 Vakgroep Gezondheidsleer [Department of Public Health.]
De Dreijen 11; P.O.Box 8128, 6700 ET Wageningen.

NL.02.00.20 Vakgroep Grondbewerking (Soil Tillage Laboratory).
Diedenweg 20; 6703 GW Wageningen.

NL.02.00.21 Vakgroep Bosteelt (Department of Silviculture).
Generaal Foulkesweg 64; P.O.Box 342; 6700 AH Wageningen.

NL.02.00.22 Vakgroep Hydraulica en Afvoerhydrologie [Department of Hydraulics and Catchment Hydrology.]
Nieuwe Kanaal 11; P.O.Box 9101 HB Wageningen.

NL.02.00.23 Vakgroep Industriële Bedrijfskunde (Department of Business Administration).
Hollandseweg 1; P.O.Box 8130; 6700 EW Wageningen.

NL.02.00.25 Vakgroep Landbouwplantenteelt, Grasland– en Onkruidkunde – (Department of Field Crops, Grassland husbandry and Weed science).
Haarweg 33; P.O.Box 9101; 6700 HB Wageningen.

NL.02.00.26 Vakgroep Landbouwtechniek [Department of Agricultural Engineering.]
Mansholtlaan 12; P.O.Box 9101; 6700 HB Wageningen.

NL.02.00.27 Vakgroep Ontwikkelingseconomie (Department of Development Economics).
Hollandseweg 1; P.O.Box 8130; 6700 EW Wageningen.

NL.02.00.28 Vakgroep Landmeetkunde [Department of Surveying.].
Hesselink van Suchtelenweg 6; P.O.Box 339; 6700 AH Wageningen.

NL.02.00.29 Vakgroep Landschapsarchitectuur – [Department of Landscape Architecture.].
Wilhelminaweg 1; P.O.Box 9101; 6700 HB Wageningen.

NL.02.00.30 Vakgroep Huishoudkunde (Department of Home Economics).
Ritzema Bosweg 32A; P. O. Box 8060; 6700 DA Wageningen.

NL.02.00.31 Vakgroep Levensmiddelentechnologie –

[Department of Food Science.]
De Dreijen 12; P.O.Box 8129; 6700 EV Wageningen.

NL.02.00.32 Vakgroep Marktkunde en Marktonderzoek – [Department of Marketing and Marketing Research.].
Hollandseweg 1; P.O.Box 8130; 6700 EW Wageningen.

NL.02.00.33 Vakgroep Microbiologie – [Department of Microbiology.].
Hesselink van Suchtelenweg 4; P.O.Box 8033; 6700 EJ Wageningen.

NL.02.00.34 Vakgroep Moleculaire Fysica (Department of Molecular Physics).
De Dreijen 6; P.O.Box 8091; 6700 EP Wageningen.

NL.02.00.35 Vakgroep Natuurbehoud en Natuurbeheer – [Nature Conservation Department.]
Ritzema Bosweg 32A; P. O. Box 8080; 6700 DD Wageningen.

NL.02.00.36 Vakgroep Natuur- en Weerkunde – [Department of Physics and Meteorology.].
Duivendaal 1 and 2; 6701 AP Wageningen.

NL.02.00.37 Vakgroep Nematologie – [Department of Nematology.].
Binnenhaven 10; P.O.Box 8123; 6700 ES Wageningen.

NL.02.00.38 Vakgroep Organische Chemie (Department of Organic Chemistry).
De Dreijen 5; P.O.Box 8026; 6700 EG Wageningen.

NL.02.00.39 Vakgroep Pedagogiek en Didactiek (Department of Education).
Hollandseweg 1; P.O.Box 8130; 6700 EW Wageningen.

NL.02.00.40 Vakgroep Planologie (Department of Urban and Regional Planning).
Wilhelminaweg 1; P.O.Box 9101; 6700 HB Wageningen.

NL.02.00.41 Vakgroep Plantenfysiologie (Department of Plant Fysiology).
Arboretumlaan 4; 6703 BD Wageningen.

NL.02.00.42 Vakgroep Plantenfysiologisch Onderzoek – [Department of Plant Physiological Research.]
Generaal Foulkesweg 72; P.O.Box 8041; 6700 EL Wageningen.

NL.02.00.43 Vakgroep Taxonomie van Cultuurgewassen en –begeleiders (Department of Taxonomy of Cultivated Plants and Weeds).
Haagsteeg 3; 6708 PM Wageningen.

NL.02.00.44 Vakgroep Plantensystematiek en –geografie (Department of Plant Taxonomy and Plant Geography).
Generaal Foulkesweg 37; P.O.Box 8010; 6700 ED Wageningen.

NL.02.00.45 IvP–Instituut voor Plantenveredeling (Department of Plant Breeding).
Lawickse Allee 166; 6709 DB Wageningen.

NL.02.00.46 Vakgroep Plantencytologie en –morfologie – (Department of Plant cytology and morphology).
Arboretumlaan 4; 6703 BD Wageningen.

NL.02.00.47 Vakgroep Pluimveeteelt – [Department of Poultry Husbandry.].
Prinses Marijkeweg 40; P.O.Box 338; 6700 AH Wageningen.

NL.02.00.48 Vakgroep Psychologie – [Department of Psychology.].
Hollandseweg 1; P.O.Box 8130; 6700 EW Wageningen.

NL.02.00.49 Vakgroep Rechts– en Staatswetenschappen van de Westerse Gebieden (Department of Jurisprudence and Political Science of the Western Countries).
Hollandseweg 1; P.O.Box 8130; 6700 EW Wageningen.

NL.02.00.50 Vakgroep Rechts– en Staatswetenschappen en Landbouwcoo peratie en –kredietwezen van de niet–Westerse Gebieden (Department of Jurisprudence and Political Science of the non–Western Countries; Department of non–Western Agricultural cooperatives and finance).
Hollandseweg 1, P.O.Box 8130, 6700 EW Wageningen.

NL.02.00.51 Vakgroep Sociologie Westerse Gebieden (Department of Sociology).
Hollandseweg 1; P.O.Box 8130; 6700 EW Wageningen.

NL.02.00.52 Vakgroep Staatshuishoudkunde (Department of Economics).
Hollandseweg 1; P.O.Box 8130; 6700 EW Wageningen.

NL.02.00.53 Vakgroep Textiel – [Department of Textiles and Their Use.].
Prof. Ritzema Bosweg 32A; P.O.Box 61; 6700 AB Wageningen.

NL.02.00.54 Vakgroep Theoretische Teeltkunde (Department of Theoretical Production Ecology).
Bornsesteeg 65; 6700 EN Wageningen.

NL.02.00.55 Vakgroep Toxicologie – [Department of Toxicology.].
De Dreijen 12; P.O.Box 8129; 6700 EV Wageningen.

NL.02.00.56 Vakgroep Tropische Plantenteelt (Department of Tropical Crop Science).
Prof. Ritzema Bosweg 32; P.O.Box 341; 6700 AH Wageningen.

NL.02.00.57 Vakgroep Tuinbouwplantenteelt – [Department of Horticulture.].
Haagsteeg 3; P.O.Box 30; 6700 AA Wageningen.

NL.02.00.58 Zoo technische Vakgroepen (Department of Animal Production).
Marijkeweg 40; 6709 PG Wageningen.

NL.02.00.59 Vakgroep Veevoeding (Department of Animal Nutrition).
Haagsteeg 4; 6708 PM Wageningen.

NL.02.00.60 Vakgroep Vegetatiekunde en Plantenecologie – [Department of Plant Ecology.]
De Dreijen 11; P.O.Box 8128; 6700 ET Wageningen.

NL.02.00.61 Vakgroep Virologie – [Department of Virology.].
Binnenhaven 11; P.O.Box 8045; 6700 EM Wageningen.

NL.02.00.62 Vakgroep Humane Voeding (Department of Human Nutrition).
De Dreijen 11; P.O.Box 8128; 6700 ET Wageningen.

NL.02.00.63 Vakgroep Voorlichtingskunde – [Department of Extension Education.]
Hollandseweg 1; P.O.Box 8130; 6700 EW Wageningen.

NL.02.00.64 Vakgroep Waterzuivering – [Department of Water Purification.].
De Dreijen 12; P.O.Box 8129; 6700 AB Wageningen.

NL.02.00.65 Vakgroep Weg– en Waterbouwkunde en Irrigatie (Department of Civil Engineering and Irrigation).
Nieuwe Kanaal 11; P.O.Box 9101; 6700 HB Wageningen.

NL.02.00.66 Vakgroep Wiskunde – [Department of Mathematics.]
De Dreijen 8; P.O.Box 8003; 6700 EB Wageningen.

NL.02.00.67 Vakgroep Wonen (Department of Ecology of Habitat).
Ritzema Bosweg 32A; P.O.Box 8065; 6700 EPWageningen.

NL.02.00.68 Vakgroep Moleculaire Biologie(Department of Molecular Biology).
De Dreijen 11; P.O.Box 8128; 6700 ET Wageningen.

NL.02.00.69 Vakgroep Wijsbegeerte – [Department of Philosophy.]
Hollandseweg 1; P.O.Box 8130; 6700 EW Wageningen.

NL.02.00.70 Vakgroep Luchthygie ne en Verontreiniging(Department of Air Pollution).
Binnenhaven 12; 6709 PD Wageningen.

NL.02.00.71 Vakgroep Experimentele Diermorfologie en Celbiologie (Department of Experimental Animal Morphology and Cell Biology).
Prinses Marijkeweg 40; P.O.Box 338; 6700 AH Wageningen.

NL.02.00.72 Biologisch station "Wijster". [Biological station "Wijster".].
Kampsweg 27; 9418 PD Wijster.

NL.03.00.00 Faculteit der Diergeneeskunde van de Rijksuniversiteit te Utrecht (Faculty of Veterinary Medicine of the State University of Utrecht).

NL.03.00.01 Vakgroep Functionele Morfologie (Department of Functional Morphology).
Bekkerstraat 141; 3752 SG Utrecht.

NL.03.00.02 Vakgroep Bacteriologie – [Department of Bacteriology.].
Biltstraat 172; 3572 BP Utrecht.

NL.03.00.03 Vakgroep Farmacologie en Toxicologie – [Department of Pharmacology and Toxicology.]
Biltstraat 172; 3572 BP Utrecht.

NL.03.00.04 Vakgroep Parasitologie – [Department of Parasitology.].
Yalelaan 7; 3584 CL Utrecht.

NL.03.00.05 Vakgroep Pathologie – [Department of Pathology.].
Biltstraat 166/172; 3572 BP Utrecht.

NL.03.00.06 Vakgroep Tropische Diergeneeskunde en Protozoologie (Department of Tropical Veterinary Science and Protozoology).
Biltstraat 172; 3572 BP Utrecht.

NL.03.00.07 Vakgroep Virologie – [Department of Virology.].
Yalelaan 1; 3584 CL Utrecht.

NL.03.00.08 Vakgroep Voedingsmiddelen van Dierlijke Oorsprong – [Department of Food Science and Technology.].
Biltstraat 172; 3572 BP Utrecht.

NL.03.00.09 Vakgroep Zoo techniek(Department of Animal Husbandry).
Yalelaan 17; 3584 CL Utrecht.

NL.03.00.10 Vakgroep Algemene Heelkunde en Heelkunde der Grote Huisdieren (Department of General Surgery and Applied Surgery of Large Animals).
Yalelaan 12, 3584 CM Utrecht.

NL.03.00.11 Vakgroep Inwendige Ziekten der Grote Huisdieren (Department of Internal Medicine of Large Animals).
Yalelaan 16; 3584 CM Utrecht.

NL.03.00.12 Vakgroep Geneeskunde van het Kleine Huisdier (Department of Small Animal Medicine and Surgery).
Yalelaan 8; 3584 CM Utrecht.

NL.03.00.13 Vakgroep Verloskunde, K.I. en Voortplanting(Department of Obstetrics, A.I. and Reproduction).
Yalelaan 7; 3584 CL Utrecht.

NL.03.00.14 Vakgroep Fysiologie – [Department of Physiology.]
Alexander Numankade 93; 3572 KW Utrecht.

NL.03.00.15 Vakgroep Biochemie – [Department of Biochemistry.].
Biltstraat 172; 3572 BP Utrecht.

NL.03.00.16 Vakgroep Radiologie – [Department of Radiology.].
Yalelaan 10; 3584 CM Utrecht.

NL.03.00.17 Vakgroep Bedrijfsdiergeneeskunde en Buitenpraktijk – [Department of Industrial Farming and Ambulatory Clinic.].
Yalelaan 20; 3584 CM Utrecht.

NL.03.00.18 Werkgroep Immunologie – (Working group Immunology).
Biltstraat 172; 3572 BP Utrecht.

NL.04.00.00 Overige Overheidsinstellingen – [Other Government Institutes.].

NL.04.00.01 Laboratorium voor Toegepaste Entomologie –

Universiteit Amsterdam (Laboratory for Applied Entomology – University of Amsterdam).
Kruislaan 302; Amsterdam

NL.04.00.04 Phytopathologisch Laboratorium "Willie Commelin Scholten" (Phytopathological Laboratory "Willie Commelin Scholten").
Javalaan 20; 3742 CP Baarn.

NL.04.00.06 KNMI – Koninklijk Nederlands Meteorologisch Instituut (Royal Netherlands Meteorological Institute).
Wilhelminalaan 10; P.O.Box 201; 3730 AE De Bilt.

NL.04.00.07 RIJP–Rijksdienst voor de IJsselmeerpolders(IJsselmeerpolders development Authority).

Zuiderwagenplein 2; P.O.Box 600; 8200 AP Lelystad.

NL.04.00.10 RAAD – Rijks Agrarische Afvalwater Dienst (Government Service for Agricultural Sewage).
Kemperbergerweg 67; Arnhem.

NL.04.00.11 RIV–Rijksinstituut voor de Volksgezondheid(National Institute for Public Health).
Sterrenbos 1; 3511 ES Utrecht

NL.04.00.12 KIT–Koninklijk Instituut voor de Tropen, afdeling Agrarisch Onderzoek (Royal Tropical Institute, department of Agricultural Research).
Mauritskade 63; 1092 AD Amsterdam.

NL.04.00.13 Onderafdeling der Wijsbegeerte en Maatschappijwetenschappen–TH(Department of Philosophy and Humanity–University of Technology).

NL.05.00.00 Organisatie voor Toegepast Natuurwetenschappelijk Onderzoek – TNO (Organization for Applied Scientific Research in the Netherlands – TNO).

NL.05.01.00 Centrale Organisatie TNO (Central Organization TNO).

NL.05.01.01 Proefboomgaard "De Schuilenburg"(Experimental Orchard "De Schuilenburg").
4041 BK Kesteren.

NL.05.01.04 OCI – Organisch Chemisch Instituut TNO – [Institute for Organic Chemistry TNO.].
Croesestraat 79;Po Box 5009;Utrecht.

NL.05.02.00 Nijverheidsorganisatie TNO (Organization for Industrial Research TNO).

NL.05.02.08 Proefstation voor de Aardappelverwerking (Experiment Station for the Utilization of Potatoes).
Rouaanstraat 27; 9723 CC Groningen.

NL.05.02.10 Hoofdafdeling Maatschappelijke Technologie–TNO(Division of Technology for Society–TNO).
Complex Zuidpolder; Schoemakerstraat 97; P.O.Box 217; 2600 AE Delft.

NL.05.02.20 Hoofdafdeling Bouw en Metaal–TNO(Division for Building and Metal Research–TNO).
Lange Kleiweg 5; P.O. Box 107; 2280 AC Rijswijk (Z.H.).

NL.05.02.30 Hoofdafdeling Industrie le Produkten en Diensten–TNO(Division of Industrial Products and Services–TNO).
Complex Zuidpolder; Schoemakerstraat 97; P.O.Box 288; 2600 AC Delft.

NL.05.03.00 Voedingsorganisatie TNO (Organization for Nutrition and Food Research TNO).

NL.05.03.01 CIVO – Centraal Instituut voor Voedingsonderzoek TNO (Central Institute for Nutrition and Food Research TNO).
Utrechtseweg 48; P.O.Box 360; 3700 AJ Zeist.

NL.05.03.02 IGMB – Instituut voor Graan, Meel en Brood – TNO (Institute for Cereals Flour and Bread TNO).
Lawickse Allee 15; P.O.Box 15; 6700 AA Wageningen.

NL.05.03.03 IVP – Instituut voor Visserijprodukten TNO (Institute for Fishery Products TNO).
Dokweg 37; P.O.Box 183; 1970 AD IJmuiden.

NL.05.03.04 NIBEM – Nationaal Instituut voor Brouwgerst, Mout en Bier TNO (National Institute for Malting Barley, Malt and Beer TNO).
Utrechtseweg 46; P.O.Box 109; 3700 AC Zeist.

NL.05.03.05 Stichting ILOB – Instituut voor Diervoedingsonderzoek (Foundation ILOB, Institute for Animal Nutrition Research).
Haarweg 8; P.O.Box 9; 6700 AA Wageningen.

NL.05.04.00 Gezondheidsorganisatie – TNO. [Organization for Health Research – TNO].

NL.05.04.01 IMG – Instituut voor Milieuhygiene en Gezondheidstechniek – TNO (TNO Research Institute for Environmental Hygiene).
Complex Zuidpolder; Schoemakerstraat 97; P.O.Box 214; 2600 AE Delft.

NL.06.00.00 Particuliere Onderzoekinstellingen (Private Research Institutes).

NL.06.00.02 NIZO – Nederlands Instituut voor Zuivelonderzoek (Netherlands Institute for Dairy Research).
Kernhemseweg 2; P.O.Box 20; 6710 BA Ede.

NL.06.00.03 IRS–Instituut voor Rationele Suikerproduktie(Sugar Beet Research Institute).
Van Konijnenburgweg 24; P.O.Box 32; 4600 AA Bergen op Zoom.

NL.06.00.04 CLO – Instituut voor de Veevoeding "De Schothorst" (Institute for Animal Feeding "de Schothorst").
Meerkoetenweg 26; 8218 NA Lelystad.

NL.06.00.05 "Grontmij" Grondverbetering en Ontginningsmaatschappij N.V.
"Houdringe"; de Bilt

NL.06.00.06 N.V. Heidemaatschappij Beheer
Lovinklaan 1; Arnhem

NL.06.00.07 Hybried Pluimveefokkers coöperatie "Hypeco"

u.a. (Hybrid poultry breeders coöperation).
Rijksweg 16; Nuland.

NL.06.00.08 Centraal Stikstof Verkoopkantoor – [Central Nitrogen Selling Agency.].
Thorbeckelaan 360; Den Haag

NL.06.00.09 Centraal Veevoeder Instituut – [Central Institute for Animal Nutrition.]
Middelweg 85; Leersum.

NL.06.00.10 Stichting Gezondheidsdienst voor Pluimvee.
Oude Rijksstraatweg 43, P.O. Box 43, Doorn

NL.06.00.11 Stichting Gezondheidsdienst voor Dieren in Noord–Holland.
Helderseweg 8, P.O. box 88, Alkmaar.

COMMISSION OF THE EUROPEAN COMMUNITIES

COMMISSION OF THE EUROPEAN COMMUNITIES

XE.06.01.00 Commission of the European Communities, GD
vi. 84 Rue de la Loi, Bruxelles, Belgium.

XE.06.01.01   Joint Research Centre.
Ispra (Varese) Italy.

LIST OF SCIENTISTS

Arrigoni, C. - Traldi, G. 20921
Arrivo, A. 15818
Arrowsmth 13992
Arthey 18702, 18703, 18704, 18705, 18706, 18707, 18708, 18709, 18710, 18711, 18712, 18713, 18714, 18715, 18736
Arthur 7576, 7577, 11896, 11897
Artmann, R. 10698, 15421, 15422
Arts, W.B.M. 15528
Aru, A. 316
Arvieu, J.C. 195, 196, 477, 478, 634, 635, 8152, 8156
As, H. van 20080
Asche, H. 18213
Ash 12512, 13327
Ashworth 663, 664, 5601
Asjes, C.J. 9923, 9925
Askar, A. 16093
Aslyng, H.C. 24, 2611
Asplin 14846
Assandri, G. 330
Asselin, C. 7393
Asso, J. 13721, 14206, 14455, 14691
Ast, K.J. van 17461
Astier, R. 20138
Astier, S. 7790, 7890
Aston 12094
Atanasiu, N. 133, 134, 135, 445, 3072, 4362, 5014, 5428
Atger, P. 8164, 8735, 8736
Athari, S. 4743
Atherton 18723, 18724, 18726, 18731
Athias, C. 1034
Atkinson 28, 29, 4079, 4175, 4181, 4182, 4217, 4246, 4298
Attaby, H. 20569
Attonaty, J.M. 2753, 4337, 17397, 17400
Aubert 4811
Aubert, D. 17503, 18097
Aubry, J. 13716, 13720
Aubury 6763
Auclair, D. 4800, 4801
Audemard, A. 8741
Audemard, H. 8679, 8680, 8681, 8719, 8720, 8737, 8742, 8743, 8817
Audsley 920, 19906
Auerhammer, H. 15367
Auernhammer, H. 17197, 17198
Auffray, P. 12497, 12716
Aufhammer, W. 3014
Aufsess, H.von 4863, 15951, 20652
Auge, P. 2563, 7791, 10347
Augustin, H. 16399
Augustini, C. 11159, 15805, 19168
Augustinussen, E. 3264, 15905
Auhagen, A. 1851
Aukema, S. 18177
Aumaitre, L. 11188, 11189, 11685, 12460, 12475, 12477, 12479, 12483, 12488, 12489, 12857
Aureli, G. 13136, 13871
Auriau, Ph. 6343, 6344, 6346
Aurousseau, B. 11704, 11838, 12073, 12075, 12076, 12308, 12309, 12311

Aussenac 16436
Aussenac, G. 4789, 4791, 4792, 4794, 4795, 4798, 4974, 5992
Aust, H.-J. 9018, 9342
Austenfeld, F.-A. 10334
Austin 2935, 2936, 2937, 2939, 5668, 7897, 12315, 14000, 14001, 14477
Autran, J. C. 6374
Autran, J.C. 6111, 6361
Auweck, F. 1803
Auxilia, M.T. 10616, 10861, 10863, 12799, 12800, 12801, 12802, 12803, 12804, 13126, 13447
Avancini, D. 9218
Avanzi, S. 4209
Avanzo, E. 4995, 4996, 4999, 7701, 7702, 7703, 7704, 7705
Avella, T. 16361, 16362, 16363
Averdunk, G. 13004, 13020
Averna, V. 10381
Aversano, B. 6003, 7737, 16304, 17021
Avery 5783, 5784, 5831
Aveyard 8288
Avigliano, M. 7733, 7736, 7737, 8947, 10052, 10054
Avolio, S. 4901, 4902, 4904, 20899, 20901
Avril, P. 1783
Avronsard 1449
Awad, G. 3983
Aydin, B. 15559
Aydin, I. 4932
Ayglon, D. 20772
Ayla, C. 16423
Aynaud, J.M. 14691, 14693, 14694
Ayres 10191
Azzam, A.M. 16591

Baackmann, W. 14132
Baader, W. 17042, 17057, 17067
Baarveld, W.C. 15317
Baath, C. 20230
Babel, U. 731
Baccetti, B. 13528
Bach, Aa. 6523, 15836
Bach, E. 8979
Bach, H. 17817
Bach, P. 17370
Bachacou 9993, 20339
Bachacou, J. 2732, 4808, 20341, 20756
Bache 64, 65, 66
Bacher, E. 3844, 3941, 3942, 15837
Bacher, R. 15497
Bächmann, K. 19424
Bachmann, O. 4380, 4381, 7377, 20683
Bachmann, P.A. 14651, 20661
Bachmann, U. 16097
Bachthaler, E. 4598, 4599, 4682
Back, W. 17218
Bäcker, G. 1362, 4406, 8803, 15326, 15370, 15504
Backus, H.C.S. 18450, 18455
Bacon 10526, 11893, 18487
Bacou, F. 10431, 11504
Badawy, N. 12652
Bade, R. 18047, 18296

Bäder, G. 7389
Bader, H. 10573, 10575, 10576, 10577, 10578, 10650, 10654, 15931
Bader, S. 2424
Baderschneider, F. 15746
Badia, J. 247, 11426, 11430, 19834
Badings, H.T. 16950, 19393
Badino, M. 3961, 6868, 6906, 6907, 7077, 7135, 19102
Badir, S. 20731
Badis, M.F. 17180, 20824
Badoux, D.M. 21076
Baer, E.von 16768
Baert, L. 431, 982, 5121
Baeteman, M. 19592
Baeten, H. 104, 1835, 10333, 10338
Baeumer, K. 2684, 2685, 3065, 3259, 5149, 5426, 10086
Baeyens, L. 107, 2539, 4733, 5945
Bagdadhi, N. 5133
Bagger, O. 9566
Baggott 14231
Baglioni, T. 13138, 20921
Bagnaresi, U. 4482
Bagni, N. 4259, 4260
Bagnoli, B. 20890, 20891, 20892
Bähr, H.-G. 17942
Bahr, R. 1373, 5201
Bahramian 5385
Bailey 4073, 8005, 8006, 10282, 14992, 15202, 15203, 15204, 15205, 15206, 15207, 15609, 18521, 18523, 18528
Bailleux, P. 5059, 5981
Bain 48
Bainbridge 9151, 9663
Baines 11617, 17077
Baird 11719
Baitsch, B. 1218
Baize, D. 185, 244, 631
Bakel, J.M.M. van 7882, 7885, 7886, 9616, 9617, 9618, 9645, 9646
Bakel, P.J.T. van 1634, 1641, 1644
Bakels, F. 12889
Baker 2776, 7793, 8748, 8840, 12095
Baker, G. 1768
Baker, K.P. 13946, 20882
Bakermans, W.A.P. 2877, 5319
Bakheit, H. 14060
Bakkendrup-Hansen, G. 10072, 10259, 10281, 10295
Bakker, H. 13167, 13465, 13466
Bakker, H. de 374, 20428
Bakker, Ij.T. 13172
Bakker, J.J. 6605, 6723, 6740
Bakker, J.W. 860, 1610, 15729
Bakonyi, E. 17995, 17997, 18278
Balancon, M. 14817
Balazs, A. 2068
Balch 12102
Baldelli, G.P. 3116, 3117
Baldelli, R. 13864, 13866, 19574
Baldi, G. 6290, 6291, 6292, 6293, 6294, 6295, 6297, 6301, 6302, 10200, 16058
Baldini, E. 4112, 4113, 4114, 4115, 4116, 4473, 4474
Baldini, P. 16211, 16212, 16213, 16708,

Baumeister, G. 4990, 7656, 7657
Baumeister, W. 10334
Baumer, M. 6168, 6169, 6170, 6171, 6172, 6804
Baumgarten, K. 17905
Baumgartl, C. 10577
Baumgartner, A. 1304, 1305, 1306, 1962, 4751, 10321, 20647
Baumgärtner, G. 10669
Bäumler, W. 1960, 13641, 20646
Baurant, R. 1826, 1827, 7956, 8034
Bausch–Goldbohm, R.A. 19774
Bausch, R. 2908
Bauwens, A.L.G.M. 2453, 2454, 2457, 2458, 2459, 2460, 2462, 18179, 18181, 18189, 18190, 18353, 18354, 18355
Bavo, L. 20770
Baxter 13337, 13338, 15543, 15544, 15546
Bayer, U. 20098
Bayer, W. 12974
Bayonove, C. 7407
Bayrle, H. 11372
Bazier, G. 1110, 1111, 1787
Bazin, G. 18077, 18108
Bazzocchi, R. 15615
Bazzoffi, P. 96, 274, 276
Bb35 7938
Bd 4314
Beal 10973
Beale 14212, 14701, 20837
Beall, E. 11404, 11407, 11426
Bean 3435, 3436
Bean, J. 6697
Beatty 12926, 12927, 12928, 13088
Bech Andersen, B. 13031
Bech–Andersen, J. 16278
Bech–Andersen, S. 10494, 11674
Bech, K. 8984, 9708, 9709
Becher, H.H. 149, 825, 881, 882, 1307, 2424
Bechet, G. 12302, 12303
Bechmann, G. 13980
Beck, D. 2407, 2733
Beck, G. 20527
Beck, H. 10470, 11983, 12826, 18853
Beck, T. 165, 469, 469, 2440, 5066, 5188, 16327, 16328, 19677
Becker 15834
Becker, D. 19466, 19501
Becker, F. 18244, 18912
Becker, G. 17249, 18256, 20693
Becker, H. 4368, 20274, 20275, 20307
Becker, H.–J. 18277
Becker, H.F. 16421
Becker, K. 15998, 16467, 17117
Becker, K.H. 19624
Becker, M. 191, 2097, 4805, 4806, 4867, 4977, 17807, 20140, 20141
Becker, N.J. 4392
Beckerich, J.M. 20819
Beckers, B. 1844, 4733, 5945
Beckers, J. 10627, 14073
Beckhoff, J. 3529, 15904
Becking, J.H. 5435
Beckmann, H. 118, 573, 819

Beckmann, R. 2540
Bedeneau 10389
Bedford 18706, 18707, 18708, 18709, 18710, 18711
Beech 18551
Beek, A.van–der 17922
Beek, C.P. van der 8423
Beek, G. van 16022, 20422
Beek, J. 714
Beek, P. van 19955, 19956
Beekman, A.G. 379
Beel, E. 4492
Beemster, A.B.R. 9339
Beer 10181
Beers, A. van 11777
Beese, F. 138, 139, 823, 1237
Beeskow, H. 7387
Beetsma, J. 11489, 11490, 11491
Beever 11866, 11867, 11868, 11869
Begheijn, L.T. 407
Beghelli, V. 11054
Beghi, B. 530, 10336
Begon, J. C. 493
Begon, J.C. 181, 184
Begtrup, J.W. 9085, 9286
Behaeghe, T. 3307, 3507, 11546
Behler, H. 16374, 20560
Behme, G. 17715
Behre, K.–E. 20706, 20707, 20708
Behrend, M.C. 3378
Behrendt, W. 10640
Behrens, D. 13924
Behrens, R. 20182
Behrens, V. 7634
Behring, J. 5143
Behringer, H. 5204, 5473
Beier, J. 20662
Beier, M. 1293
Beijersbergen, J.C.M. 4561, 9922
Beinhauer, R. 750, 3163, 10375, 14161
Beintema, A.J. 2191, 2199, 2213, 2227, 2229, 2230, 2232, 2233
Beirens, P. 18472
Bejer–Petersen, B. 2094, 7984, 7985
Bejer, B. 8917
Bekaert, H. 11092, 12378, 12379, 12381, 17162, 19140
Bekendam, J. 20448, 20457, 20458
Bekendam, M. 13252
Bekkum, J. G. van 14410
Bekkum, J.G. van 13949, 14063, 14409, 14411, 14412, 14413, 14577, 14578, 14769, 14770, 14771, 20488
Bel, F. 18137, 19813
Belcher 15179
Belderok, B. 16061, 16542, 19017, 20514
Belford 2759
Belgraver, W. 7592
Belitz, H.–D. 16531, 18918, 19093, 20702
Bell 8019, 8533, 10810, 14318, 16008, 19558, 19936
Bell, A. 680
Bellanger, J. 11693
Bellegem, T.M. van 16626, 16629,

19071
Bellenand–Mayeur, P. 5039
Bellenand–Mayeur, P. 195, 196, 5040, 5240, 17044, 17045, 20760
Belli, G. 9190, 9943
Bellia, F. 1500
Bellini, E. 2864, 4267, 6072
Bellini, P. 1769
Bellitti, E. 11754, 13244
Bellon, Nicole 478
Bellucci, V. 17587
Belmans, C. 1191
Belouin, A. 7259
Belshaw, B.E. 15061
Belt, A.H.M. 17187
Bem, Z. 17141
Bembenek, M. 10115
Bemelmans, J. M. H. 16717
Bemelmans, J.M.H. 18971
Bemmer, P. 1245
Ben Harrath, A. 818, 867
Benad, A. 18249
Benaguid, T. 449
Benassi, A. 18163, 18164
Benassy, C. 8125, 8766, 8768
Benazzi, P. 13868
Benckiser, G. 141
Benda, I. 7386, 16613
Bendall 18530
Bender, L. 5726
Bender, R. 17921
Benders, G.A. 15447, 19714
Bendixen, E.O. 18291, 18292
Bendixen, P.H. 13506, 14193, 14194
Benecke, P. 1240
Benedikz 9300
Benetti, M.P. 9581, 10021
Benfatto, D. 4351, 8771, 8772
Benhamou, N. 9113, 9115, 20777
Benians 77
Benini, G. 305, 1508
Bennema, J. 409, 412, 413
Bennet 6113, 18714
Bennett 2933, 6114, 6118, 9295, 9755, 9773, 9774, 9810, 18705
Bennetzen, F. 5379
Benning, G. 426, 1701, 1707, 1714
Benoit, F. 3692, 7297
benschop, M. 4558, 4569
Bensink, J. 21032
Benthin, G. 10932, 10933, 10934
Bentz, A. 818, 866, 867
Benvenuti, A. 6481, 6508, 6509
Benvenuti, B. 18383
Beny, G. 13716
Benzian 5388
Beran, N. 4410, 20709, 20710
Beranger, C. 10586, 10745, 10746, 11709, 11825, 12045, 12064, 12065, 12068, 12072, 12310, 17501
Beranger, G. 17500
Berben, J. 107, 2539, 4733, 5945
Berbigier 11192
Berbigier, A. 6183, 6184, 6314
Berbigier, P. 11502
Berchu, L. 20822

Biemans, J.M. 18353
Bienfait, J. 5527
Bienfait, J.M. 11544, 11545, 11548, 11549, 11550, 11551, 11552
Bienfet, V. 10628
Bierg, N. 4880, 4985
Bierhals, E. 1940, 2412
Bierhuizen, J.F. 3786, 3787
Biering–Sørensen, U. 14686
Bierl, J. 19302
Biermann, U. 19092
Biermans, V. 431, 982, 5121
Biernaux, J. 7745, 7746, 11412
Biersteker, K. 19724, 19773, 20992
Biesterfeld, H. 2425
Biewenga, W.J. 14433
Biezen, J.B. 20968
Biezen, J.B. van 2218
Bigelli, G. 6980
Biggs 14894
Bij de Vaate, A. 2350
Bijkerk, C. 2479
Bijl, R.S. 20397, 20437
Bijlsma, I.G.W. 14415, 14774
Bijlsma, S. 391
Bijsterbosch, B.H. 20497
Bilio, M. 11436
Billard, M. 11431
Billard, R. 11375, 11378, 11379, 11380, 11381, 11382, 11383, 11384
Bille, N. 14683, 14684
Bille, S.W. 5227
Biller, R.H. 15287, 15288, 17178
Billiard, B. 11386
Billib, H. 1269, 1273
Billing 9745, 9747, 9975
Billington 15771
Billot, C. 2644, 3174, 3327, 3662, 3897, 4066, 4067, 4233, 6449, 6657, 6658, 20337
Bina, L. 11189
Binark, M. 1292
Binazzi, A. 8916, 8922
Bindseil, E. 14685
Bines 10770, 12102
Bingham 6375
Bini, G. 2862, 2865
Binkhorst, G.J. 13964
Binnerts, W. T. 12242
Binnerts, W.T. 11782
Binns 8668, 13990, 14698, 15728
Binot, R. 2596
Binzel, R.–M. 11639, 16185, 20213
Biométrie : Badia 20188
Biondi, A. 284, 1178
Biondi, G. 3241, 4365, 10374, 16151
Bird, G. 2996
Birker, F. 12976, 12977, 12978
Birkkjær, H.E. 16238, 16790, 16795, 16798, 16799, 16800, 16803, 16804, 16807, 16808, 16809, 16810, 16811, 19347
Birmie, J. 21005
Biront, P. 13485, 14612
Birot 4811
Birot, Y. 4868, 7646, 7647, 10000

Birse 68
Bischof, F. 10091, 10095
Bischoff, T. 11140, 15557, 15643, 15644, 16036, 17196
Bischofsberger, W. 15720, 15721, 17055, 17056, 19637
Bisgaard, M. 13496
Bisgård, K.M. 20060, 20061
Bishop, N.I. 20717
Bisiach, M. 9189, 9193
Bismarck, C.von 16366
Bisping, W. 13588, 13589, 13590, 13591, 14630, 15024, 17094
Bissinger, E. 11096
Bisson, J. 7401
Biston, L. 15796
Biston, R. 1202, 2898, 3630, 11563, 15896, 18809
Bitoun, B. 17506
Bitsch, I. 13578, 18847, 19510, 19513, 19762
Bitsch, R. 19466, 19468, 19501
Bittante, G. 10875, 12203, 12204, 12357, 12776, 13141, 13142, 13143, 13144, 17516
Bjergskov, T. 18825
Björkman, N. 14681
Bjældager, P. 15115
Blaak, G. 5054
Blab, J. 1991, 1996, 1997, 20685
Black 11008, 13760, 18428
Blackett 2983, 5390, 6141, 7762
Blackman 6377
Blackwell 2942
Blaeser, M. 8043
Blagden, P. 510, 5501, 5611
Blago, A. 5048, 7735
Blähser, S. 1883
Blaich, R. 4380, 4381, 7378, 9045, 9847, 20684
Blair 10129
Blaisinger, P. 2106, 7908, 8181, 8583, 8747
Blanc 12751
Blanc, D. 5696
Blanc, D. Mme 479, 5039, 5101, 5102, 5241, 5697, 5900, 8139, 20760
Blanc, D.Mme 5240
Blanc, J.M. 13424, 13425
Blanc, M. 10435, 10958, 10960, 10966, 17395, 18231, 18305
Blanc, M.R. 10952
Blanc, P. 1135, 1142, 1446
Blanchard, M. 2741, 10126
Blanchet, R. 227, 650, 837, 3175, 3633, 5109, 5110, 5464, 20798
Blanckenburg, P.von 17903, 17904, 18241, 18417, 20194
Blanco, A. 6408, 6409, 6410
Blanco, V.V. 1549, 3873, 3932, 3971, 3972, 5733, 7191, 10275
Blandini, G. 15376
Blank 4887
Blank, H.–G. 4131, 15930, 16117
Blankholm, E. 6860
Blanz, P. 20111, 20676

Blaschke, H. 15950, 20644
Blasdale 8458, 8595
Bleasdale 3212, 17302
Blecken, F. 2417, 2418
Bleckmann, E. 14118, 14121
Blendl, H.M. 11161, 11162, 13285, 14661, 14662, 19680
Blérot, P. 4912
Blesenkemper, L. 10665
Bless, R. 1994, 20685
Bleve Zacheo, T. 8368
Bleymüller, H. 4884
Bligh 10525, 10526
Blight 8279, 8939
Blobel, H. 13567, 20102, 20578, 20579, 20580
Bloch, B. 20327
Bloch, G. 19659
Blöchinger, J. 11150
Block, C. 3088
Block, H. 20198
Block, K. 3015, 3086
Blohm, B. 10444
Blok, I. 9646, 9647, 9695
Blokker, K.J. 18415
Blokland, A.J. 9340
Bloksma, A.H. 11789, 16544, 16551, 19016, 19019, 19022
Blom, J.Y. 15538
Blomme, R. 4492
Blondel, L. 4338, 4339, 5241, 7331
Bloomfeld 36, 509
Bloomfied 37
Blosser–Reisen, L. 18276, 18438, 18439, 18440
Blum, W.E. 125
Blume, E. 5151
Blume, H.–P. 108, 109, 110, 438, 439, 869, 990, 991, 1203, 20525
Blumenbach, D. 19801, 19803, 19806, 20283, 20284, 20286
Blundstone 18720, 18737
Blüthgen, A. 11994, 15112, 19312, 19313, 19316, 19317, 19322, 20247
Blyth 17311
Boa 3712, 17103
Boag 8617
Boatman 3549
Boatto, V. 17419, 17516, 18351
Boccard, R. 10746
Boccardo, G. 10035
Boch, J. 1977, 13628, 13629, 13632, 13633, 13927, 13928, 14145, 14654, 14655, 15034, 15035
Bochkoltz–Maufroid, C. 3822
Bock, E. 1309
Böckenhoff, E. 11140, 17782, 17784, 17989, 17990, 17991
Böcker, R. 1847, 2398
Bockor 1912
Bockstaele, L. 3008, 3136, 3137, 3790, 5032, 7846, 7955
Bockstedte, W. 4061
Bocquet, G. 13048, 13365, 20137
Bodart, A. 10439
Bodart, C. 10440, 11093, 11566, 11808

Carneiro, J.G.A. 4736
Carnovale, E. 16954
Caron, J.E.A. 8858, 10300
Caroppo, S. 16848, 16851, 16853
Carotenuto, R. 5049, 7740
Carow, B. 4526, 4527, 4595, 4596, 4597, 15941
Carpenter 7614, 15541
Carr 7834, 7864, 9316, 9562, 9571, 9572, 13206, 13207, 18546, 18548, 18550, 19562, 19563, 20007
Carravetta, R. 303, 1504
Carré, J. 15737, 15739, 15740, 15826
Carre, M. 7915
Carre, S. 8235, 8601
Carrere, G. 17499
Carruthers 15201, 15208, 15212, 16281
Carruthrs 15214, 15215
Carstens, W. 14642
Cartechini, A. 7440, 7654
Cartia, G. 9902
Carton, O. 2130
Cartwright 9779, 14743, 14746, 14747, 14748, 14751
Cartwrigt 14749
Caruso, A. 2834, 4345, 4346, 7338, 8772
Caruso, P. 3761, 3762, 3965, 7090
Casalicchio 5504
Casalicchio, G. 685, 790, 20370
Casarini, B. 3284, 3285, 3925, 5503, 5763, 6603, 7078, 7080, 7142, 9530, 9531, 10364
Casas, J. 18092, 20807
Casati, D. 18157
Casati, M. 12947
Casati, Marta 13137
Caserio, Giuseppe 19285, 19379
Casey 7065
Casey, C. 5746
Casier, J. 6321, 17733, 18460, 18461, 18462, 18463, 19965, 19966
Casimir, J. 3630
Casini, E. 2868, 3160, 4092, 4102, 7457
Casolari, A. 16015, 16016, 16159, 16661
Casper, R. 9067, 9068, 9447, 9584, 9720, 9787, 20281, 20282
Caspers, N. 1213
Cassaniti, S. 6411
Cassidy, J.C. 3746, 3748, 3749, 3811, 3813, 3814, 3815, 3816, 7876, 9614, 12174
Cassin, J. 231, 5836, 7332
Cassini, R. 6236, 9289
Castagnola, M. 783
Castagnoli, M. 8485, 8486
Casteels, M. 11092, 12378, 12379, 19140
Castellain 4235
Castellani, C. 4825, 4906, 6736, 17861, 20368
Castellani, E. 9904, 9905
Castelli, G. 15380
Castelvetri, F. 16016
Casteran, P. 4415

Castiglione, A. 15472, 17089
Castino, M. 16998, 16999, 17000, 17001, 17005, 17008
Castle 10772, 12111, 12112
Castrucci, G. 14379
Casu, B. 16302
Casu, S. 13186
Catalano, M. 16065, 18801, 18939
Catalano, V. 4433
Catara, A. 7341, 7342, 7476
Cataudella, S. 11443, 11455, 11456
Cate, J.A.M. ten 348, 349, 385, 392
Catherall 6125, 9318, 9555
Catizone, P. 297, 10224
Catroux, G. 1027, 2107, 8184, 17087, 19696, 20801
Catt 34, 35
Cattabiani, F. 14382, 14947
Cattaneo, M. 6398, 6399, 6401
Cattaneo, P. 16014, 16206
Caubel, G. 6762, 8221, 8446, 8448, 8449, 8512, 9096
Cauchy, L. 14809, 14811, 14813, 14814
Cauderon, Y. Mme 6106, 6345, 6946
Causse, R. 8158, 8159, 8170, 8171, 8678, 8681, 8720, 8743, 9105
Caussin, R. 7748
Cavailhes, J. 18081, 18082, 18083
Cavalcaselle, B. 4895, 4896, 8309
Cavalchini, L. 11363
Cavalchini, L.G. 11361, 11362
Cavallari, L. 783
Cavallero, A. 3365, 3366
Cavazza, L. 789, 1487, 1488, 1489
Cavelier, M. 7829
Cavenel, B. 3098
Cawthorne 14283
Cayley 7847
Cayrol, J.C. 8142, 8145, 9096, 9100, 9104, 9711, 20763, 20765
Cayrol, R. 8159, 8160, 8211
Cazemier, W.G. 2274, 2280, 2282, 2286, 2288
Ceccarelli, S. 6207, 6208, 6415, 6721, 7152
Cecchi, V. 6146, 6395, 6400, 6403
Celestre, P. 1521, 1522
Celi, R. 11050
Cengel, M. 996
Cenni, B. 11057, 12818, 12819
Centurier, C. 13630, 13631, 13637, 13638, 13641, 14657
Centurier, H. 13537
Ceoloni, C. 6394, 6395, 6397
Cera, M. 15382
Cerato, C. 3237, 6555, 6597, 6603, 9530, 9531
Ceretto, F. 13875
Cerletti, P. 11761
Cermak 15260, 15262, 15548, 19569
Cerny, G. 15999, 16000
Ceruti, A. 544, 1076, 1077, 1078, 1079, 1080, 1081, 4031, 4032, 4835, 4836, 4837, 4838, 20949, 20950, 20951
Cerutti, F.M. 10866, 10867
Cerutti, G. 19456, 19457, 20075

Ceruzzi, B. 11753
Cervato, A. 7088
Cervelli, S. 319
Cervenka, L. 17053
Ceselli, P. 12203
Cetin, M. 19988
Ceulemans, E. 6024
Ceustermans, N. 3692
Ceusters, A.M. 101
Cevaal, P.K. 3128, 10175, 17462, 17468, 17470, 17630
Chabert, J.P. 17502, 17846, 18120
Chabouis, Mme 765, 5256
Chacon, J. 438
Chadwick 6139
Chadwick, K.H. 20961
Chae, Y.A. 6334
Challa, H. 3768
Challice 2779, 3211
Challice, A.H. 4081
Chalmers 10609, 11213
Chamberlain 18743
Chambers 9823, 9910, 9911
Chambon, J.P. 8112, 8444, 8594, 8676, 20737
Chambonnet, D. 7721
Chamboredon, J.C. 18325
Chambroy, Y. 16992
Champagne, P. 18325, 18326, 18329
Champeroux, A. Mme 5539
Champion, R. 6044, 6494, 10126
Champredon, C. 10599, 11636, 11835, 11836, 12307, 20006
Chan 18481
Chancellor 2763, 10131, 20843
Chancelor 10241
Chandler 13836, 14219, 14715
Chang, A. 15116
Channon 9333, 9639, 9640, 9678
Chansigaud, J. 8226
Chapa, J. 7461, 7471
Chaplin 15163
Chapman 4137, 4243, 4245, 6380, 6381, 6382, 8763, 13772, 13941, 14332, 14912, 17877, 18577, 18578
Chappell 15139, 18537
Charbonniere 12669
Charbonnieres, R. 16616, 16617
Chardon, P. Renard, C. 14688
Charlemagne, D. 13716, 13717
Charles 3432, 3561, 6689, 6690
Charles, M. 20738, 20740, 20745
Charles, P.J. 2100
Charlet-Lery, G. 11186, 12453, 12456, 12665, 12834
Charlet, G. 12496, 12666, 12668, 12669
Charlet, M.G. 12835
Charley, B. 14690, 14691
Charley, J. 13711, 13713
Charlier, J. 19131
Charpenteau 492, 20188
Charpenteau, J.L. 2746, 15125, 19834
Charriaut, C. 13697, 13704
Charrier 6043
Charrier, J. 10949
Chartier, P. 2385

Coelingh, J.P. 13170
Coen, R.C. 5990, 7948
Coeterier, J.F. 2574, 2575, 2579, 2580
Cogan, T. 12176, 12178, 16829, 19364
Coggins 18549
Cognie, Y. 10953
Cohat, J. 4532, 7505
Cohen–Stuart, M.A. 20986
Cohen, J. 13714, 14207, 14208
Coiro, M. 8366, 8369
Colagrande, O. 17025, 19450, 19451
Colapietra, M. 1479
Colas, G. 10954, 10964, 15938, 15939
Colatruglio, P. 13245
Colbourn 841
Cole 6530, 6531, 6532, 16052
Cole, A. 10555, 12170
Coleman 10971, 11850, 11851, 11852, 20834
Coleman, D. 5502
Coleno, A. 8223
Coles, C.L. 9792
Colesanti, F. 1472, 5452, 6248
Colin, L. 12786
Colin, M. 12788, 12789, 12791, 12792, 12793
Colleau, J.J. 12061, 13053, 13057, 13058, 13060, 13064, 13071
Collier 5706
Collignon, G. 10633
Collins 15144, 16009
Collins, D.P. 3590, 10838, 12159, 17104, 17508
Collins, H.–J. 1223, 1224, 1225, 1226, 19604
Collins, J.D. 10592, 12186, 14371, 19704, 19705
Collins, J.K. 13860, 18934, 19270, 20156
Collins, T. 2126
Collombel, B. 18341, 18924
Colombani, B. 10619
Colon, F.J. 15764
Colon, M. 1134, 1140, 1141, 1143
Colorio, G. 1482, 4193, 4256, 4330, 5786, 15799
Colson, F. 18235, 18408
Colucci, R. 287, 289, 524, 526, 527, 683
Colzani, G. 1470, 12187, 15294, 15332, 15348, 15373, 15374, 15471, 17190, 19708
Combaud, J.F. 10750
Combe 16994
Combe, E. 10948
Comberg, G. 10458
Combettes, S. Mme 8145, 8150, 9096, 9104, 20765
Comerford, P.J. 11622, 15551, 20880
Comi, G. 18946
Commins, P. 2394, 18236
Compagnucci, M. 14860
Compère, R. 1825, 3063, 11538, 11539, 11808, 12280, 17439
Conan, J. 14817
Concaret, J. 206, 642, 649, 836, 913, 1428, 1447, 1762, 2107

Conchie 10412
Concilio, L. 7189
Conesa, A. 488, 641, 2104, 2105, 5492, 16436, 20343
Congiu, F. 11085, 11764
Coni, V. 14763
Connaughton, J. 15438
Connaughton, M.J. 20358
Connell 10764, 15174, 16333
Connolly, J. 12181
Connolly, J.F. 2830, 16830, 18831, 19359, 19360, 20362, 20879
Connolly, V. 2134, 6714, 6734, 6747, 6771, 7866
Connolly, W. 17408
Conrad, A. 15932
Conrad, R. 20590
Conroy, N.J. 10221
Conry, M.J. 261, 262, 2989, 3032
Conte, L. 15472, 17089
Conti, M. 7447, 9370, 9382, 9700, 9701
Conti, S. 6980, 7146, 7189
Continella, G. 4353, 20909
Contini, A. 13878
Contrepois, M. 10742, 12305, 14197
Convent, B. 19575
Convertini, G. 526, 785, 3115, 7473, 11750
Conway, D.A. 13942, 14367
Conway, M. 11435
Cooianchi, D. 7287
Cook 6585, 8454, 8581, 11864, 14915, 14917, 14918, 14921, 15912, 17848
Cooke 41, 7798, 7800, 8563, 8564, 9459
Cooke, B.M. 9369, 9408
Coolen, W. 7961, 8040, 8041, 8820, 9906
Coolen, W.A. 7958, 8821
Coop 14012
Cooper 2814, 2816, 3636, 3952, 8477, 9557, 10143, 14823, 14824, 14825, 14895, 14896, 14897, 15539
Coosemans, J. 985, 8038
Copelli, A. 17422, 17870, 17871, 18163, 18164
Copeman 3575, 6732, 6733
Copin, A. 1107, 1109, 7748
Coppenet, M. 497, 3405, 3406, 3906, 5258, 5517, 5542, 5543, 5544
Coppens d'Eeckenbrugge, G. 1787
Coppens, R. 15969, 16181, 18457, 19024
Coppola, S. 20927
Coppolino, F. 6259
Corazza, L. 9409
Corbel 13838, 14049, 14315, 14317
Corbellini, M. 6392, 6399, 6401
Corberi, E. 20918
Cordesse, R. 10431
Cordiez, E. 5527, 10628, 11544, 11545
Cordonnier, P. 17561, 18109, 18110
Coretti, K. 16270, 19189
Corfield 3105
Corino, L. 4445, 7424, 8814
Corke 4080, 9732, 9758
Corleto, A. 3348, 3349

Cormack 4305, 4306, 4307, 4308
Cornaglia, E. 13527, 14933
Cornelisse, J.L. 12566, 14579, 20173
Cornelius, H. 12403
Cornelius, R. 20100
Cornet, D. 980
Cornford 18746
Cornic, J.F. 2100, 8940
Cornillon, P. 198, 2657, 2658
Cornu 4811
Cornu, A. 6240, 7564, 7565, 7568, 7569, 7570, 7572, 7573, 20803
Cornuet, J.M. 11479
Cornuet, P. 7790, 7890, 9458, 9710
Cornwell, S. 150
Corpet, D. 12861
Corradini, C. 16861
Corrall 3332, 3334, 3409
Corrao, A. 16961
Corre, J. 2753
Correia, M. 8236
Corrias, A. 14385
Corrigall 13794, 14591, 14722
Corring, T. 11187, 12457, 12460, 12465, 12467, 12468, 12471, 12503, 12505, 12506, 12507, 12785, 12856, 12864
Cors, F. 3132
Corsico, G. 19212
Corsten, L.C.A. 19950, 19951
Corstiaensen, G.P. 19742
Corte, A. 9881, 10247
Corteel, J.M. 11078
Corteel, M. 10436
Corten, A.A.H.M. 2263, 2269
Corthier, G. 14690, 14692, 14693, 14694, 14695
Corti, R. 3148, 4981
Cortot, J. 18313, 18315
Corvalan 13736
Cosentino, E. 13245, 13448, 16707
Cosmo, I. 7418, 7422
Cosse, V. 17738
Cossu, P. 14763
Costa 13239
Costa, G. 7217
Costacurta, A. 2637, 4434, 4436, 4437, 4444, 7418, 7419, 7420, 7421, 8815
Costamagna, L. 12881, 16244, 16871, 16879, 17046, 20933
Costantini, A. 283
Costantini, F. 13406, 13407
Costello, M. 19365
Costelloe 6548
Cotteleer, C. 13480, 13481, 13482, 13483, 14075, 14611, 15827
Cotten 8298, 8299, 8763
Cottenie, A. 430, 2639, 19578, 19579, 19580, 19581
Cotter, J. 20156
Cotton 15609
Cotton, J.C.F. 1768
Cotton, M. 8727
Cottyn, B. 11947, 11948, 11950, 11951, 11952, 11954, 11955, 11957, 11960, 15895

14374
De Marzo, L. 8323
De Meuter, F. 14611
De Mey, L. 17039, 20192
De Meyer, E. 11953, 11958
De Michele, A. 4354, 7365
De Michele, Andrea 7347
de Montalembert 11379, 11380, 11452
De Montard, F. 5593, 11825, 18076
de Montard, F. X. 204, 1126, 1421, 2102, 3540, 5591
de Montard, F.X. 3541, 5590, 5592
de Montard, F.X. (Agro) 18075
De Montis, S. 15440, 15672, 15673
De Moor, A. 13972, 14067, 14068
De Moor, H. 18459, 19286
de Nancy 19810
De Nayer, J. 1843
De Paepe, G. 6321, 17733, 18460, 18461
De Philippis, A. 4982
De Ranieri, M. 4551, 4656, 4657, 5907, 5908, 7579, 10296
De Ravignan, F. 18067
De Rijcke, R. 14602
De Robertis, A. 10207
De Roo, R. 6640
De Rossi, C. 519
De Sanctis, F. 7417, 9875
De Schutter, F. 19848
De Stefanis, E. 3116, 3118, 6404, 6406
De Temmerman, L. 104, 1835, 10332, 10333, 10338
De Vaubernier, E. 1447, 2924, 3545, 3546, 12086
De Vecchi, E. 7660
de Verneuil, B.H. 18070
De Verneuil, H. 18068
De Vilder, J. 16228, 16747, 18467, 19292
De Vincentiis, M. 13261
De Vita, M. 4553, 4554, 5876, 10296
De Vleeschouwer, D. 5680, 17047, 17048
De Volder, E. 1787
De Vos, N. 11946
De Vos, R. 19967
De Wever, L. 17734
De Wilde, J.J.F.E. 21040
De Wilde, R. 12377
De Wulf, F. 99
De Zutter, L. 19584
Deanesley 12913
Debaere, R. 11547
Debailleul, G. 17503, 18128
Debecg, J. 13477
Debecq, J. 13486
Debergh, P. 6856
Debets, F.M.H. 21053
Debil, J.P. 3905
Debois, J.-M. 17930
Debouche, C. 19781, 19845, 19847
Debouck, P. 14608
Debruyckere, M. 15532, 15637, 15638
Debuisson, J. 2538, 7524, 7525
Debus, H. 16447

Decaen, C. 12056, 12370, 19355
Decallonne, J. 8035
Decau 491, 492
Decau, J. 1434, 3633, 3900, 3901, 5463, 5467, 6485, 20812
Deckelmann, B. 1964, 4887
Deckers, J. 1190, 4211, 4223, 4224, 4274, 5802
Deckers, T. 3824, 3938, 4275, 4576
Deckert, T. 13500
Decleire, M. 1114, 10074
Declerck, D. 16217, 19262
Decourt, M. 18076
Decourt, N. 2110, 2111, 2751, 4794, 10389, 11844, 18136, 18217, 18340
Decourtye, L. 7242, 7243, 7609, 7610, 7611
Dedryver, C. 8449, 8450
Dedryver, Ch. 8130, 8219, 8633, 8660, 20811
Deegen, E. 10579
Deelder, C.L. 2262, 2268
Defares, P.B. 19777
Deffontaines, J.P. 2386, 2387, 18068, 18070, 18299, 18303, 18423, 19864
Defosse, L. 7830
Defraigne, J.P. 17895
Defrance, H. 483, 1416, 1760, 1761, 4067, 4234, 4235, 4476, 4477, 5810, 20348
Degand, J. 17344
Degani, J. 8357
Degenhardt, H.C. 10565
Degert, G. 18090
Degkwitz, E. 20561
Degle, I. 19333
Degras, L. 6621
Degrijse, E. 6160
Deidda, M. 6432
Deidda, P. 4356, 7350, 7351, 7446
Deil, U. 1863
Deinema, M.H. 19727, 19729
Deinum, B. 3062, 3505, 3625, 3675
Dejardin, D. 8234
Dejou, J. 203, 835, 911, 912, 3540, 5592
Dekhuijzen, H.M. 5749, 6826
Dekker, A. 10058, 10620, 12281
Dekker, J. 9240
Dekker, L.W. 855
Dekker, P.H.M. 3879, 3937
Dekker, T. 14391
Dekock 5288, 5289, 5290
Del Bene, G. 8312, 8313
Del Favero, R. 4947, 17861
Del Monte, M. 685, 20370
Del Re, A. 10348, 16660, 16859, 19109, 19448, 19449, 20031
Del Vecchio, C. 283
Del Zan, F. 5454, 7097, 7192
Delabraze, P. 10032, 10309, 10387, 10388, 10390
Delaere, L. 5061
Delage, J. 12371, 12372
Delanghe, F. 15405, 15406
Delannay, J. 6024

Delarbre, F. 20338
Delas, J. 484, 638, 759, 761, 4417, 5246, 5461, 5659, 5859, 5860, 5861, 9869
Delatour, C. 9993, 10000, 20754
Delattre, P. 19769
Delaude, A. 3640, 3895
Delbeke, R. 16746, 18469, 19293
Delbut, J. 7731
Delcarte, E. 1829, 2639, 10318, 10319, 10327
Delecour, F. 5944
Deleu, R. 1107
Delfini, C. 17009, 17011, 19443
Delforno, G. 6635
Delhaye, J.P. 11544, 11545
Delhaye, R. 3299, 6273, 10204, 10401, 20521
Delhey, R. 3195, 3196, 3300, 3301, 15940
Delincee, H. 18895, 18896
Dell'aendla, G. 792
Dell'Aquila, S. 11039, 11040, 11042, 11043, 11044, 11045, 11046, 11047, 13240, 13243, 14574
Dell'Orto, V. 10869, 10870, 10871, 13137
Della Giustina, W. 8113
Della Lucia, D. 1510
Della Strada, G. 7287, 7288
Della, Giustina, W. 8445
Dellacecca, V. 528, 3963
Dellaert, L. 6077
Dellagatta, C. 7082
Dellaglio, F. 20911, 20912
Deller, B. 594
Dellman, H.D. 15116
Dellweg, H. 16446, 16447, 16448, 17040, 20528
Delmas, A. 20734
Delmas, J. 4001, 4002, 4003, 4004, 4005, 4008, 4010, 4015, 4809, 5768, 5770, 5772, 5773, 20796
Delmas, J.M. 6454, 7484
Delobel, B. 8228, 8229, 8230, 8232
Delogu, G. 6202, 6204, 6206
Delord, B. 17579, 18085
Delord, F. 7463, 7464
Delorme, A. 20603
Delort, F. 4478, 7469
Delouis 10753
Delouis, C. 10426, 10430
Delpech, P. 4004
Delpech, P. et Desmoulin, B. 11683
Delpech, S. 10963, 10964
Delphin 488, 2104
Delrio, G. 8500
Deltour 15969
Deltour, J. 6004, 15531, 19960
Delvaux, J. 4732, 4853, 4854
Delventhal, H. 1937, 11370
Delver, P. 396, 1556, 1577, 5787, 5788, 5789, 5790, 5816, 5817, 5818, 5819
Demarne, Y. 11627, 11630, 11681, 11684, 11686, 11689, 12851, 12863
Demarquilly 3541
Demarquilly, C. 6627, 6807, 11697,

Droege, H.P. 1119
Droege, P. 20221
Droeven, G. 2601, 5063, 11562
Drommer, W. 10577, 14622, 14623, 14624, 15092
Dronne, Y. 11710, 17840, 19410
Druart, C. 1136
Druart, J.C. 1129
Druart, Ph. 6857, 7209
Drysdale 18632
Du Crehu, G. 5679, 6836
Du Merle, P. 2100, 8902, 8911
Duben, J. 9268, 9270
Dubenkropp, G. 15459
Dubois, J. 6323
Dubois, J. P. 5084
Dubois, J.P. 1144
Dubois, M. 10434, 11381, 11431, 20818
Dubois, M.P. 10958
Dubois, P. 19697
Dubos, F. 11688, 12846, 12850, 12853, 12859, 15003
Dubos, J. 17838
Dubourguier, H.C. 10742, 12305, 14197
Dubray, G. 13697, 13699, 13704
Dubuisson, J. 2602, 4491, 7296, 7298, 7724
Ducatelle, R. 14604
Duchêne, M. 16181, 19131
Ducluzeau, R. 11630, 11688, 12483, 12494, 12495, 12664, 12665, 12845, 12846, 12850, 12852, 12853, 12854, 12855, 12856, 12857, 12858, 12860, 12861, 12862, 12863, 12864, 14687, 15003, 15119, 15120, 15121, 19490, 20752
Ducrey, M. 4791, 4793, 4796, 4797, 4974
Duden, R. 3704, 16612, 18444, 19089, 19090
Dudley 9594, 18594
Duee, P. 12474, 12485, 12487, 12490, 12492
Dufey, V. 15272, 15273, 15274, 15320, 15552, 15767, 17163
Duff, C. 5302
Duffus 2977
Dufour, J. 10957
Dufresne, S. 15119, 15120
Duggan, J.J. 3233, 6144, 7765, 7766, 7767, 7850, 7851, 7852, 7888, 7919, 8568, 8569, 8841
Dührssen, E. 20563
Duin, H. van 19392
Duineveld, T.J. 9922
Dulieu, H. 7570, 7573, 7731, 20803
Dullemen, E. van 15269, 15306
Dulor, J.P. 10949
Dulphy, J.P. 6627, 11697, 11698, 11701, 11705, 12070
Dumas de Vaulx, R. 7362, 7722
Dumas, J. 11405, 11426, 11428
Dumas, Y. 764, 3710, 3902, 6625
Dumas, Y. - Clairon, M. 5514
Dumay, C. 11629, 11690, 11824

Dummel, K. 17260, 17655
Dümmler, H. 990
Dumont 15508, 16699, 19901, 19902, 19904
Dumortier, B. 10346, 20735, 20736, 20738, 20740, 20741, 20742, 20743, 20744, 20745, 20746
Dumstorf, H. 17362
Dun 12620
Duncan 4549, 4650, 9158, 9821, 11291, 11332, 11895, 13368, 13403, 14892, 20027
Dunlop 15866
Dunn 3583, 6050, 7996, 8649, 20854
Dunne, B. 3113, 7840, 9172, 9335
Dunne, C. 14371, 19704, 19705
Dunne, R.M. 7768, 7889, 8619, 8764
Dunne, W. 2995
Dunning 7802, 8262, 8565, 8566, 8567, 9362
Dunwell 7508
Dupire, S. 104
Dupont, E. 13422
Dupont, J. 11379
Dupont, Y. 17506
Duprat, F. 16616
Dupuis, M. 224, 225, 644, 645
Dupuy, G. 13048
Duquesne, J. 7484
Durand, E. 14973, 18922, 19134
Durand, G. 11627, 11631, 11633, 11682, 12473
Durand, J.H. 201, 762, 1417, 1419, 1806
Durand, M. 11629, 11690, 11824
Durand, P. 10589, 13695
Durand, R. 232, 654
Durand, Y. 8467
Duranti, A. 1503, 2843
Duranti, G. 4819
Duranti, M. 11761
Dureau, Mme 5247
Dureau, P. 1418
Dureau, P. Mme 484, 760, 3847, 5698, 17060, 17086
Durey, P. 5667
Düring, H. 4375, 4377
Duron, M. 7612, 20758
Dürr, G. 17900
Dürr, R. 1869
Durrant 670, 1808, 5497
Durwen, K.-J. 2430
Dusaussoy, G. 2100, 8912, 8915
Dussaigne, A. 20822
Dusseldorp, D.B.W.M. van 18367
Dusseldorp, D.W.B.M. van 18374, 18375
Duthion, C. 1422, 2107, 5107
Duthoit, J. L. 8173, 8176
Duthoit, J.L. 19767, 19768, 20774, 20775, 20776, 20780, 20781, 20782, 20783, 20787, 20791
Dutil, P. 655, 1442, 5448
Dutrecq, A. 6021, 6033, 9266
Dutzler-Franz, G. 992
Duval, L. 497, 653, 5515, 5516, 20823

Duval, Y. 500, 5259, 5518, 12847, 15122, 15123
Duvekot, W.S. 16163, 16169, 17136, 20965
Düvel, D. 19616
Düx, A. 13266
Duym, J. 7825, 7826, 7827, 8429, 9341, 10179
Duyvendak, R. 6080, 6150, 6723
Dwivedy 8269
Dwyer, E. 6143, 16536, 18829, 19014
Dyer 2783, 6389, 10522
Dyson 2810, 14561
Dzapo, V. 13270
Dziedzic, A. 20750
Dziengel, A. 16583, 16590

E J 6671
E White P 8668
Eades, J.F.K. 7769, 18830, 19098, 19204, 20030
Eadie 10981, 10983, 10984, 11900, 11901
Eagles 6630, 6631, 6632, 6674
Eales 10977
Early 18777, 18778
Easson 2960, 2961, 3027, 3682, 15863
Easterbrook 8702, 8730, 8759
Ebbels 9874
Ebben 9677, 9693, 9895, 9896, 9939
Ebbesen, F. 19523
Ebbinghaus, R. 7545, 7550, 7554
Eben, C. 14
Eberhard, D. 6230, 6232
Eberle, G. 15022
Ebert, A. 4056
Ebing, W. 7781, 7782, 7783, 8070
Ebner, R. 20230
Eccher Zerbini, P. 4199, 16127, 16128, 16130, 16132, 16133, 16135, 17128, 19103
Eccher, A. 4819, 4873, 4874, 4892, 4894, 7665, 7666, 7667, 7668, 7669, 8921, 20885
Eccher, T. 4094, 4205, 4208, 7290, 7323
Echard, G. 13316, 13318, 13446
Echaubard 19699
Eck, G. van 15706, 15724
Eck, J.H.H. van 14940
Eckardt, C. 16194
Eckehardt, H. 9020
Eckstein, D. 15832, 20680
Eckstein, K. 12391
Ectors, F. 10400, 10627, 14073
Eddie 11219
Edelman–Vlam, A.W. 352
Edelsten, D. 16825, 16826, 19523
Edema, J.M.P. 18390, 19532
Edenharder, R. 19763
Edens, F.J. 17191, 17229
Eder, H. 20048, 20101, 20180
Edgar 3341
Edmonds 67
Edney 9753, 9754
Edokwe, G. 13542

Flink, J.M. 16472, 16473, 16474, 16614, 16643
Flint 12916, 12917, 12918, 12945, 18926
Floate 46, 5603, 10981
Flock, D. 14877, 14879
Flock, D.K. 11122
Flohn, H. 1222
Floor, H. 10337
Florence 20013
Florent, A. 7180
Florenzano, G. 543, 1075, 2176, 16565, 18832, 19105, 19454, 20948
Flores, C. 10580
Flori, P. 7813, 9880
Fluhr, R. 14442
Fluit, J. 20387, 20391
Flurl, H. 4758, 4970, 19854
Flynn, A. 18931
Flynn, A.V. 5523, 11622, 12168, 12172, 17104, 17329, 19570
Flynn, P. 18931
Flynn, V. 5450
Focant, M. 12283
Focardi, P. 273
Fockedey, J. 12637, 19964
Foeken, D. 20404
Foerst, K. 4927
Foerster, E. 2073
Fogh, H.T. 5215
Fogher, C. 6264
Fogliani, G. 7943
Föhr, C. 2374
Foing, J.J. 8932
Foissner, W. 1948, 1949
Foldager, J. 12029, 14192
Foldi, I. 8225
Foldø, N.E. 6524
Folkers, D. 20264
Folkers, G. 18849
Folkerts, J. 11354
Follet 13376
Folstar, P. 16487, 17036, 18953
Fölster, E. 3700, 3984, 5754
Fölster, H. 1241, 5996
Folving, S. 2445
Fontaine, G. 11555, 11557, 12639, 12640, 12641
Fontana, F. 1478, 3284, 5503, 5505, 6603
Fontana, P. 9186, 16660
Fontanazza, G. 3157, 6468
Fontaubert, Y. de 10751
Fonteyne, R. 15405, 15406
Foot 12330, 12332
Forbes 3422, 5555
Forceille, M.J. 5527, 5528, 11544, 18834
Forche, E. 20565
Forche, S. 9343
Forchthammer, L. 4607, 4609, 5887
Ford 2936, 5796, 5797, 11586, 18584, 18585
Formanek, H. 12837
Formisano, M. 16870
Forneris, G. 12734

Forni, C. 6146, 6394, 6397
Forni, E. 18937, 19104
Forsingdal, K. 16809, 16810, 16811
Förster, E. 3379
Forster, H. 4885, 5204, 5210, 5473, 9271
Förster, M. 13009, 13013, 13464, 20233
Förster, U. 18432
Forstner, M.J. 13640, 14446, 14447, 15036, 19631
Forsyth 12961, 12962
Fortini, S. 3119, 6255, 20894
Fortune, A. 2993, 2994, 3277, 3485, 15329
Fortusini, A. 7437
Fos, A. 9117, 9728, 9870
Foschi, S. 1082, 2840, 7819, 7820, 8377, 8570, 8753, 9418, 10216, 10267, 15936, 15937, 18948
Fossati Galli, E. 20919
Fossati, G. 3074, 6293, 6294, 16058
Fosset, J. 14982, 19767, 20773
Foster 6197, 7935, 8301, 8833, 12697, 13221, 13405, 18628
Fostier, A. 11375, 11377, 11381, 11385, 11386, 11400, 11402, 11404, 11420, 11421
Foti, S. 1550, 3243, 3353, 3354, 3767, 3874, 3875, 3991, 6981, 6982, 7086, 7095, 7155
Fottrell, P.F. 11410
Fouarge, G. 6007, 8963
Foucault, M. 8114
Fouet, J.P. 17721, 18098
Fouilloux, G. 7052, 7053, 7054, 7055, 7056, 7057
Foulhouse, I. 18103
Foulhouze, I. 18126, 19887
Foulkes 13099, 13101, 13103, 13104, 13108
Foulley, J.L. 13052, 13056
Foully, J.L. 13049
Foulon, M. 18833
Fourbet, J.F. 229, 651, 2923, 3022, 3101, 5541
Fourbet, J.P. 230, 5257
Fournier, J.C. 208, 763, 1025, 1029, 1030, 20142, 20802
Fournier, N. 16563
Foury, C. 6949
Fowler 10070, 11211, 11212, 12521, 12526, 12528, 12529, 13328
Fox 9465, 9466
Fox. P.F. 18931
Fox, M. 15979
Fox, P.F. 16841, 17106, 18933, 19365, 19366, 19367, 19370
Foxe, M. 9824
Foxell 16342
Foy 1151, 1152
Fraakinet, M. 818
Frädrich, G. 10654
Frahm, H. 18814
Frahm, J. 7843
Frahm, K. 10679, 10683, 13002, 13006
Fraipoint, L. 16359

Fraipont 16360
Fraipont, L. 15734, 16358
Fraiture, A. 7729
Frame 3587, 3588, 3589, 3651, 10612, 11618
Francani, 16850
Francani, R. 15814, 16242, 16843, 16846, 16847, 16849
France 11871
Franceschetti, U. 7190, 7581
Francia, U. 3084
Franck, E. von 5164
Francke, R. 16649
Francke, W. 8062, 8881, 8882, 20104, 20611, 20612
Franco, E. 8135, 8138, 8735, 8766, 8768
Franco, V. 6815
Franco, W. 1241
François, E. 5436, 10439, 11093, 11565, 11808, 11961, 12280, 12284
François, M. 5089
Frandsen, F. 13681
Frandsen, K.J. 6653, 6654
Frank, A. 18889
Frank, H. 13558, 15014
Frank, H.K. 5183, 9719, 18872, 19085
Frank, W. 1941, 1942, 1943, 1944, 1945, 1946, 11518, 13610, 15029, 15099, 15100
Franke 2033
Franke, E. 4686
Franke, W. 3697, 4493, 4585, 15830, 20540, 20541
Franken, H. 121, 122, 123, 441, 1729
Franken, M. 462
Franken, P. 13906
Franken, S. 7167
Franken, W. 462
Frankinet, M. 1108, 6006, 7831, 10204
Frankowski, J.P. 8145
Franz, C. 2695, 5028, 5033, 5035
Franz, F. 1964, 4753, 4754, 4756, 4757, 4758, 4885, 4886, 4887, 4970, 17643, 19854
Franz, G. 20636
Franz, J.M. 20120
Franzusky, D. 15041
Frascati, F. 270, 276
Fraselle, J. 7748, 8950, 10058, 10620, 12281
Fraser 9593, 10989, 12124, 13723, 14018
Frassine, W. 17800
Frat, Y. 17844
Fraticelli, A. 519
Fratteggiani Bianchi, R. 13246, 13247
Frau, A.M. 7483
Frauen, M. 3656, 6785
Frawley, J. 17410, 18343, 18410
Frebling, J. 13056, 13059, 13062
Frech, P. 16396, 17116
Freddi, G. 16438
Frede, E. 16769, 16771, 16772
Frederiks, J. 15355
Frederiksen, J. 5383

Heger, K. 826, 3261
Hegewald, B. 2674
Hegner, D. 13621, 15103
Heide, B.A. 19755, 19756
Heidemann, F. 18826
Heidhues, T. 17772, 17773, 17969
Heidler, G. 1321, 9060, 20271
Heidrich, H.J. 10565, 13544, 13975, 14086, 14874
Heidt, H. 20526
Heij, G. 3779
Heij, W. 17274
Heijbroek, W. 6618, 6833, 7859, 7860, 8576, 8577, 8578, 8579, 8580, 20516
Heijning, J.J. 15315, 15402
Heil, G. 11279, 11280, 13358, 13360, 13392, 13393, 13394
Heilenz, S. 19973
Heilinger, F. 3201, 6619, 10365
Heilmeier, A. 14805
Heimann, W. 16499, 18849, 19425, 20637
Heimeshoff, B. 15746, 15747, 16392, 20228, 20235, 20236
Hein, A. 9031, 9036, 9040
Heindl, M. 10049
Heine, K. 19329, 19662, 20059
Heinemann, G. 10940
Heinemeyer, O. 1005, 5174
Heinlein, R. 9540
Heinrich, F. 17781
Heinrich, I. 17487, 17488, 18023
Heinrich, W.-D. 2400, 2402
Heins, H.G. 16019
Heins, H.H. 2687
Heintsberger, G. 810
Heinze, W. 4226, 5881
Heise, P. 15933
Heiseke, D. 11657
Heisig, G. 1873
Heisterkamp, S.H. 19825, 19946
Heitefuss, R. 9267, 9344, 9365, 9654, 10088, 10181, 10211, 10358
Heitmann, J. 13592
Heitzhausen, G. 17367
Heitzman 10536, 19546
Heizmann, A. 16388
Heland, M. 11406, 11426, 11427, 11428, 13426
Held, J.J. den 2537
Held, W. 16446
Helder, J.F. 11354, 11773, 12626, 12627
Helder, Th. 14962
Helfferich, B. 5530, 11973
Helle, W. 8425, 8427, 14437, 21087
Hellemann, C. 15932
Hellemans, R. 18394, 19783
Hellemond, K.K. van 11941, 12265, 12267, 12272, 12572, 20177
Heller, H. 1891
Heller, P. 19790
Heller, R. 11519
Helleris, A. 2138, 20881
Helles, F. 2443, 18064
Hellings, A.J. 1572, 1573, 1591, 1592,

1593, 1616, 1617, 1618
Hellriegel, T. 991
Hellrigh, B. 4949
Hellweg, W. 15700
Hellwig, I. 19157
Helm, H.-U. 5791
Helming, G. 16759, 17148
Helming, W. 2554
Helms, W. 12395
Helweg, A. 20, 10076
Helwig, A.J.G. 20386
Hemington 10519
Hemkes, O.J. 11772
Hemmer, K. 15564, 15567, 16381
Hemminga, H. 13255
Hemminga, M.A. 21001
Henderickx, D. 11945
Henderickx, H. 11944, 12376, 16679, 19500
Henderson 7806, 8293, 8294, 8586, 11902, 15913
Hendi, A. 8046
Hendrick, E. 1461, 5963
Hendrickx, G. 7958, 8041, 8820, 8821, 9906
Hendriks, H.J. 15064
Hendriks, L. 8
Hendriks, M. 1682, 19957
Hendriks, T.H.B. 19954
Hendrix, A.T.M. 17221, 17224, 17227, 17230, 17231, 17240, 17241, 17242, 17243, 17244, 17245, 17629
Hendrix, W.M.L. 15050
Hengeveld, A.G. 12227, 15925, 15926
Henin, S. 621
Henke, H. 2557, 2558
Henkel, H. 12408, 12409, 12410
Hennaux, L. 11808, 12280
Henne, A. 2442, 2562, 4770, 4771, 5984, 15789, 17665, 17666
Hennerty, M. 9824
Hennerty, M.J. 4192, 5304, 5813, 5814
Henning, W. 19079
Henninger, W. 20530
Henningsen, K.W. 3892, 6029, 20717
Hennlich, W. 15929
Henrard, G. 7524
Henrichfreise, A. 1998, 1999, 2008, 2009
Henrichs, J. 5145
Henrichsmeyer, W. 17346, 17746, 17914, 17915, 17916, 17917, 17918
Henrici, D. 20228
Henriet, J. 7753, 19073, 19970, 19971
Henriksen, Aa. 16641
Henriksen, H.A. 4786
Henriksen, J. 10734, 12041
Henriksen, K. 3977
Henriksen, P.S. 15075
Henriksen, S.Aa. 13494
Henrion, I. 451
Henry, J.B. 18228, 18316
Henry, Y. 11191, 11627, 12474, 12476, 12478, 12479, 12480, 12482, 12484, 12486
Hens, J. 12643

Henseler, K.L. 1748
Hensen, K.J.W. 7595
Hensey, M. 2145
Henshall 18716, 18717, 18718, 18719, 18735
Henstra, S. 20430
Hentgen, A. 493, 2644, 3326, 3402, 17201
Hentig, W.-U.von 4612, 4613, 4614, 4615, 4616, 4617, 4619, 4619
Hentschel, G. 5093, 5159, 20055
Hentschel, H. 16112, 19085
Henze, A. 17783
Henze, J. 4045, 4368, 16096, 16097, 16098, 16631
Hepherd 15170, 15171
Heppel 3340
Hepting, L. 6234
Herbert 12956, 13209
Herbert, J. 5263, 18257
Herborn, A. 4042
Herbst, M. 14641
Herdlitschka, P. 13612
Heredia, N. 7416
Herfs, W. 20276, 20277
Herinckx–Pirlot, J. 17173, 17894
Hering 1272, 1276
Heringa, A. 2310
Heringa, J.W. 2870
Heringa, R.J. 6799, 6800, 6801, 6802
Herken, A. 13924
Herlihy, M. 5298, 5501, 5522, 5559, 19368
Herlyn, D. 15104
Hermanin, L. 4832
Hermann, A. 19307
Hermann, J. 17918
Hermann, L. 7256, 7259
Hermann, P.-G. 2403
Hermanns, W. 14636
Hermansen, J.E. 6105
Herms, A. 15650
Herms, U. 459
Hermsen, J.G.T. 6561, 6565, 6571, 6572, 6575
Hermsen, J.G.Th. 6574, 6576
Heron 12508
Herregods, M. 16092
Herring 14016, 14017
Herrington 7797, 7800
Herrington, B.K. 7798
Herriot 3572, 3573
Herrmann, C. 19160
Herrmann, G. 19420
Herrmann, H. 2061
Herrmann, H.H. 11075, 11107
Herrmann, K. 18283, 19078, 19079, 20621
Herrmann, S. 7631, 7635
Herrmann, W. 10567
Herröder, W. 15563, 15568
Herschel, A. 13599
Hertel, J. 19523
Hertampf, B. 13608
Hertsch, B. 10580, 13918, 13921, 13922
Hertveldt, L. 7873

LIST OF SCIENTISTS

Höfner, W. 2901, 5144, 5417, 5418, 5841, 9372
Hofschreuder, P. 19741
Hofstee, E.W. 18381, 18382
Hofstee, J. 20041
Hofstetter, N. 20305
Hogenboom-Kesler, B.E.Th.A. 18451
Hogenboom, N.G. 7160, 7161, 7162, 7163, 7196, 7197
Högner, W. 20108
Hohenadl 1963
Hohls, F.W. 19151
Höhm, H.-P. 17806, 18039
Hohmann, B. 5027
Hohmann, K. 18049
Hohn 19608
Höhn, H. 1887, 13566, 14092
Hokse, H. 16628
Holden 2785, 2788, 6542, 9404, 15162
Holding 1157, 3648
Holdsworth 18732, 18733
Hole 3910
Holla, M. 18047, 18293, 18294
Hollaender, J. 18920
Holland 19895, 19897
Hollatz, R. 20658
Holle 1267
Höller, H. 12290
Holliman 11204, 13739
Hollings 9126, 9127, 9128, 9713, 9899, 9900, 10056
Hollins 9299, 9301
Hollmann, P. 18022, 18025
Hollomon 9361
Holloway 2776, 4299, 4301, 7793
Hollwich, W. 14662, 19132
Holm, E. 4972
Holm, F. 11581, 16276, 16290
Holm, S.N. 3165, 6750, 13431, 13432
Holmes 2630, 2631, 2980, 2981, 3030, 3474, 3574, 9505, 9558, 14834, 14916
Holmsgaard, E. 2077, 4993
Holsheimer, J.P. 12630, 12712, 20410
Holst, A.F. van 379, 554, 1562
Holst, H. 8051
Holsteijn, G.P.A. van 3781, 3782, 15635
Holstein, B. 18854
Holsten, H.H. 10579
Holsteyn, C.W.M. van 13891
Holt 15166, 15507, 16225, 18607, 18608
Holterman, H.J.W. 17682, 17882
Holtz, E. 2542
Holtz, W. 10647, 10648, 10649, 10929, 11075, 11106, 11107, 11108, 11109, 11110, 11111, 11112, 11113, 11114, 11414
Holtzbauer-sharman 10506
Holz, B. 9854, 9855, 9856, 9858
Holzbauer-sharman 10507, 10512, 10514
Homagk 1266
Homberg, E. 16557, 19042, 19485, 19486, 19487
Homble 7035, 7037
Homrighausen, E. 1876, 1877, 5140

Hondelmann, W. 2697, 2698, 6037, 19796
Honig, H. 3391, 15779, 15899, 15900
Honikel, K. 13490, 20261
Honing, Y. van der 11774
Hood, D.E. 16706, 17853
Hoof, H.A. van 9224, 9515
Hoog, C. de 18385
Hoogendoorn, H. 12768
Hoogerbrugge, A. 11265, 13381
Hoogerkamp, M. 1554, 3620, 10165, 10248, 10283
Hoogeterp, P. 4560
Hoogkamer, W. 9340
Hoogmoed, W.B. 978
Hoopen, A. ten 1090
Hooper 8010, 8661, 15151
Hoorens, J. 14602, 14603, 14604, 14605
Hoorn, W.G. van 21060
Hoornweg, J. 17530
Hooydonk, M.J. 16483, 16675
Hooydonk, M.W. 16482
Hope Cawdery, M.J. 13521, 14366, 14570, 14571
Hopgood 4133
Hopkins 14841, 14925
Hopkins, L. 19702
Hopmans, P.A.M. 3788
Hopp, H. 9081
Hoppe, H.H. 9654, 10087
Hoppe, R. 17709
Hoppmann, D. 1327, 4384, 4385, 4386, 4387, 4388
Hoque, R. 5442
Horak, P. 16761, 19306
Hörchner, F. 13540, 13541, 13542, 13973, 14614, 15083, 15084, 20090
Horgan, V. 19206, 19270
Horn, G. 17941
Horn, R. 439, 869
Horn, W. 4528, 5883, 7486, 7487, 7488, 7489, 7490, 7544, 15425
Hornby 2950, 9305, 9306, 9308
Horne 9130, 16744, 20010, 20851
Horney, G. 20307
Hörnicke, H. 10581, 11499, 20626
Hornig, B. 10664
Horsburgh 17315, 17322
Horst, P. 10445, 10636, 11314, 12882, 13386, 13461
Hörsten, H.von 13613
Horsten, J. 9269
Hörstgen, G. 20601
Horstink, A.R.M. 13164
Horstmann, K. 2607, 8935
Hörtig, W. 20046
Horzinek, M.C. 13905, 15054
Höschele, K. 2372, 19626
Hosie 16007
Hoste, G. 20088
Hottes, K. 17745, 18244
Hötzel, D. 19464, 19465, 19466, 19501, 19602
Hötzel, H.-J. 1950
Houba, V.J.G. 5361

Houben, J.H. 16711
Houben, J.M.M.T. 1770
Houben, J.M.M.Th. 856
Houdard, Y. 2387, 2729, 2730, 3102, 3661, 3894, 17391, 17392
Houdayer, M. 14696
Houdebine, L.-M. 10430
Houdebine, L-M. 10429
Houee, P. 18230
Houlier, G. 13051
Houseman 11608, 12526
Houssin, Y. 10946, 10947
Houston 20072
Houte de Lange, S.M. ten 2253, 2504
Houtmeyers, J. 19583
Houwing, H. 15766, 16223, 16734, 16737, 19280
Houwing, J.F. 960, 3250
Houx, N.W.H. 398
Houx, W.N. 7824
Hovart, P. 15405, 15406, 16216, 16217, 16218, 17277, 18466, 19262, 19592
Hovell 12522, 14724
Hoven, T.J.J. van den 20498
Hovmand, M.F. 619, 2726
Howard 4141, 4692, 4693, 6533, 13733, 13736, 13741, 14255
Howell 7848, 8527, 12525
Howes 8268, 14905
Hoyer, R. 17647
Hoyle 15177, 15178
Hoyningen-Huene, J.von 1008, 1009, 1325, 2653, 2704, 2705, 2706, 20185
Hoyoux, J.M. 7777
Hradetzky, J. 17663
Hruschka, H. 19637
Hsueh-err, C. 16639
Huang, P. 2069
Hubbard. 18929
Hübel, K. 2061, 19682, 19682
Huber, F. 4978
Huber, J. 8074, 8080, 8717
Hubert, J. 14589
Hübler, K. 14155, 15816
Hübler, K.H. 2542
Hübner, R. 3659
Hudalla, B. 19268, 19488, 20265, 20689
Hudault, S. 12858, 20752
Huebschle, O. 13659
Huet 2923
Huet, Ph. 651, 3100
Hugard, J. 4241
Huge, P. 988
Hugenroth, P. 446, 1238
Huger, A.M. 8076, 9073, 9075
Huges 3103, 6378
Hughes 38, 3426, 3427, 3559, 3643, 5074, 8992, 11331, 14224, 14246, 14257, 18509, 20860
Hughet, L. 12312
Huglin, P. 4419, 4422, 4423, 5862, 7401, 7406
Huguet, C. 1760, 1761, 4067, 4234, 4476, 5808, 5809, 5810
Huguet, C. Mme 5779, 5807, 5869

Kanning, K. 10645
Kant, N.F. van de 17283
Kanters, F.M.L. 6874, 7008
Kanters, H.L. 709
Kaplan, Y. 2303
Kapol, F. 864, 2032
Kappus, W. 19521, 19522
Karanth, N. 20115
Karg, H. 10692, 10694, 10695
Karl, I. 11126
Karl, J. 1345, 1346
Karlsen, P. 2722, 5100, 5118, 5731, 10406
Karnatz, A. 4038, 4039, 4278, 7234, 10287
Karnbaum, B. 11999
Karnop, G. 16219, 16220
Karo, M. 20213
Karras, K. 12996
Karrasch, A. 13973
Karrenberg, H. 1372
Karssen, C.M. 10178
Kasbohm, C. 11515, 13546, 13547, 13548, 15012
Käser, H. 6786
Kassam, K. 16593
Kassanis 9150, 9152
Kassemeyer, H. 4391
Kassner, H. 19601
Kasteren, H.W.J. van 20077
Kaswalder, F. 9581
Kat, M. 2275
Katan, M.B. 18956, 19492
Kathol, G. 17560
Kato, F. 1926, 17636, 17640
Katona, Ö. 12941
Kattein, U. 16981
Katz, M. 14656
Kau, M. 3520
Käufer, I. 13557, 13558, 14788
Kaufholz, H. 5231
Kaufmann, F.von 11106, 11111
Kaufmann, R. 15105
Kaufmann, U. 6029, 7678, 20717
Kaufmann, W. 11819, 11979, 11993, 15903, 19308
Kaule, G. 1983, 2373
Kaulen, H.A. van 6082
Kausch, H. 1873, 11397, 11398
Kausch, W. 4367, 20535, 20536, 20537
Kaushik 12730, 12752
Kauss, W. 15489
Kavanagh A.J. 10841, 15667
Kavanagh, J.A. 8993, 9173, 9508, 9615, 16057
Kavanagh, T. 4325, 4329, 6707, 7316, 7317, 7318, 7320, 7911, 7918
Kay 10988, 11080, 11081, 11742, 11913, 11914, 12122, 12123, 12150, 12530, 12965
Kayser, R. 17051
Keane, G.P. 3486
Keane, M.G. 12165, 12171, 12173
Keane, T. 20358
Kearney, A. 11026, 14569
Kearney, P.A. 12545

Keating, J. 18430
Keay 18759
Kechel, H. 10031, 10393
Keding, K. 16983
Keenan, L.R.J. 14760
Keep 7306, 7307, 7613
Keerthisinghe, D.G. 581
Kees, H. 10187
Keesen, H. 11413, 12741
Kehoe, H.W. 3236, 6551, 6552
Keiding, H. 7687, 7715, 7716
Keiding, J. 20715
Keijbets, M.J.H. 16084, 16620, 16621, 16622, 16623, 16624, 16664, 19070
Keil, W. 2062, 2063, 2064, 2066, 8104, 8893
Keilen, K. 124, 579, 19605
Keim, A. 1942
Keimer, B. 2400, 2402
Keinhorst, A. 15980
Keitel, K. 16296
Keizer, A. de 20984
Keizer, M.G. 715
Kejwal, K. 15555, 15556
Kekem, A.J.T. van 17552, 17555, 17556, 20442
Kelami, A. 10444
Kelany, J.M. 1880
Kell, R. 9540
Kelleher, D.L. 13237
Keller, H. 10568, 13910, 13912, 13913, 13914
Keller, P. 10727, 11176, 12448, 15652, 15654, 16432, 17112, 17560
Keller, R. 1231, 4962, 4963, 4973, 4977, 10032, 15751
Kellermann, B. 2547
Kellermann, M. 1319
Kellert, M. 20290
Kellner, P. 10447, 10448, 10449, 11368, 13177, 13271
Kellner, R. 11655
Kelly 11885, 15256
Kelly, A. 18936
Kelly, J. 2825, 4652, 4654, 4710, 4711, 4712, 4715, 4716, 7942, 20880
Kelly, P. 16832, 16835, 16836, 16837
Kelly, T. 10422, 11391, 11392, 11453, 17278, 17279
Kelly, T.A. 10857
Kelly, W.R. 14762
Kelman 15245
Kelso 3797, 16648
Keltjens, W.G. 5362
Kemmers, R.H. 1637, 1639, 1653, 1654
Kemp 10971, 11849, 20834, 20835
Kemp, A. 5632, 11772, 12212
Kempen, G.J.M. van 12564, 12565
Kempeneers, L. 6094, 6328, 6936, 6937
Kemper, K. 19421
Kempf, W. 16600, 16602
Kempton 7930, 7996, 19914, 19915, 20854
Kendall 2780, 14050, 14732
Kendlbacher, R. 9347
Kenk, G. 4765, 4768, 4769

Kennedy 2119, 9819, 9820, 11409
Kennedy, L. 10852, 10853
Kennel, E. 4755
Kenneweg, H. 7946
Kenny, T.A. 3750, 16263
Kent 10417, 11389, 14212, 14701, 18765
Kenworthy 70, 15293
Keogh, M.K. 16832, 16840
Keogh, R. 20359
Keppel, N. 603
Keppens, L. 13383, 13384
Kerber, E. 3835
Kerboeuf, D. 13708, 13709, 13986
Kerck, K. 8870, 8933, 8934
Kerkenaar, A. 9263
Kerkhof, J.A. 15442
Kerkhof, P. 16554, 16930, 16943
Kerkhoff, M.A.T. 2267, 19271, 19272, 19273
Kerkstra, K. 2586, 2588
Kerling, W. 19980
Kermarrec, A. 1049, 1050, 8222, 8223, 8940, 19769, 20764
Kern, H. 893, 1339, 1340, 1742, 1743, 2609
Kern, K.-G. 1730
Kerner, H. 4750
Kerry 8258, 13732
Kerschner, K. 16184
Kersjes, A.W. 13955, 13957, 14426
Kersken-Bradley, M. 16392
Kersten, L. 17796
Kersten, U. 15022, 15024, 15096
Kersting, K. 2226, 20171
Kesavan, R. 2
Keser, M.H. 1260, 1263
Kess, U. 17758
Kesselschläger, J. 17284, 17707, 17912
Kessler, H.G. 16459, 16460, 16461, 16759, 16760, 16761, 16762, 17148, 19306, 19632
Kett, J.J. 12184, 15917
Kettenis, D.L. 19826
Kettern, W. 19433
Kettmann, R. 14066
Keuffel, W. 17637, 17638, 17639, 19789
Keulemans, J. 4223, 4224, 4274
Keulen, H.A. van 6875
Keulertz, K. 10568
Keuls, M. 6091, 19952
Keuning, D.H. 1675
Keuning, J.A. 10907, 10924
Keuning, J.H. 12231, 12278
Keuskamp, J.W. 20982
Kevany, J.P. 18936, 19527
Keydel, F. 6339, 6342
Keymeulen, H. 18240
Keys 20857
Khafaga, E. 3186
Khairia el Sayed 5059
Khalil, H. 13012
Khan 1265
Khanna, P.K. 7, 137, 584
Kibler, E. 142

Madsen, A. 3017, 10274, 11167, 11169, 12418, 12422, 12423, 12430, 12438, 12439, 15846
Madsen, E. 3165, 20323
Madsen, H.B. 472
Madsen, H.E.L. 19993, 19997
Madsen, J. 11823
Madsen, N.P. 12447, 15290, 15371, 15432, 15469, 15654
Madsen, P. 14167
Madsen, P.P. 1018
Madsen, T.L. 1396, 4866, 4941
Maene, L. 6856
Maenhout, C.A.A.A. 2882, 2884, 3128, 8573, 8606
Maertens, C. 492, 1435, 2745, 5108, 5111, 5252, 5253, 5254, 5255, 5464
Maertens, D. 7960, 19592
Maertens, L. 11553, 11554
Magagnoli, P. 13868
Magaldi, D. 275, 276, 681, 1815
Magema 1826, 7956
Mager, A. 13022
Maggioni, A. 5112
Maggiore, T. 3054, 6250, 6251, 6252, 6257, 6260, 6813
Magherini, R. 10155, 15819
Magini, E. 5012, 7652, 7653, 7707
Magistretti, A. 17204, 17422, 17871, 18163, 18164
Maglio, D. 6303
Magnus, L. 15488
Magro, P. 9197
Maguire, J. 17106
Maguire, M.F. 3342, 10839, 11621, 14356, 16334, 16344
Mahe, L. 17394, 17837
Maher, M.J. 2827, 5303, 5708, 5762, 7169, 17167
Mählhop, R. 19685
Mahlmeister, K. 20241
Mahmut 9998
Mahnel, H. 17054, 19629, 19630
Mai, D. 17770, 17957
Mai, E. 6162
Maïa, E. 2733
Maia, N. Mme 17044
Maichle, I. 14786, 14787
Maier, D. 1329
Maier, H.G. 16451, 18845, 19423, 20205, 20547
Maier, J. 5038, 7726, 10051, 19679
Maier, K. 18002
Maier, M. 17908
Maillard, J. 8233, 8234, 8594
Mainie, P. 2388, 17201, 17393, 18113
Mainil, J. 7729
Mainland 17323
Mainsant, P. 16701, 17844, 20822
Maiorana, M. 13260
Mair 15242
Maison 8809
Maison, P. 8745, 8746, 9116
Maizonnier, D. 7564, 7565, 7568, 20803
Majohr, M. 17349

Major, F. 12882
Makagon, S. 18101
Makboul, H. 585
Maklad, M. 18979
Malato, G. 8223
Malausa, J.C. 8224
Maldonado, P. 20819
Malestein, A. 12250
Maletto, S. 10882, 12210, 12709, 13151, 13879
Maliani, C. 6433
Malisch, R. 1986
Malissiovas, N. 5842
Malkomes, H.–P. 10103, 10106, 10113
Mall, B. 4734
Mallet, S. 14204, 14205
Malone 9321, 9322, 9563, 9574, 9994
Malone, D. 19702
Malossini, F. 10860, 11751, 11752, 11927, 12188, 12189, 12354, 13242
Malphettes, C.B. 4977
Malquori, A. 321, 322, 541, 542, 697, 698, 699, 700, 2175
Malterre, C. 12064, 12065, 17500, 17501
Maltini, E. 16134, 16141, 16303, 16654, 16655
Mammerickx, M. 14074
Mammi De Leo, M. 15760
Mamoun, M. 19811, 20824
Mamy, J. 623, 628, 629, 833, 1401
Manby 15155
Manca, M.G. 5456, 5457, 6461
Mancini, A. 11443
Mancini, F. 325, 326, 327, 2168
Mandelli, G. 13873
Mändl, B. 16974, 16975, 16976, 16977, 19426, 19636
Manegold, D. 17804, 18033, 18034
Manenti, G. 3121
Manenti, S. 6400
Manfredi, E. 15266
Manfredini, M. 11240, 11755
Mang, P. 19982
Mangan 11853, 11854, 11855, 11861, 11862
Mangold, H.K. 16271, 18908, 18909, 19048, 19268, 19488, 20265, 20268, 20269, 20689
Mangoni, L. 8343
Mangstl, A. 2693, 9388
Manhardt, J. 13600
Manichon, H. 496, 2749, 3047
Manig, W. 17962
Manikarnika, R. 1309
Manil, G. 5944
Manker, L. 18263
Manmana Novaro, P. 6147, 6251, 6255, 6396, 19839
Mann 10515, 11712, 11899, 13514, 13770
Mann, C. 6231
Mann, G. 1298
Mann, J. 11073
Mannewitz, U. 14654
Manning 4016, 5870, 13086, 14829,

14830, 14831, 18587
Mannion, J. 17203, 17413, 18344, 18345, 18411, 18429, 18430, 18431
Mannipieri, P. 540
Manos, G. 6002
Mansat, P. 6660, 6663, 6664, 6666, 7624, 7625, 7626, 7627
Mansfield 10602
Mansfield, E. 19369
Manshard, W. 17351
Manson 11891, 16741, 18611
Manston 10757, 10758, 13082
Mante, W. 17741, 17742
Mantel, K. 1870, 2368, 4738, 20554
Manteufel, C. 2067
Mantovani, A. 13864, 13868, 13869, 17511, 19574
Mantovani, G. 3289
Manunta, G. 10884
Manwell 9973
Manz, D. 2067
Manzari, R. 4828, 4829
Manzo, P. 1482, 4087, 4198, 4256, 4330, 5786, 7251, 8750
Manzocchi, L.A. 5453
Maracchi, G. 1495
Maracchi, G.P. 3120
Maraite 9008
Marangoni, B. 4089
Marani, F. 9000
Marano, B. 689, 2845
Marano, G. 5468
Marasas, W.F.O. 19182
Marboutie 8719, 8742
Marboutie – Iperti, G. 8741
Marboutie, C. 8682
Marboutie, G. 4068
Marcelis, W.J. 17337, 17338, 17340
Marcelle, R. 4122, 4123, 8698, 8710, 8754
Marchal, J.L. 1828 •
Marchal, M. 8190, 8193, 8199, 8200
Marchal, P. 1189
Marchant 15173
Marchant Shepperson 15165
Marchesi, G. 7148, 7149
Marchesini, A. 19101
Marchesini, C. 17516
Marchou, M. 7462, 7465, 7467, 7468
Marchoux, G. 7180, 9106, 9112
Marcilloux, J.C. 12497, 12716
Marcoen, J. 569
Marcotrigiano, G. 11928
Maréchal, R. 6510, 6511, 7035, 7037
Marenaud, C. 9727, 9886, 9887
Marenaud, Cl. 7277, 9726, 9771, 9795, 9885, 9888
Marenco, G. 18158
Mares, D. 3199
Mariana, J.C. 10432, 10591, 10957
Mariani, B.M. 6147, 6255, 6257, 6396, 19839
Mariani, G. 1472, 3052, 5452, 6247, 6248, 6249, 6261, 6320, 9380
Marie, R. 2734, 3073, 6287, 6288, 6289, 8473

17487, 17488, 18020, 18021, 18022, 18023, 18024, 18025, 18026, 18027, 18028, 18029, 18030, 18031
Meire, R. 20089
Meisel, H. 11270
Meisel, K. 1999, 2003, 2005, 2006, 2010, 2433
Meissler, M. 13571
Meisterjahn, R. 12979
Meiwes, K.J. 7
Meixner, R. 10581, 13427
Mejer, G.-J. 15418, 15499, 19645
Mekers, O. 4489, 4577, 4580, 4581, 4582, 7530, 7538, 7539, 7540
Melchionna, M. 10153
Melchior, G.H. 7631, 7635, 7636, 7638, 7639
Melgård, P. 17387
Melis, G. 8312
Melisenda, I. 309, 310
Melle Chesneaux, M.T. 3098
Melling, H. 19298, 20214
Mellini, E. 8327, 8328
Mellins, J. 15117
Mellon, K. 18149
Mellor 10977, 12327, 12328, 14520
Melotti, P. 2159, 2160, 2163, 2165
Melzer, R. 9802
Menden, E. 11640, 19476, 19477, 19512, 19515, 19974
Mendes Pereira, E. 12865, 12867
Mendes-Pereira, E. 11703
Mendgen, K. 9023, 9652, 9653
Meneghini, A. 10317, 11485, 12774
Menegon, S. 3613
Menegonx, S. 3614
Menet, M. 486, 760, 910, 1418, 5461, 5659
Meneve, I. 4579, 4676, 4677, 4678, 7526, 7527, 7532, 7535, 7536, 7537
Meng, W. 2439, 17658
Mengarelli, A. 6504
Mengel, K. 581, 2902, 5141, 5142, 5143, 5842
Menger, A. 16516, 16517, 16518, 16523, 19001
Menguzzato 4901, 20901
Menhennet 4505
Menissier, F. 13049, 13052, 13055, 13056, 13059, 13061, 13062
Menke, G. 9674
Menke, K.H. 11132, 11647, 11648, 11977, 12401, 12402, 12403
Mennella, V. 15439, 20932
Mennessier, P. 2729, 2730, 3330, 3661, 3894, 3897, 6449, 17391, 17392
Menniti, A.M. 3241, 9509, 9761, 10363, 16152
Menoux Boyer, Y. 3138
Mentges, A. 20541
Mentz, I. 18838
Meppelink, E.K. 16541
Merat, P. 13400, 13401, 13402
Mercer 508, 658, 10128
Mercier, J.C. 10497, 13050
Merck, C.C. 10442, 13974, 13975,

14082
Mercurio, R. 20898
Meregalli, A. 13128, 13129, 13130, 13131
Mergeay, P. 9009
Mergenthaler, E. 20701
Mergner, W. 8888
Mergui, G. 17401, 17723, 18114, 19835
Meriaux, S. 5667, 5742, 5835
Merkel–Gottlieb, A. 11076
Merkel, D. 172, 5489, 19683
Merkel, K. 17911, 18049
merkus, G.S.M. 16710
Merlanti, M. 13867
Merle, P.L. 2103
Merletto, F. 11488
Merlini, L. 9195, 9196
Merlo, M. 2395, 17516
Merne, O. 11360
Merten, D. 19408
Mertens, P. 17895
Mertins, K.-H. 15490
Mesdag, J. 6434, 6435, 6436, 6437, 6438, 6439, 6440, 6441, 6442
Meshref, H. 439
Meske, C. 12736, 12737, 13419
Mesken, M. 6609, 6610, 6613, 6615, 6616
Meskens, J. 19965
Meslin, J.C. 10424, 12466, 12848, 12860
Mesnier, Y. 4237, 7270, 7279, 16645
Mesquida, J. 3166, 6478, 11480
Messem 1452, 3268
Messer 12097
Messner, M. 19634
Messori, F. 17510
Messow, C. 11164, 13584, 13919, 20622
Mettauer, H. 641, 1425, 2104, 3329, 5492
Metz, J.H.M. 10915, 10916
Metz, M. 20194
Metz, S.H.M. 20962
Metzger, F. 590
Metzger, J.J. 14691, 14696, 14697
Metzger, U. 14130
Metzner, C. 15335
Metzner, R. 15651
Meuleman, J. 15403
Meulemans, G. 13479, 14783, 14784, 14785
Meulen, U.ter 11642, 11972
Meulenberg, M.T.G. 17888
Meunier, A. 18341, 18924, 20829
Meunier, M. 8113
Meurens, M. 11543, 11809, 15894
Meurs, W.T.M. van 15626
Meuser, F. 16492, 17050, 18841, 19010, 20197
Meuther, R. 17196
Mey, G.J.W. van der 10594, 13171
Meÿaard, D. 17876
Meyboom, F.W. 10891
Meyer, A. 2013
Meyer, A.M.T. 21066

Meyer, B. 446, 447, 1237, 1238, 1735
Meyer, D. 16511, 16512, 18995
Meyer, E. 2403, 8063, 8513, 8629
Meyer, F. 4757, 12982, 12983, 13551, 14089, 20571
Meyer, G. 1229, 18254
Meyer, G.–U. 136
Meyer, H. 6309, 10658, 11796, 11797, 11975, 12397, 12398, 12399, 14630, 14634, 14636, 17761, 17940, 20607
Meyer, H.von 17932
Meyer, J. 1270, 1424, 1426, 1427, 8035, 9008, 9926, 13001, 13279, 13462, 15593
Meyer, J.-M. 14129
Meyer, J.-N. 13272
Meyer, J.L. 1424, 1425, 1426, 1427
Meyer, J.P. 7405, 16475
Meyer, S. 7373
Meyer, U. 18244
Meÿer, W.J.M. 3499, 3501
Meyerweissflog, C. 8876
Meyling, A. 13493
Meymerit, J.C. 8721, 8745, 8746
Meynadier, G. 8172, 8173, 8176, 8178, 8192, 9107, 9110, 9114, 14982, 20773, 20774, 20776, 20778, 20780, 20791, 20792
Meynet, J. 7561, 7562
Meys, C.C.A.M. 1676
Mezzetti, A. 9181
Michael, B. 5375
Michael, G. 2906, 3042
Michalet–Doreau, B. 11700, 11701, 15908
Michalsky, F. 15798
Michaux, C. 12870, 12969, 12972
Michel 14319
Michel M.France. 4802
Michel, G. 19865
Michel, H. 8067
Michel, M.C. 14970, 14971
Michelesi, J.C. 7258
Michels, P. 1190
Michels, Th. 2465, 2475, 2485, 2489
Michie 11305, 11340, 11341, 11342, 11343, 12701
Michieli, G.A. 9221
Micieli, De Biase, L. 8510
Micieli, L. 8691
Mick, W. 19080
Mickwitz, G.von 10459, 11121, 17140, 17158
Miclaus, N. 267, 274, 282, 516, 782
Miclet, G. 18094, 18225, 18315
Micol, D. 17501
Middelboe, V. 19992, 20724
Middelhoven, W.J. 20996, 20998
Middelkoop, J.H. van 11346, 11365, 13414
Miedema, P. 6087, 6271
Miedema, R. 405, 966
Mielke, H. 9278, 9281, 9282
Miert, A.S.J.P.A.M. van 14596
Miesel, G. 13001
Mieskes, G. 9025
Mifliin 5411

Moore, J. 2130
Moore, J.F. 5079
Moosmayer, H.U. 4764, 4888
Morand–Fehr, P. 10600, 11843, 12373
Morand, J.C. 7612
Morandi Cecchi, M. 19820
Morandini, R. 7949, 20366
Morcale, A. 1109
Mordenti, A. 11756, 12549, 12550, 16307
More O'Ferrall, G.J. 11025, 13114, 13119, 13121, 13122, 13227, 13228, 13229, 13230, 20876
More, E. 3406, 5517, 5542, 5543, 5544
More, J. 13687, 13693
Moreale, A. 1831
Moreau, C. 12854, 12855, 12864
Moreau, J.P. 8033, 8443, 8517, 9088
Moreau, M.C. 12852, 12860
Moreels, A. 11541, 15822
Morel, R. 230, 765, 5256
Morelet, M. 10034
Moreton 8264, 8266
Moretti, G. 4432, 4438, 4439, 4442
Moretti, R. 1466
Morfaux, J.N. 19698
Morgan 2937, 5668, 6053, 7797, 7932, 12959, 20841, 20850
Morgan–Jones 11304
Morgan, J.V. 3959, 3960, 4655, 5732, 5834
Morgan, K.G. 7901
Morgan, M.A. 1768
Morgan, P. 10858
Morgante, D. 16850
Morganti, L. 13864, 13866
Morgenstern, K. 6312
Morgner, F. 3660, 9552, 20317
Morice, J. 5459, 6475, 6476, 6477, 7732
Moriondo, F. 10007
Morisot, A. 479, 20761
Morizet, J. 835, 5104, 5105, 5106, 5250
Morlat 5384
Morlat, R. 192, 757, 5858
Morley–Jones 19894
Morlry 17156
Mormede, P. 13686, 14689
Moroney, S. 15967, 15968
Morrice 2984, 6142, 6855
Morris 13820, 13821, 19913
Morrison 3476, 3550, 3554, 5599, 12115, 12116, 12339, 14526
Morrison, M. 6705
Morrissey, P. 19270
Morrissey, P.A. 18932
Morrow 13782
Morrow, A. 13946
Mortelmans, J. 14077, 14613
Mortensen, B. 15662
Mortensen, B.K. 16797, 16802, 19338, 19339, 19345, 19346, 19353, 19354, 20713
Mortensen, G. 3165, 6175, 6750, 10123, 10198
Mortensen, H.P. 12439
Mortensen, J.V. 5067, 5427

Mortensen, S. 10585
Morvan 4235
Mörzer Bruyns, M.F. 2311
Mörzer, Bruyns, M.F. 2309
Mosca, G. 6269
Mosch, W.H.M. 8382
Moschini, E. 3765, 3967, 6722
Moseley 12321, 12322
Moser, E. 1285, 2692, 5163, 7954, 15365
Moser, L. 2660, 7228
Moskophidis, M. 20613
Mosonyi, E. 1288, 1289, 1290, 20226
Moss 2930, 11214
Mosse 2799
Mosse, J. 20064
Mossel, D.A.A. 19224
Motoulle, A. 19853
Mott, N. 5584
Motta, E. 9428, 10021
Mottar, J. 16229, 19294
Mottet, A. 15734, 15735, 16358, 16359, 16360
Motto, M. 6250, 6251, 6257, 6260, 6813
Mottram 18529, 18531
Mouchet, C. 17504
Mould 14005, 14006, 14482, 14483, 14484, 14485
Moulinier, H. 194, 479, 481, 5040, 5899, 5901
Mount 10509, 10516, 11202
Mouras, A. 7213, 7280
Mourgues, J. 16647
Mourguet, A. 12085
Mourot, J. 12506
Mousa, S. 14801
Mousain 4868
Mousain, D. 20796
Mousain, D. (Clermont Fd.) 4809
Mousdale 4074, 4077
Moussa, J.B. 20784
Mousset, C. 6811
Moustafa, A. 4655
Moustafa, A.T. 3960
Moustgaard, J. 11680, 12450, 13299, 14172
Mouton, B. 15372, 17266, 18071, 18073
Moutonet, M. 16995, 19096
Moutous, G. Mme 9870
Mowat 8618, 9910, 9911, 13779, 13780, 13781
Moyaert, I. 10623, 12968
Mpakagiannis 8060
Mrass, W. 2377, 2380, 2555, 2559
Msaye, B.M. 179
Mtetwa, J. 18248
Mucci, F. 5050, 6413
Mückenhausen, E. 113, 114, 115, 117, 870, 1209
Mudd 8256
Mue Delmas 13440
Mueller, F. 20969, 20970
Muermans, L. 17737, 17739
Mugniery, D. 8221, 8446, 8512, 8561

Mühlbauer, F. 17783
Mühlbauer, W. 16035
Muhle, O. 4774, 7691
Mühlhäusser, G. 13, 889
Muhs, H.–J. 7630, 7632, 7633, 7636
Muir 7909, 8941, 8942, 10773, 18599, 18600, 18604, 18605, 18606
Muiswinkel, W.B. van 21069
Mukherjee, K.D. 16271, 16558, 19047, 19049, 20266, 20267, 20690
Mülder, D. 1926, 1927, 4747, 4748, 4955
Mulder, E.G. 1100, 1101, 5654
Mulder, H. 16895
Mulder, I. 14064, 21085
Mulder, R.W.A.W. 11355, 11773, 19217, 20164
Mulder, W.P. 15627, 15684
Muljadi, S. 1203
Mull, R. 1264, 1265, 1266, 1278
Mullany, M. 18346
Mülle, G. 2900
Müller–Darss, H. 17258
Müller–Haslach, W. 4611
Müller–Hohenstein, K. 1863, 17924
Müller–Peddinghaus, R. 14621, 15020
Müller–Schlösser, F. 10455
Muller, A. 862, 2369, 5365, 5471, 5574, 5575, 6647, 10238, 11801, 11833, 12057
Müller, A.von 5490
Müller, B. 17716, 18042
Müller, D. 591
Müller, F. 4371, 7843, 8780, 9039, 10097, 10098, 10182, 10183, 10315
Müller, G. 4740, 17936
Müller, H. 1874, 10085, 15087, 16115, 18258, 18881, 20212
Müller, H.–M. 15887, 15891, 16033, 16034
Müller, H.L. 10689
Müller, H.P. 16632
Müller, I. 4059
Muller, J. 767, 1441, 1951, 5446, 8557, 9078, 9079
Müller, K. 1344, 3090, 4394, 5439, 5508, 5854, 7924, 16069, 16592, 19062, 19426, 20216
Müller, K.B. 11104
Müller, L. 4608, 5688
Müller, M. 13539
Müller, P. 2431, 20674, 20675
Muller, P.J. 20435
Müller, R. 12649
Müller, S. 13, 606, 889, 2912, 4765
Müller, U. 16231, 17943
Müller, W. 572, 1365, 1366, 1367, 1368, 1748, 13611, 13612, 13613, 17158, 17937, 18418, 19621, 20107, 20663
Müller, W.E.G. 20642
Müllerstael, H. 4763
Mülling, M. 10441, 14083, 14084, 14085, 14615, 20193
Mulqueen, J. 1171, 1172, 1173, 1174, 10846
Mulsant, Ph. 13317, 13318, 13319
Multon, J.L. 16476, 19012

Opitz–von–Boberfeld, W. 2680, 3309, 3376, 3509, 3510, 5137, 5138, 5569, 5570, 5571, 5572, 6644
Opitz, M. 11516, 13545, 13547, 13548, 15011, 15012, 15013
Opletal, L. 18445
Oppenländer, K.H. 18052
Oppenoorth, F.J. 8393, 8397
Oppenoorth, W.F.F. 20461, 20462, 20464, 20467
Oppermann, W.H. 13537
Oppitz, K. 6340, 6341
Orand, A. 1134, 1138, 1141, 1143
Orban 12873
Orbeck, K. 15368
Ord 778
Ordonez, M.T. 9344
Orfei, Z. 14388
Orgis, K. 2075, 20711
Orlandi, D. 3608, 5045
Orlovins, I. 13416
Orlovius, I. 6886, 6940, 6941
Orpin 11845, 11846, 11858, 12314
Orsi, S. 2841, 3359, 3360
Orskov 11912, 11914, 12337, 12338
Ort, W. 17931, 17932
Ortavant, R. 10958, 10965
Orth, H.W. 15463, 16268, 17042
Os, E.A. van 15352
Osborne 8535, 8650, 10134, 10135, 10136, 10137
Osbourn 11585, 12093
Osinga, A. 10912, 10914
Oskam, A.J. 18198, 18200, 18201
Osmers 8059
Ossard, H. 18127
Osterkorn, K. 10684, 13001, 13002, 13003, 13004, 13005, 13279, 13462, 14137, 14650
Osty, P. 17495, 18086, 18303
Osty, P.L. 2387, 18069, 18134
Oswald 3547, 8015, 20844
Oswald, H. 4790, 4804, 4973, 4974, 4976, 4977
Oswin 18704
Other 13075
Othman, M.bin 5429
Otoul, E. 7035, 7037
Ott 2368
Ott, E.C.J. 2326, 2328, 2332
Ottaro, J. 10641
Ottaviano, E. 6816, 9878
Otte, E. 18209
Otte, I. 19660
Otten, A. 19958
Ottenwaelter, M. 5780, 7397, 7398, 9868
Otterbach, G. 16453
Otto, A. 1322, 1324
Otto, C. 5696, 11282, 11283, 11324
Otto, C.Mme 5102
Otto, C.Mme 5101
Otto, H. 11143
Otto, R. 18811
Ottogalli, G. 16862, 16863
Ottolini, P. 8332

Ottorini, J.M. 4790, 4794, 4803, 4804, 4890, 4973, 20339
Ottow, J.C.G. 141, 456, 585, 586, 732, 8861, 20629
Ottoy, J. 19849
Ouden, H. den 6910, 8625
Ouden, J.H.B. den 18371
Ouhayoun, J. 13440, 13443, 13444
Outen 11866, 11869
Outer, R.W. den 21044
Ouwerkerk, C. van 938, 944, 946, 948, 949
Ouwerkerk, E.N.J. van 15581
Ouwerkerk, J. van 1723
Over, A.M. 18103
Over, H.J. 14402, 14405
Overal, M. 1827
Overbeck, G. 5183, 16112, 19084
Overbeck, H. 8631
Overdieck, D. 10341, 20525
Overdulve, J.P. 13903
Overeem, J.C. 9264, 9265
Overgaauw, J.G.A. 18177
Overvest, J. 12227, 15926
Oving, R.K. 17184, 17703
Owen 9363, 10814, 10816, 11014, 12343, 13224, 15198, 15200, 15510, 18770, 18771, 20019, 20020
Owyer 15508
Özgür, F. 8048
Özmen, T. 16116
Öztek, L. 19297

Pabst, K. 13018, 19309
Pabst, W. 12991
Pace, V. 11751, 11752, 11927, 12354
Pacher, J. 1871, 2369, 19788, 20209, 20554, 20555, 20556, 20557
Pacini, E. 20940
Pacucci, G. 6306, 6777
Padula, P. 13877
Paeffgen, D. 20544
Paelinck, H. 19406
Pagano Toscano, G. 12822, 13150
Pagano, Toscano, G. 12821
Paganucci, L. 4832
Page 5278, 5703, 5704, 14914
Pagel, B. 15010, 20523
Pagella, M. 2450, 17869
Paglialunga, S. 19412
Paglietta, R. 7219
Pahlen, H.–D.von der 17139
Pahlke, K. 111, 1205
Pahlow, G. 9030
Pahlow, R. 5155, 20616
Paiboon, P. 5431
Pain 2781, 2782
Pain, J. 11473, 11474, 12770
Painvin, R.M. 18229
Paizs, L. 12999
Paksoy, S. 19046
Pakzad, R. 11495
Palenzona, M. 4033, 5013, 7660
Palland, C.L. 1721
Pallauf, J. 11638, 11967, 12390, 12391
Pallavicini, C. 17028, 20032

Pallotta, U. 17022, 17023
Palludan, B. 12449, 12451, 14672
Palm, G. 7224
Palm, R. 19844
Palmer 4174, 4297, 10543, 15209, 15218, 15219, 15220, 15221, 15222, 15511
Palmer, E. 10591
Palmer, J. 16443, 16838, 19572, 19573
Palmia, F. 16211, 16212
Palmieri Pedrini, E. 17182
Palmieri, F. 688, 689, 1184
Palmieri, G. 689, 2845, 5468
Palmieri, S. 3962, 20895
Palti, J. 9018
Paludan, N. 9591, 9907
Paluschka, S. 7300
Pampiglione, S. 13866, 13867, 13869, 19574
Panaro, V. 15797
Pancaldi, D. 9579
Panconesi, A. 9215
Panebianco, F. 14383
Panella, A. 6208, 6638, 6721, 6784, 6818, 6819
Panero, V. 18238
Panholzer, J. 17715
Pani 14903, 14904
Panicucci, M. 274, 849, 1176
Panis, A. 8126, 8127, 8824, 20762
Pannelli, G. 3141, 3142, 3146, 3147
Panouille, A. 6242, 6243, 6809
Panten, I. 19511
Pantusa, M. 1481, 20897
Panzer, K.F. 1989, 19797
Paoletti, R. 3492, 3686
Paolim, K. 19603
Pape, J.C. 363, 368, 376, 394, 557, 1564, 2184
Papendick, K. 11570, 12000
Papenhagen, A. 4496, 4497, 15602
Papetti, G. 16852, 19374
Papineau, J. 6792, 6793
Papparella, V. 14859
Paquay, R. 11963
Paquignon, M. 10433, 11194, 11196
Paquot–Gasia, M.C. 19960
Paquot, M. 16359, 16360, 18456, 18457
Paradisi, F. 6400
Paradisi, U. 6980
Paraf, A. 13716, 13718
Parameswaran, N. 10025, 15957, 16397, 20681
Parde, J. 4804
Pardini, G. 3364, 6780
Pardon, P. 11694, 13702, 14452, 15127
Parente, G. 3613, 3614, 11771
Parenzan, P. 8322
Parez, M. 13061
Parfeit 20522
Parfitt 5785, 14733
Parigi Bini, R. 12811
Parigi–Bini, R. 12812
Paris, P. 3355
Parisini, P. 11756
Park, B.H. 3377

LIST OF SCIENTISTS

Seynaeve, R. 7210
Sfalanga, M. 270, 278, 281, 1175
Sgrulletta, D. 3119, 6255, 20894
Shabbir, S.M. 11965
Shahen, A. 19612
Shan 19195
Shanahan, R.M. 16841
Shand 20024
Shannon 12672, 12677, 12681, 12689, 12718, 18532
Sharifi, M.R. 5529
Sharma 9487
Sharma, T.R. 6096
Sharman 10506, 10508, 10510, 10517, 11200, 11712, 11909, 12932, 13728, 13793, 14029, 14591, 14720
Sharp 7857, 9168, 9366, 9613, 9912, 11296, 13370, 13371, 13372, 13374, 14505, 15166, 15347
Sharpe 11873, 18555
Sharples 2615, 18533, 18537
Shaw 7810, 9171, 13078, 14322, 15252
Shawki, E. 875
Shea 15180, 15181, 15182
Sheahan, B.J. 14373
Sheard 15142
Shearer 16054, 16055
Sheehan, W. 11025, 11925, 12348, 12351
Sheehy 3417, 3642
Sheldrick 3333, 3409, 3551, 3680
Shepherd 4141, 4693, 5294, 8012
Shepperson 15162, 15853
Sheridan, J.J. 16706, 19203
Sherlock 7999
Sherrington, J. 3281, 16124
Sherry, C. 1814
Sherwood, M. 848, 5650, 19701
Shewry 5411
Shiach 15253
Shiel 4548, 4549
Shipton 9170, 9331, 9332, 9367
Shirlaw 11303
Shirley 14908
Shone 2756
Short 9135, 9136, 9137, 16702
Shreeve 13822, 14040, 14314
Siaens, M. 2595
Sibbald 19922
Sibbing, F.A. 2333
Sibi, M. 7113
Sibma, L. 2872, 3498, 5633
Sibomana, G. 19145
Sicherer-Roetman, A. 21019
Sickel, E. 11148
Sicker, W. 19799, 19805
Siddons 11867, 14279
Sidibe, M. 11074
Siebeck, O. 1352
Sieben, W.H. 1685, 1691, 1697, 1702, 2594
Siebenbürger 2608, 17659
Sieber, J. 2641, 4610
Siebert, D. 1865
Siebert, G. 10463, 11125, 11126, 16953, 19478, 20627, 20628

Siebert, H. 4934, 4989, 5954, 5985, 8894
Siebert, W. 1320, 19265, 20251
Sieder 20699
Siegert, E. 1326, 5186
Siegert, M. 15011, 15013
Siegfried, R. 18872
Siegmann, O. 13488, 14883
Siegmund, H. 10075
Sieker, F. 1271, 1275
Siems, H. 18837, 19142
Siepman, A.H.J. 20398
Siepmann, R. 9999, 10002, 10014, 10015
Sieverding, E.G. 5051
Sievert, M. 1958
Siewert, E. 19133
Sigersted, E. 16810
Sigmund, U. 1944
Signoret 9115
Signoret, J.P. 10438, 10967, 10969
Sigwalt, C. 6790
Sijtsma, R. 10197, 10234
Sikora, R.A. 8555
Silfhout, C.H. van 9423
Siller 11287, 14891
Silva-Montenegro, M. 5428
Silva, S. 530, 2167, 5309, 10336, 20910
Silver 10524, 10601, 14718
Silvestri, B. 3147, 6455, 6458, 6459
Sim 11601
Simatupang, M.H. 16400, 16401, 16403
Simeone, A. 4193
Simesen, M.G. 14676
Simmermacher, W. 20100
Simmonds 6126
Simmonds, M. 19613
Simoens, X. 19967
Simoes 12507
Simon, D. 12884, 12978, 12979, 12980, 13269, 17477
Simon, F. 13698
Simon, G. 183, 187
Simon, G. Mme 5237, 5741
Simon, G.Mme 1019
Simon, I. 19465, 19501
Simon, M. 3098, 6361, 6450
Simon, P. 4122, 4123, 8698, 8710, 8754
Simon, S. 1020
Simon, U. 1877, 1878, 1879, 3312, 3377, 3378, 3379, 3511, 3512, 6622, 6645, 11811, 14146
Simonic, T. 13872
Simonic, T. – De Luca, A. 20921
Simonin, G. 13716, 13718
Simonis, M.T. 7908, 8559, 8747
Simons, D. 15524
Simons, P.C.M. 11353, 13411, 14934, 19405
Simons, W. 8415
Simonsen, H.B. 11286
Simpson 3651, 6051, 6115, 11481, 11482, 11618
Sinatra, M.C. 11757
Sinclair 676, 3410, 14053

Sine, J. 1188
Sine, L. 818, 866, 867, 1109, 1189
Sinell, H.–J. 16182, 18837, 18838, 18839, 19141
Sinell, H.J. 19142
Singer 11746, 18785, 18786, 18787, 19568
Singer, B.D. 7801
Singh, A. 9012
Singh, K. 16042
Sinke, J. 3002
Sinn, H. 1285, 2692
Sinnaeve, J. 347, 20034
Sinner, H.–U. 17645, 17646
Sinner, M. 16403, 16407
Sioli, H. 1328, 2034, 5052
Siragusa Campanello, N. 20887
Sirks, J.L. 14439
Sissingh, H.A. 337, 701, 702, 5349
Sissons 12108
Sisto, D. 9417
Sital, J.T. 17473
Sittler, B. 1868
Six, R. 5669
Sixdenier, G. 6790, 7059
Sjardijn, R.C. 553
Sjöberg, A. 11511
Sjut, N. 3939
Ska, P. 12282
Skaer 8272
Skatulla, U. 7945, 8883, 8886
Skene 2765, 4138
Skerrett 7795, 7796
Skibsted, L.H. 19996, 20725
Skinner 1059, 7995, 10540, 13778
Skirde, W. 1233, 2548, 5086
Skomroch, W. 17349, 17350, 17922, 20546
Skorupka, A. 14951
Skou, J.P. 9355, 9356
Skovborg, E.B. 11571, 11572, 12014, 15842, 15843
Skovgaard, N. 18825
Skriver, K. 5383, 5443, 5645
Skröppa, T. 20324
Skudder 18579
Slack 3948, 3978
Slade 13999
Slager, S. 420
Slamka, P. 4400
Slangen, L.H.G. 17685, 18202
Slappendel, R.J. 14600
Slater 2765, 12328, 13215
Slavekoorde, S.M. 7944, 9984
Slee 10503, 10504, 13208
Sleurink, J.P. 17276
Slight 18657, 18658, 18659, 18660, 18661, 18662, 18663
Slijkoord, F. 1725
Slim, P.A. 2220, 2221
Sloet van Oldruitenborgh, C.J.M. 2301, 2313
Slogteren, D.H.M. van 9229, 9918
Slope 9302, 9305, 9306
Sluijs, P. van der 354, 369, 856, 1560, 1566, 1770

LIST OF SCIENTISTS

Vazzana, C. 10152

Vear, F. 6042, 6490, 6492, 6493

Vecchio, V. 10208

Vecchione, G. 19381

Vecchiotti Antaldi, G. 10593

Vedel, H. 2082, 2084, 2085

Veen, B.W. 2878, 2879, 3134, 15878

Veen, H. 4662, 20959

Veen, J. van de 19061

Veen, J.A. van 707

Veen, J.C. van 2291

Veen, J.F. de 2264, 2265

Veer, A.A. de 384, 393, 2565, 2566

Veer, J. de 17434, 18352

Veerkamp, C.H. 16215, 17138

Veerman, A. 8426

Veerman, G.J. 1595

Vehlow 15965

Veierskov, B. 2722, 2723

Veil, L.B. 13560, 13562

Veirskov, B. 2721

Velde, H.A. te 2881, 3375, 3671, 3672

Velde, J.J. van de 20388, 20389, 20390

Veldeman, R. 9716, 9931, 10030

Velden, N.A. van de 13458

Velden, N.A. van der 13456, 13459, 13460

Veldhuyzen, C.J. 17676, 17681, 17683

Velghe, G. 19581

Velghe, K. 19576

Velicogna, E. 17586

Velsem, A.F.M. van 19736

Veltmann, E. 14624

Veltsistas 1963

Vemmer, H. 11660, 11662, 12415

Ven, W.S.M. van de 21050

Vender, C. 3240, 6555

Veneau, G. 633

Venekamp, J.T.N. 335

Venezian Scarascia, M.E. 288, 1469, 3051, 3345, 5421, 5451, 20893

Venezian, M.E. 3487, 3638, 3652, 6812

Venge, O. 15044, 15069, 15070, 15072, 15073

Veniale, F. 694

Venker, A.J. 15058

Vente, J.M. 15629

Venter, F. 3702, 5682, 5683, 5685

Venturi, G. 6515

Venturo, R. 10203

Verachtert, H. 16265, 16266, 19582, 19583, 20520

Verbaan, A. 17280

Verbeck, B. 11458

Verbeke, N. 17897

Verbeke, R. 19138

Verbiese, R. 1787

Verboon, M.C. 13163

Verbruge 20348

Vercauteren, R. 13533, 14069

Verdonck, O. 1, 436, 1200, 1201, 1788, 4488, 4490, 5680, 10383, 17047, 17048

Verdooren, L.R. 19959, 21062

Verduin, B.J.M. 9007

Vereecke, D. 10624

Verf, M.M. 9241

Verga, M. 10871

Vergara, S. 8192, 8196

Verger, J.M. 13696, 13698, 13702, 14452

Vergnes, A. 4426, 7408, 7412

Vergniaud, P. 2613, 3793, 3896, 6894, 7121, 17219

Vergos, S. 4881

Vergouwe, A.A. 722

Verhaegh, A.P. 17609

Verhoeff, J. 12255, 14436

Verhoeff, K. 9260

Verhoeyem, M. 7750

Verhoyen, M. 8964, 9926

Veri, G. 1074, 2667, 2833, 5113, 20943

Veringa, H.A. 19394

Verità, P. 12820

Verite, R. 11834, 12059

Verkaik, A.P. 20392, 20401

Verkerke, D.R. 5749

Verkley, F.N. 9690

Verloo, M. 19581

Vermeiren, L. 1188

Vermeulen, H. 7729

Vermi, P. 206

Vermorel, M. 10747, 10748, 11703, 11704, 11705, 11839, 12052, 12072, 12073, 12074, 12308, 12309, 12310, 12311, 12868

Verna, M. 3084, 11751, 12187

Vernon 11728, 11892

Versailles, S.E.I. 1764

Verschaege, L. 3184

Verschraeghe, L. 6511

Verschuren, P. 13898

Versteeg, M.N. 2873

Versteegh, H.A.J. 11773

Verstegen, M.W.A. 11262

Verstraeten, C. 1828

Verstraeten, L. 434, 5061

Verstraeten, M.J. 433

Vertessen, J. 17739

Vertregt, N. 20033

Vervack, W. 11541, 11542, 15928, 18833

Vervaeke, I. 12376

Vervoort, M.J. 17537

Verweij, E.J. 2478, 2493

Verwey, J.A. 7693, 7709, 7710, 7714

Verworn, H.-R. 1271

Verworn, W. 1273

Vestergaard Thomsen, K. 20128

Vestergaard, K. 10495, 11184

Vestergaard, P. 2088

Vesth, B. 13027

Vetter, G. 12644

Vetter, H. 5196, 5197, 19684, 19685

Vetter, R. 19625

Vey, A. 7988, 8175, 14982, 20767, 20770, 20772, 20781, 20782, 20784, 20785

Veyrunes, J.C. 8178, 19768, 20774, 20782, 20783, 20785

Vezinhet 12311

Vezinhet, A. 10949

Vezinhet, F. 20809

Viaene, N. 13472, 14873

Viallow, J.B. 18080

Vianello, G. 685, 20370

Viau, C. 17847, 18131

Vibe–Petersen, G. 13934

Vible, J.C. 6237, 6238

Vicente, M.A. 756

Vicini, E. 16159

Victor 11890

Vidal, A. 6483, 6484

Vidano, C. 8030, 8356, 9203, 11487

Vieweg, G. 19972

Viggiani, G. 7923, 8341, 8342

Viggiani, P. 10224

Vigneron, P. 10431, 11504

Vigoureux, A. 3253, 7855, 15134, 15828

Vigouroux 4235

Vila, J.P. 19869, 19871, 19876, 19878, 19879, 19880, 20138

Villa, B. 6295, 6296, 6298, 6299, 6300, 6302, 8474, 10200, 10201, 10202

Villejoubert, C. 14204, 14205

Villemur, P. 4170, 6454

Villers, S. 7862

Villiers, P.A. 20139

Villumsen, J. 5929

Vilsmeier, K. 154

Vincent 2621

Vincent, A. 6579, 6580

Vincent, J.J. 8200

Vincke, H. 3185, 4361

Vincourt, P. 6626

Vink, L.W. 2484

Vink, N.H. 1678, 1679

Vink, P. 9915

Vinkenborg, C. 19060

Violante, A. 687

Violante, P. 690, 1184

Violente, P. 687

Vipond 11016, 12344, 12345

Vis, T. 372, 373

Visai, C. 4458

Visschedijk, A.H.J. 11313

Visscher, A.H. 13250, 13251

Visscher, H.R. 4035, 4036

Visser, A.C. 1818, 2468, 2569

Visser, A.J. de 17626, 17627, 17628

Visser, D.L. 7002

Visser, G.J.M. 2246

Visser, G.R. 17705, 19776

Visser, H. de 12216

Visser, J. 1695, 1711, 5801, 20977, 20978

Visser, J.H. 8405

Visser, M.B.H. 21065

Visser, S. 16909

Visser, T. 7253, 7254, 7255

Vitagliano, C. 3153, 4098

Vitagliano, M. 19375

Vite, J.P. 8025, 8864, 8907

Viticulture Bordeaux 5860

Vitlox, O. 15274, 15320

Vittori, F. 4338, 7331